SO-BYB-093

Microwave Electronic Devices

Microwave Technology Series

The *Microwave Technology Series* publishes authoritative works for professional engineers, researchers and advanced students across the entire range of microwave devices, sub-systems, systems and applications. The series aims to meet the reader's needs for relevant information useful in practical applications. Engineers involved in microwave devices and circuits, antennas, broadcasting communications, radar, infra-red and avionics will find the series an invaluable source of design and reference information.

Series editors:

Michel-Henri Carpentier, Professor in 'Grandes Écoles', France, Fellow of the IEEE, and President of the French SEE

Bradford L. Smith, International Patents Consultant and Engineer with the Alcatel group in Paris, France, and a Senior Member of the IEEE and French SEE

Titles available

1. **The Microwave Engineering Handbook Volume 1**
   Microwave components
   Edited by Bradford L. Smith and Michel-Henri Carpentier

2. **The Microwave Engineering Handbook Volume 2**
   Microwave circuits, antennas and propagation
   Edited by Bradford L. Smith and Michel-Henri Carpentier

3. **The Microwave Engineering Handbook Volume 3**
   Microwave systems and applications
   Edited by Bradford L. Smith and Michel-Henri Carpentier

4. **Solid-state Microwave Generation**
   J. Anastassiades, D. Kaminsky, E. Perea and A. Poezevara

5. **Infrared Thermography**
   G. Gaussorgues

6. **Phase Locked Loops**
   J.B. Encinas

7. **Frequency Measurement and Control**
   Chronos Group

8. **Microwave Integrated Circuits**
   Edited by I. Kneppo

9. **Microwave Tube Transmitters**
   L. Sivan

10. **Microwave Electronic Devices**
    Theo G. van de Roer

Microwave Electronic Devices

Theo G. van de Roer PhD

Associate Professor in Electrical Engineering
Eindhoven University of Technology
Eindhoven, The Netherlands

CHAPMAN & HALL
London · Glasgow · Weinheim · New York · Tokyo · Melbourne · Madras

Published by Chapman & Hall, 2–6 Boundary Row, London SE1 8HN, UK

Chapman & Hall, 2–6 Boundary Row, London SE1 8HN, UK

Blackie Academic & Professional, Wester Cleddens Road, Bishopbriggs, Glasgow G64 2NZ, UK

Chapman & Hall GmbH, Pappelallee 3, 69469 Weinheim, Germany

Chapman & Hall USA, One Penn Plaza, 41st Floor, New York NY 10119, USA

Chapman & Hall Japan, ITP-Japan, Kyowa Building, 3F, 2–2–1 Hirakawacho, Chiyoda-ku, Tokyo 102, Japan

Chapman & Hall Australia, Thomas Nelson Australia, 102 Dodds Street, South Melbourne, Victoria 3205, Australia

Chapman & Hall India, R. Seshadri, 32 Second Main Road, CIT East, Madras 600 035, India

First edition 1994

© 1994 Chapman & Hall

Typeset in 10/12 Times by Pure Tech Corporation, India
Printed in Great Britain by the University Press, Cambridge

ISBN 0 412 48200 2

Apart from any fair dealing for the purposes of research or private study, or criticism or review, as permitted under the UK Copyright Designs and Patents Act, 1988, this publication may not be reproduced, stored, or transmitted, in any form or by any means, without the prior permission in writing of the publishers, or in the case of reprographic reproduction only in accordance with the terms of the licences issued by the Copyright Licensing Agency in the UK, or in accordance with the terms of licences issued by the appropriate Reproduction Rights Organization outside the UK. Enquiries concerning reproduction outside the terms stated here should be sent to the publishers at the London address printed on this page.
  The publisher makes no representation, express or implied, with regard to the accuracy of the information contained in this book and cannot accept any legal responsibility or liability for any errors or omissions that may be made.

A catalogue record for this book is available from the British Library

Library of Congress Catalog Card Number: 94–71063

∞ Printed on acid-free text paper, manufactured in accordance with ANSI/NISO Z39.48-1992 (Permanence of Paper).

Contents

Contents

Preface

This book deals with microwave electronics, that is to say those components of microwave circuits that generate, amplify, detect or modulate signals. It is based on a course given in the Electrical Engineering Department of Eindhoven University since 1985 and on about twenty years of experience in the microwave field.

Somewhat to my surprise I found that there were hardly any textbooks that addressed the specific properties and demands of microwave devices, including vacuum devices and their interactions with circuits. Numerous books exist on semiconductor electronic devices, dealing in an excellent way with the basic device physics, but being somewhat brief on typical microwave aspects. On the other hand there are also many books that concentrate on electromagnetic theory and passive circuits, treating devices without reference to the underlying physics. In between there are some entirely devoted to a particular device, for example, the GaAs MESFET. With regard to tubes the situation is even worse: books that treat the basic principles are usually quite old and modern books often concentrate on specific devices, like high power tubes.

So it seems that there is room for a book like this one. Its aim is to provide an elementary understanding of microwave electronic devices, both vacuum and semiconductor, on the one hand in relation to the basic physics underlying their operation and on the other in relation to their circuit applications.

The book starts, after a short review of the historical developments in Chapter 1, with microwave tubes, Chapter 2, partly for historical reasons but also because this gives an opportunity to introduce the concept of transit-time, which is the one principle pervading all microwave devices.

Since semiconductors nowadays dominate the field at low and intermediate power levels, a large part of the course is devoted to a discussion of semiconductor devices, in particular diodes, Chapter 5 and transistors, Chapter 6. To make the book reasonably self-contained, this part is preceded by a review of semiconductor physics, Chapter 3, which is kept short assuming that students have gone through a semiconductor course previously. There is also a chapter, included on modeling and noise, Chapter 4, two aspects that in my opinion are increasingly important but often tend to be neglected.

The final application-oriented part is formed by the discussion of amplifier and oscillator circuits, Chapters 9 and 10. Again, for completeness' sake, these chapters are preceded by a discussion of passive circuitry, especially microstrip. It is kept brief since there are many excellent books treating this subject in depth. In this chapter (7) attention is also paid to microwave measurement techniques. The book closes with a short discussion of monolithic microwave integrated circuits, Chapter 11.

I would like to thank the people of Chapman and Hall, who guided me through the difficult process of producing a coherent manuscript in a limited time, and to Bradford Smith, whose comments greatly helped to improve Chapter 2. Thanks are also due to my colleagues Hugo Heyker and Mathieu Kwaspen, who contributed significantly to Chapter 7. I also want to thank all members of the group of Electronic devices for creating a work atmosphere that allowed me to concentrate on this book and last but not least my wife Elly who had to miss me so many evening and weekend hours.

This book is not perfect and nobody knows this better than the author. Therefore I would strongly encourage my readers to send me their comments and suggestions for improvement.

Theo G. van de Roer
Eindhoven, 1994.

1
History

1.1 INTRODUCTION

The development of microwave electronics has always been driven by the demands of short wavelength radio (and later radar) systems. The history of microwaves started already with the first experiments of Heinrich Hertz around 1887. Hertz used a spark transmitter that produced signals in a very large frequency band and he selected from these a band around 420 MHz with an antenna that measured half a wavelength at this frequency. The receiving antenna had the same dimensions. Hertz also used parabolic mirrors and lenses of dielectric material. In 1893 Lord Kelvin gave a theoretical analysis of hollow waveguides and shortly afterwards Oliver Lodge demonstrated waveguides at frequencies of 1.5–4 GHz.

Following Marconi's impressive results in long-distance communication the development of radio moved towards the very large wavelengths, first because with these greater distances could be bridged and second because at low frequencies it was easier to generate narrow band (sinusoidal) signals. Only radio amateurs occupied themselves with the short waves.

In the 1930s a new interest arose for the ultrashort waves, mainly because of the rise of radar, an acronym of 'radio detection and ranging'. The first radars used wavelengths of a few meters but it was realized from the start that shorter wavelengths would yield better directivity of the radar beam as well as stronger reflections from small objects such as airplanes. The imminent threat of new wars in Europe and East Asia made it all the more urgent to develop the short wavelength field. Among others Schelkunoff, Southworth, Slater and Hansen did important theoretical and experimental work on waveguides, resonators and antennas. However, the main things needed were generators and detectors for these high frequencies.

1.2 ELECTRON TUBES

1.2.1 Early history

In 1919 H. Barkhausen and K. Kurtz [1] discovered high frequency oscillations in a triode transmitter tube. They ascribed these to the up and down movements of electrons in the region between cathode and grid. This led to the first practical high frequency sinusoidal generators. Two years later A. Hull developed the coaxial magnetron, as a low frequency amplifier and oscillator. Via the split-anode magnetron this led to the invention of the cavity magnetron (section 1.2.3).

W. E. Benham in 1931 [2] and I. Müller in 1933 [3] found that the transit time delay in vacuum diodes should lead to a diode impedance with a negative real part at frequencies corresponding to the inverse of the transit time, but this did not lead to practical applications. However, the principle was later applied with success in semiconductors (section 1.3).

1.2.2 The klystron

In 1937 the brothers Sigurd and Russell Varian succeeded in using the transit time of an electron beam for the amplification and generation of high frequencies. They were among the people who worried about how to detect incoming airplanes at night or in cloudy weather. Russell Varian tried out many ideas for centimeter wave generators, all unsuccessful, and while he was making a classification of his ideas to see whether something was overlooked he hit upon the idea of using a flat cavity invented by William Hansen to impart a velocity modulation on an electron beam. Further on in the beam this leads to a density modulation which they called 'bunching'. These electron bunches were used to induce an alternating current in a second cavity through which the beam passed. The tube was called the klystron [4].

As important as generating waves is to find ways to detect whether anything has been generated. For these early pioneers this was a big headache. The Varians solved it by deflecting the electron beam, after it had passed the second cavity, onto a fluorescent screen and judging from the size of the spot on the screen whether the tube was oscillating.

1.2.3 The magnetron

The invention of the klystron was published in 1939 and the news soon reached Britain where the preparations for war were much more intense than in the United States. At the University of Birmingham a laboratory was set up to do research on klystrons and related devices. Looking for a device that could produce high pulse power (necessary for radar systems) they came across the split-anode magnetron which had a very low efficiency but whose geometry allowed large input powers. In this magnetron the electrons start from a cylindrical cathode and are forced into spiral orbits by a magnetic field in the direction of the cathode axis. The split-anode magnetron was further developed by Klaas Posthumus of Philips Research Laboratories into a device with a four-segment anode [5]. This allows a rotating wave in a way that is analogous to the rotating field in an electric motor with two orthogonal windings.

The problem now was how to increase the conversion efficiency of the magnetron and Henry Boot and John Randall tried to do this by borrowing ideas from the klystron, especially the use of cavity resonators. They conceived the idea of taking the old Hertz resonator which was just a circular wire loop with a small gap. By stretching this it would become a cylindrical cavity with a slot coupling. A number of these would fit nicely around the cylindrical cathode of a magnetron and provide a circuit that would couple well with the circulating electrons. The inner diameter of the anode was made 12 mm, a wavelength of 10 cm being aimed at [6]. It worked right from the start and produced a lot of power. Neon lamps lit up at a distance from the tube and cigarettes could be lighted at the output lead. When the wavelength was measured it turned out to be 9.8 cm, remarkably close to the design goal. One of the first prototypes was shipped to Bell Laboratories in the US and soon development here began too. This was quickly followed by theoretical analysis by among other Hartree in the UK and Slater in the US.

In the meantime, completely isolated from these developments, in Japan a group led by Yoji Ito was researching on ways to improve the split-anode magnetron. This first resulted in an eight-segment anode in 1937, followed by a cavity magnetron in 1939 [7].

1.2.4 Traveling wave interactions

One of the members of the Birmingham team was Rudolf Kompfner, Austrian of birth and architect by training, a rather unlikely figure to work on military electronics for the Allies. His research was aimed at building an amplifier that would facilitate the detection of weak radar returns. His work on the klystron led him to the conclusion that its beam modulation efficiency was severely limited by the localized interaction of the electric field and the electron beam. So he looked for a distributed interaction and naturally came to the idea of making a wave move along with the electrons so their interaction would be extended over the full length of the beam. To this end the electromagnetic wave would have to be slowed down to a speed less than one-tenth of the velocity of light. He conceived the idea, genial by its simplicity, to use a coaxial line with a helically wound inner conductor, in the hope that the wave would propagate along the wire thus having a much reduced speed in the axial direction. Some simplified calculations he made convinced him that a wave could be amplified in such a structure. In his first experiments, in the autumn of 1943, he sent the beam between the inner and outer conductors of the coax, with the idea that there would be not only bunching but also sideways deflection, which could be made visible on a fluorescent screen, the same trick the Varians had applied to see whether anything was happening. This way he already reached an amplification of 8 dB at a wavelength of 9.1 cm [8]. However, he soon found out that it was much more efficient to send the beam through the inside of the helical wire, in which case he could even leave out the outer conductor altogether. This is the basic structure that is still used in many of today's traveling wave tubes or TWTs as they are called. Shortly afterwards the news reached John Pierce of Bell Laboratories, who within two months had formulated a detailed theory of the traveling wave tube and in his own experiments obtained an amplification of 23 dB at 4 GHz with a bandwidth of 800 MHz. From then on the development of TWTs moved at a much quicker pace in the US than in the UK.

Kompfner did not sit still after the invention of the TWT. The next thing he thought of was an electronically tunable oscillator. The TWT can be made to oscillate by feeding the output back to the input, but its oscillation frequency will depend only very weakly on the beam voltage. The reason is the same as the reason why it has such a large bandwidth, namely that the propagation velocity in the structure is nearly independent of frequency. So he looked for a dispersive structure and he found it in a waveguide that was folded to cross the electron beam at periodic intervals. In fact this looks more like a klystron than a TWT. Because of the dispersion of the waveguide (the changing of the phase velocity with frequency) he could now change the frequency of optimum interaction by changing the beam voltage. After some time Kompfner realized that this structure would work also with a wave traveling in the opposite direction to the beam, provided that the wave and the electrons arrived at the intersections with the right phase. This only happens for specific wavelengths. The final step was to realize that one needs a wave where the *phase* velocity is in the same direction as the beam but the *group* velocity is in the opposite direction. Then the feedback necessary for oscillation is provided by the beam transporting energy in one direction and the circuit in the other. Also, the ends of the tube must be reflectionless otherwise standing waves between the ends will determine the wavelength. He called this device the backward-wave oscillator (BWO).

This far Kompfner had come without being able to realize his ideas in practice. In 1948 he learned that one of Pierce's TWTs oscillated in a mode where the beam and the wave went in opposite directions but apparently Pierce had not realized the significance of this. On a visit to Bell Laboratories in the beginning of 1952 Kompfner asked to experiment with this tube and immediately found that the oscillation frequency was tunable by the beam voltage. At the Electron Tube Conference in June 1952 they learned that in France, guarded by military secrecy, B. Epsztein of Thomson-CSF had independently been developing a magnetron-like device which he called the carcinotron and which worked on the same principle as the BWO. So, as in the case of the magnetron, essentially the same invention was made at two different places independently.

In later years many other tubes have been invented that rely on the traveling wave principle. The most impressive of these is the *gyrotron* [9] in which an electron beam with relativistic velocity is applied. These tubes can deliver megawatts of power at frequencies of 100 GHz and higher. One of the things that one tends to do when looking back is to concentrate on the developments that have been successful and to forget about those that fell by the wayside. In the tube field one of these is the beam–plasma amplifier. The idea is that an ionized gas can sustain a variety of propagating modes, especially when it is put in a constant magnetic field. Some of these modes are slow and since it is easy to shoot a beam of electrons through a plasma it should be possible to obtain interaction between the beam and a slow wave. The nice thing about this idea is that no metallic slow wave structure is necessary and so it should make for simple constructions usable to very short wavelengths. The idea worked but unfortunately it worked too well. It turned out to be very difficult to control the many interactions that are possible in a beam–plasma system. Besides, gas plasmas have electron temperatures in the order of 10^4 K, and so they produce a lot of thermal noise. So many progress reports on the development of beam–plasma amplifiers ended with the remark that the tube could always be used as a noise generator and the whole research died out without having produced any practical applications.

1.3 SEMICONDUCTORS: PASSIVE DEVICES

1.3.1 The crystal detector

The history of semiconductors really began in 1877 when Ferdinand Braun discovered that crystals of metal sulfides showed non-linear *I–V* characteristics, in particular when the two contacts were of different size. In later years Braun occupied himself with radio transmission (he would eventually receive the Nobel prize together with Marconi) and in 1899 it occurred to him to use this effect for the detection of high frequency signals. The word semiconductor existed already but it was used for all substances with a conductivity in between those of the metals and the insulators, including wood, seawater and the human body but also stuff that we nowadays classify as semiconductors, such as the above-mentioned metal sulfides and selenium. These were normally not a subject of electrical research, the main reason being probably that their conductivities could vary enormously and in an unpredictable way.

In those days various kinds of detectors were used, including carbon microphones. In 1902 the American G. W. Pickard accidentally discovered that such a microphone could be used without batteries as a detector of radio waves [10]. He concluded that it must be the combination of a well-conducting material with a highly resistive one that did the trick and started a study of many years in the course of which he tested hundreds of different material combinations for their abilities as radio detectors. The best material turned out to be silicon contacted by a tungsten wire. Since silicon was difficult to procure in crystal form most people in that time used carborundum (SiC), because it was much more readily available, being fabricated in large quantities as grinding material. The detection properties of carborundum were not discovered by Pickard, but by H. C. Dunwoody, an associate of Lee de Forest, the discoverer of the vacuum diode.

The crystal detector was very popular for a while but soon it was replaced by the vacuum diode. The main reason for this was the lack of stability of the fragile construction in which a tungsten wire had to be pressed against a crystal. Also the variability of properties between different samples was not understood. People did not realize yet in that time that very small amounts of impurities could have a large influence on the electrical properties of semiconductors.

1.3.2 Crystal detectors for microwaves

However, with the advent of microwave technology around 1940 the crystal detector made a come-back. The vacuum diodes did not work at high frequencies because of their large capacitances and also because of transit time effects. In the meantime the knowledge of semiconductors had made great progress and Walter Schottky had provided a theoretical explanation of the rectifying properties of a metal–semiconductor junction. The great demand for detectors for radar equipment motivated an intensive research into the preparation of pure silicon and the chemically related germanium. This not only led to better detectors for radar systems but eventually also made the invention of the transistor possible. The availability of high quality germanium (which was an easier to handle material than silicon) stimulated the search for a solid state amplifier and, as is well known, by the end of 1947 Shockley, Bardeen and Brattain succeeded by discovering the point-contact transistor, which was quickly followed by the junction transistor. So one could say that it is from microwaves that semiconductor electronics was born. In its turn transistor technology would later give a strong impulse to the development of microwave semiconductors.

Improving the base material of detector was one thing; another that was absolutely necessary before it could be used in radar systems was a better mounting that provide a good electrical and a stable mechanical contact as well as low parasitic capacitance and inductance. In the mid-1930s a mount that conformed to these demands was developed by A. P. King and R. S. Ohl, associates of Southworth at Bell Laboratories. This mount, called the point-contact cartridge (a cross-section is shown in Fig. 8.1.2), was extensively used until well into the 1960s and is probably still in use at some places.

1.3.3 Non-linear reactances

The sensitivity of a radar system can be increased by increasing the transmitter power but also by increasing the receiver sensitivity. One way to achieve this is to replace direct detection by heterodyne downconversion followed by intermediate frequency application. So almost immediately after the crystal detector was made fit for microwave frequencies it was also used as a mixer.

The sensitivity of a receiver is determined by its noise figure, a quantity that was defined by H. T. Friis who developed the noise theory of receivers in the 1940s. Since a diode mixer always has conversion loss its noise figure cannot be very small. The only way to achieve a low system noise figure is to precede the mixer with a low noise amplifier. It was known long ago that amplification could be obtained from a non-linear element which is pumped by a high frequency signal and Jack Manley had in 1934 formulated some basic relations for non-linear devices driven by signals of different frequencies, but the significance of this was not appreciated. In 1948 Aldert van der Ziel predicted that low noise amplification could be obtained from a non-linear capacitance but still nothing happened.

In 1952 Harrison Rowe was led by some experiments at Bell Laboratories to re-examine the old Manley theory and together they expanded it into what is now known as the Manley–Rowe relations. In 1956 Marion Hines of the Western Electric plant at Allentown studied non-linear capacacitance diodes and coined the name varactor for these devices. He also obtained amplification and oscillation at about 6 GHz from a varactor driven at 12 GHz. This was the first parametric amplifier and it was quickly followed by others. After that varactors were also used for frequency multiplication. These were the first semiconductor amplifiers and signal generators for the microwave range [11].

1.4 SEMICONDUCTORS: ACTIVE DIODES

1.4.1 Transit time in semiconductors

It was going to take until the late 1960s before silicon technology had advanced so far that microwave frequencies were reached with Si transistors. In the meantime one of the inventors of the transistor, William

Shockley, who before the second world war had worked on tubes and was well aware of transit-time effects, proposed in 1954 to employ transit-time effects in semiconductor diodes for the generation of microwaves. The structure he proposed was a simple p–n–p diode and indeed, in the 1970s, it was realized under the name Baritt diode (Chapter 4).

In 1958 Shockley's coworker W.T. Read proposed to combine the transit-time delay with injection of charge carriers from an avalanching p–n junction. The realization of such a device put high demands on the technology: layer widths of a few microns had to be realized with great uniformity. Many diodes were burned out because of non-uniform avalanching. So it took until 1965 before a working Read structure was made for the first time [12]. It turned out to be an efficient microwave generator at frequencies up to 300 GHz and became known as the IMPATT diode (from 'impact avalanche transit time').

1.4.2 Intrinsic negative resistance

The discovery of the transistor spurred worldwide activity on semiconductor research and fabrication. Zener had proposed quantum mechanical interband tunneling as a mechanism of electrical breakdown in dielectrics at high electric fields and this model had also been applied to explain the reverse breakdown of p–n junctions. In 1958 Leo Esaki, working at Sony, was researching how high he could dope p–n junctions before they lost their injecting properties as an emitter–base junction. Since he wanted to write a PhD thesis he kept an attentive eye on the scientifically more interesting tunnel effects. Increasing the doping to over $10^{19}/cm^3$ on both sides he found that a region of negative differential resistance occurred in the forward direction [13] and it did not take him long to decide that this indeed was the result of tunneling. This discovery of the tunnel diode brought him the Nobel prize in physics. A diode with negative differential resistance can be used as an oscillator or amplifier and the tunnel diode was the first active component of this kind that could be used up to the microwave range.

About 20 years later Esaki did it again [14]. In the meantime he had moved to the IBM Thomas J. Watson Research Laboratory and became interested in so-called superlattices, that is thin alternating layers of different semiconductor materials. When one takes a single layer of a wide bandgap material sandwiched between two layers of narrow gap material electrons cannot cross from one side to the other because the high bandgap material acts as a potential barrier. However, when this barrier layer is very thin, 10 nm or less, electrons can tunnel through it. Now when one takes two barrier layers, separated by an equally thin layer of narrow gap material, resonant tunneling can occur, that is the transmitted and reflected wavefunctions of the electron can interfere constructively. This way at a certain bias voltage electrons can tunnel very easily while at other biases they cannot. This also leads to an *I–V* characteristic with negative differential resistance.

1.4.3 Negative differential mobility

Another line of thought that was followed by several people was to look for intrinsic material properties that could lead to an *I–V* characteristic with negative slope. It was known already soon after the discovery of the transistor that the carrier velocities in semiconductors were proportional to the electric field only at comparatively low electric fields. At higher field strengths the velocity increased less than proportionally and finally it did not rise any more but saturated at a value of about 10^5 m/s. So it was conceivable that in a suitable material the velocity would go down again at high fields. Already in 1948 Herbert Kroemer had suggested that one could use the fact that at high electron or hole energies the energy bands are no longer parabolic, meaning that the energy increases less than quadratically with momentum. This can also be phrased differently by saying that the effective mass increases as the carriers become more energetic. Indeed, it can be proved that the effective mass must go to infinity at some point in the conduction band or valence band so the velocity of these electrons or holes should go to zero. It turned out, however, that the energies at which this happens are so high that other effects such as impact ionization prevail before one could see any effect.

In 1961, however, Brian Ridley and Tom Watkins [15] pointed out that the conduction bands of most semiconductors have different branches in different crystal directions and that there may be higher-lying minima. If these are not too far above the lowest minimum where the electrons normally reside (say a few tenths of an electronvolt) and if moreover the effective mass in these minima is much greater than in the lowest one, one could expect that at high electric fields the electrons on the average would become so energetic ('hot electrons') that a substantial part of them could transfer to a higher minimum which would lead to a substantial increase of the average effective mass. This in turn could lead to a decrease of the average velocity with increasing electric field. Ridley and Watkins did not go deeply into the question of which material would be most suitable for this electron transfer mechanism, but Cyril Hilsum [16] was quick to point out that gallium arsenide had a favorable band structure for this effect.

Both groups did not pursue the matter experimentally but, not knowing of their theories, Ian Gunn at the same above-mentioned IBM laboratory was studying the behavior of hot electrons by measuring their noise properties. The idea was that when an electron population acquires more energy this can be expressed as a rise in the electron temperature and this should show up as an increase of the thermal noise. He had done noise measurements on germanium and silicon before in his home country Britain and after he moved to the US he continued this work and also took into consideration some of the more exotic semiconductors such as GaAs and indium phosphide, of which the first samples were becoming available at that time which was 1963. To his surprise [17], when he increased the voltage on a GaAs sample, at a certain bias the noise suddenly increased enormously to values that were unexplainable as thermal noise. When he did some more experiments he discovered that in some samples the 'noise' was not random at all but consisted of regular current oscillations, the period of which was related to the length of the sample indicating a transit time effect. This soon became known as the Gunn effect. This caused great excitement in the semiconductor community and many people started repeating the experiments. Still, it was not realized immediately that this was an example of the effect predicted by Ridley, Watkins and Hilsum. Several other explanations were put forward. It was finally the aforementioned Kroemer who conclusively showed that the Gunn effect indeed was caused by the transfer of electrons to higher conduction band minima.

Now it became feasible to build semiconductor microwave oscillators. The tunnel diode has an inherent limitation in the voltage range over which a negative differential resistance exists so as an oscillator it can only produce microwatts of power. With the Gunn effect this changed drastically. It put GaAs in the picture as a technical semiconductor material besides Si and Ge. A modest industry developed that concentrated on the technology of GaAs and the manufacture of Gunn diode oscillators. The development of GaAs technology was an important step forward for the microwave community because it made possible the next big development, the GaAs MESFET.

1.5 MICROWAVE TRANSISTORS

1.5.1 Gallium arsenide MESFETS

One of the interesting properties of GaAs is that the electron mobility is much higher is than that of Si. In 1966 Carver Mead made the first GaAs MESFET's which were not microwave devices because of their layout. However, he pointed out that the limiting frequencies of these transistors should be considerably higher than similar devices made using silicon technology. In the late 1960s technology had advanced so far that S. Middelhoek of the Zurich IBM laboratory could make FETs with gate-lengths of 1 µm. In 1969 the first GaAs field effect transistors with Schottky gates, called MESFETs (from metal semiconductor field effect transistors), were made with these short gates and they had already cutoff frequencies of about 10 GHz. A comparison by Baechtold *et al.* showed that MESFETS on GaAs had about four times higher cutoff frequencies than Si devices with the same layout [18] and from that moment the GaAs MESFET was on its way to becoming the workhorse of microwave technology. It is now the most used microwave component. The frequency region has in the meantime been extended to 100 GHz and noise figures are pleasantly low too.

From then on developments went at an accelerating pace. Already in 1974 the first GaAs MESFET integrated circuit was demonstrated by Liechti and coworkers [19]. Even more important was the development of new epitaxial growth techniques: molecular beam epitaxy (MBE) was pioneered by Esaki at IBM and Alfred Cho at Bell Laboratories, and metal–organic vapor phase epitaxy (MOVPE) by a group at Thomson-CSF in France. Previously liquid phase epitaxy (LPE) was the main growth technique. It produced layers of good quality but with low uniformity and on small-area wafers only. The new techniques made it possible to grow highly uniform layers with sharp boundaries on large areas.

1.5.2 New technologies, new transistors

The invention of the laser diode in 1961 also did much to stimulate the development of GaAs technology and when in 1970 the first heterojunction lasers were made it became important to develop heterostructure technology, e.g. the growth of Al_xGa_{1-x} As on GaAs. The new techniques MBE and MOVPE made it possible to make heterojunctions with very sharp transitions, even as sharp as one atomic layer. The invention of the double barrier resonant tunneling diode, described above, was made possible by MBE. Another important development that was made possible by MBE and MOVPE is the invention of improved field effect transistor structures. The most conspicuous of these is the HEMT (high electron mobility transistor) in which the channel is formed by an AlGaAs–GaAs heterojunction. This device was proposed by Ray Dingle of Bell Laboratories [20] and first realized by a Fujitsu group [21]. It has turned out to be a transistor with a higher frequency limit than the MESFET and very good noise properties. Besides a proliferation of various FET structures (Chapter 5) the new technologies also made possible the heterojunction bipolar transistor based on AlGaAs–GaAs and other III–V heterojunctions.

Although it seemed for a while that the III–V materials had taken exclusive possession of the microwave field, silicon is coming back into the race by also adopting MBE technology. Already around 1970 the Si technology had advanced so far that silicon bipolars reached the gigahertz region. After that the development did not progress very much any more and the limiting frequencies of Si devices seemed to be stuck at about 10 GHz. MBE technology made it possible, however, to fabricate heterojunctions of Si and the alloy Si_xGe_{1-x}. This could be used to make heterojunction bipolar transistors with a very thin SiGe base which have excellent high frequency properties. Cutoff frequencies of 75 GHz have already been reached. This is yet another example of the cross-fertilization of different technologies that started with the crystal detector.

Finally it should be mentioned that GaAs technology offers the possibility of integrating transistors and passive microwave components thanks to the fact that the substrate of GaAs can be made semi-insulating. This leads to so-called MMICs (microwave monolithic integrated circuits). Also, digital ICs with gigahertz clock speeds become possible. Here too silicon is not outdone yet. Si MMICs and fast digital circuits are under development and compete strongly with the GaAs technology.

1.6 MICROWAVE INTEGRATED CIRCUITS

1.6.1 Microstrip lines

Microstrip line is the name given to a waveguide consisting of a thin dielectric layer covered with a metal ground plane on one side and a narrow metal strip on the top side. It was conceived around 1950 as a lightweight and flexible transmission medium intended especially for airborne applications. A review of its development has been given in ref. 22.

The most interesting feature of microstrip is that in the top metal all kinds of patterns can be made by the use of well-known printed circuit techniques. So, it is very easy to integrate transmission lines, couplers, capacitors, inductors, resonators and filters on the same substrate. Active devices can be mounted as bare chips or packed in stripline packages. This produces a low cost manufacturing technique for microwave circuits and so its applicability is much wider than the above-mentioned one. For these hybrid circuits the name microwave integrated circuit (MIC) has been coined. By 1970 the theory of microstrip transmission line and components was well developed and so were the fabrication techniques. Then with perfect timing the GaAs MESFET appeared on the scene, a device which is much easier to mount in microstrip than in coax or waveguide. The happy marriage of these two now dominates the microwave industry, at least at frequencies up to about 40 GHz. Waveguide or coaxial systems are preferred only when low losses, high accuracy (as in measurement systems) or very high power are essential. The next logical step was to take semi-insulating GaAs as the dielectric substrate and to integrate passive components and active devices monolithically.

1.6.2 Monolithic ICs

In the early 1960s T. Hyltin at Texas Instruments was already developing MMICs on silicon, using a high-ohmic p-type substrate. These were not successful because during processing the substrates tended to convert to n type. In 1964 Uhlir proposed to use semi-insulating GaAs substrates with an epitaxial germanium layer to make the devices in. In 1968 the TI group made the first GaAs MMIC but the field really started going when Charles Liechti *et al.* at Hewlett-Packard made the first digital IC using GaAs MESFETs and in 1976 Ray Pengelly and Jim Turner of Plessey fabricated the first monolithic X-band MESFET amplifier. Since then great progress has been made both in analog and digital MMICs. Notable achievements are complete integrated receivers, wide band amplifiers (2–20 GHz) and microprocessors.

REFERENCES

1. Barkhausen, H. and Kurtz, K. (1920) Die Kürzesten, mit Vakuumröhren herstellbaren Wellen. **21**, 1–6.
2. Benham, W. E. (1928, 1931) Theory of internal action of thermionic systems at moderately high frequencies. *Philos. Mag.*, **5**, 641–62; **11**, 457–517.
3. Müller, I. (1933, 1934) Elektronenschwingungen in Hochvakuum. *Hochfrequenztech. Elektroakust.* **14**, 156–67; **43**, 195.
4. Ginzton, E. L. (1975) The $100 idea. *IEEE Spectrum*, **12**, 30–9.
5. Burns, R. W. (1988) The background to the development of the cavity magnetron, in *Radar Development to 1945*, Peregrinus.
6. Boot, H. A. H. and Randall, J. T. (1976) Historical notes on the cavity magnetron. *IEEE Trans. Electron Devices*, **23**, 724–9.
7. Nakajima, S. (1992) Japanese radar in WW II. *Radioscientist (URSI Mag.)*, **3**, 33–7.
8. Kompfner, R. (1976) The invention of traveling wave tubes. *IEEE Trans. Electron Devices*, **23**, 730–8.
9. Flyagin, V. A. and Gaponov, A. V. (1977) The gyrotron. *IEEE Trans. Microwave Theory Tech.*, **25**, 514–21.
10. Douglas, A. (1981) The crystal detector. *IEEE Spectrum*, **18**, 64–67.
11. Hines, M. E. (1984) The virtues of non-linearity: detection, frequency conversion, parametric amplification and harmonic generation. *IEEE Trans. Microwave Theory Tech.* **32**, 1097–104.
12. de Loach B. C. (1976) The Impatt story. *IEEE Trans. Electron Devices*, **23**, 657–60.
13. Esaki, L. (1976) Discovery of the tunnel diode. *IEEE Trans. Electron Devices*, **23**, 644–7.
14. Chang, L. L., Esaki, L. and Tsu, R. (1974) Resonant tunneling in semiconductor double barriers. *Appl. Phys. Lett.*, **24**, 593.
15. Ridley, B. K. and Watkins, T. B. (1961) The possibility of negative resistance effects in semiconductors. *Proc. Phys. Soc.*, **78**, 293–304.
16. Hilsum, C. (1978) Historical background of hot-electron physics (a look over the shoulder). *Solid State Electron.*, **21**, 5–8.
17. Gunn, J. B. (1976) The discovery of microwave oscillations in gallium arsenide. *IEEE Trans. Electron Devices*, **23**, 705–13.
18. Baechtold, W., Daetwyler, K., Forster, T., Mohr, T. O., Walter, W., Wolf P. (1973) Si and GaAs 0.5 μm-gate Schottky-barrier field-efeect transistors. *Electron. Lett.*, **9**, 232–4.
19. Van Tuyl, R. L. and Liechti, C. A. (1974) High-speed integrated logic with GaAs Mesfet's. *IEEE J. Solid-State Circuits*, **9**, 269–76.
20. Dingle, R., Störmer, H. L., Gossard, A. C. and Wiegmann, W. (1980) Electronic properties of the GaAs–AlGaAs interface with applications to multi-interface heterojunction superlattices. *Surf. Sci.*, **98**, 90.

21. Mimura, T., Yoshin, K., Hiyamizu, S. and Abe, M. (1981) High electron mobility transistor logic. *Jpn. J. Appl. Phys.*, **20**, L598.
22. Howe, H. (1984) Microwave integrated circuits – an historical perspective. *IEEE Trans. Microwave Theory Tech.*, **32**, 991–6.

2

Vacuum electron devices

2.1 INTRODUCTION

From its invention in 1906 until about 1950 the vacuum tube has been the key element of electronics. Then it started to be replaced by semiconductor devices. From our everyday experience with consumer electronics we are inclined to conclude that the days of the vacuum tube are over. Nothing is less true, however. Many vacuum devices have been replaced by cheaper, smaller, more reliable and less power-consuming semiconductor devices. The low power consumption of semiconductors, which is essential for applications such as computers and electronic telephone exchanges, becomes their weak point in applications where high power is needed. Then tubes are difficult or downright impossible to replace by semiconductors. Also, at the very high frequencies, say above 300 GHz, semiconductor

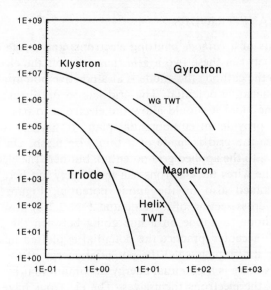

Figure 2.1 Power against frequency for microwave tubes.

devices are scarce and tubes will continue to be used for quite some time to come. Figure 2.1 gives an impression of the power output that can be obtained from microwave tubes. In short, despite the enormous development of semiconductor technology, it is much too early to declare the vacuum tube a fossil of a past period. In fact, developments in the tube field have never stopped and still continue. Also, as is shown in the last section of this chapter, the tube field is able to generate new development by borrowing from semiconductor technology.

For a time the demand for higher frequencies could be satisfied by shrinking the dimensions of triodes and more carefully designing the connecting leads, thus reducing the transit time as well as the parasitic inductances and capacitances. However, this soon reached its limits: fabrication tolerances and the heat produced by the cathode prevent the cathode–grid distance from being reduced indefinitely. It became necessary to invent other ways of amplification or oscillation and thanks to the creative imagination of many people this has been achieved. Between the invention of the klystron in 1937 and that of the gyrotron around 1970 many ingenious ideas have been applied to overcome the limits of frequency and power.

In the following sections a treatment of the most important vacuum devices will be given aimed at understanding their principles rather than a detailed comprehensive treatment. In Appendix 2.A some theorems from electromagnetic theory are presented which are useful in the analysis of vacuum devices (but also of semiconductors) and which can be applied to explain why triodes let us down at high frequencies.

2.2 MICROWAVE TRIODES

A triode consists of a *cathode* emitting electrons, and *anode* collecting the electrons, and between these two a *grid* that controls the electron current. With respect to the cathode the anode is at a positive potential and the grid usually is at a negative potential. The grid has to fulfill rather conflicting demands: on one hand it should allow the electrons to pass freely; on the other it has to provide an equipotential plane transverse to the electron flow. In practice the grid is made of a thin wire mesh. On the wires the potential is equal to the applied grid potential, but here the electrons cannot pass. Between the wires the electrons can pass freely but here the potential is partly determined also by the anode potential. Figure 2.2(a) shows schematically a cross-section of a triode and Fig. 2.2(b) how the potential varies from cathode to anode on a line going between the grid wires. In accordance with accepted practice the potential is plotted upside down, so that it gives (on a different scale) the potential energy of the electrons. As the figure shows there is a potential energy maximum that is created by the space charge of the electrons themselves. The electrons have to overcome this potential barrier to reach the anode. When the anode voltage is

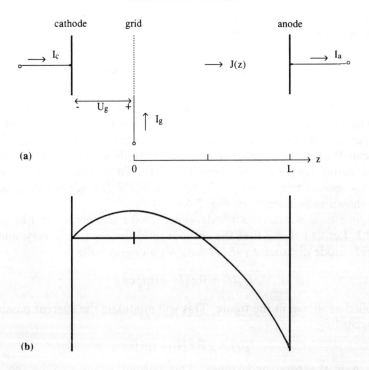

cathode   grid   anode

I_c

$\longrightarrow$ J(z)

I_a

U_g

I_g

(a)

0   L   z

(b)

Figure 2.2 (a) Schematic picture of a triode and (b) potential diagram.

increased this maximum is reduced and eventually disappears. This is a situation one should avoid for the following reason: the electron current emitted from the anode fluctuates owing to various causes. When the emitted current increases the space charge increases also with the effect that the potential barrier becomes higher. This makes it more difficult for the electrons to reach the anode, so a negative feedback occurs that attenuates the effect of the emission fluctuations on the anode current. Thus it is important to operate triodes (as well as other vacuum electron devices) in the so-called *space-charge limited regime* where a potential maximum in front of the cathode exists. Otherwise electron tubes would be very noisy.

The vacuum triode has an inherent frequency limit for two reasons: in the first place the interelectrode capacitances combined with the lead inductances form a low-pass filter and secondly there is the finite transit time of the electrons between grid and anode. When this transit time is greater than the period of the signal, the electrons which passed through the grid in the positive half of the period have not yet reached the anode when the grid is already in the negative half of the period. In this case *charge density waves* propagate between grid and anode and the anode current does not truly follow the grid voltage. The effects of lead inductances can be illustrated by the following argument. Consider the well-known formula for the inductance of a wire over a ground plane:

$$L \text{ (H/m)} = \frac{\mu_r \mu_0}{2\pi} \cosh^{-1}(h/a) \tag{2.1}$$

where a is the wire diameter and h the height over the ground plane. A wire of 0.5 mm diameter and $h = 2$ mm has, with $\mu_r = 1$, an inductance of about 4 nH per centimeter of length, which at 1 GHz gives an impedance of about 25 Ω/cm. With a capacitance to ground of 10 pF we have a low-pass filter with a cutoff frequency of about 0.3 GHz. To minimize these parasitic elements special triode designs have been made for microwaves, such as those shown subsequently in Fig. 2.4.

Suppose now we have a triode which looks schematically like that in Fig. 2.2. Let us assume that the cathode–grid distance L_1 is very small but the grid–anode distance L_2-L_1 is not. An a.c. grid voltage

$$U_g(t) = \text{Re}\,[U_1 \exp(j\omega t)] \tag{2.2}$$

is applied as shown in the figure. This will modulate the current passing the grid according to

$$I_g(t) = g\,\text{Re}\,[U_1 \exp(j\omega t)] \tag{2.3}$$

where g is the *transconductance*. This current is injected in the anode space. For simplicity we now assume that the anode space is field free so that the electrons drift here with constant velocity v_0. At $z = 0$ the alternating current is given by equation (2.3). The electrons that pass at a point z were a time z/v_0 earlier at $z = 0$, so the alternating current at the point z is

$$I_1(z, t) = I_1(0, t - z/v) = -g\,\text{Re}\,\{U_1 \exp[j\omega(t - z/v_0)]\}. \tag{2.4}$$

The a.c. component of the anode current (exclusive the capacitive current) is according to the theorem of Ramo (see Appendix 2.A) an average of this modulation over the grid–anode distance:

$$I_a(t) = \frac{1}{L}\int_0^L I_1(z, t)\,dz = -g\,\text{Re}\left[U_1 \exp(j\omega t)\,\frac{\exp(-j\omega L/v_0) - 1}{-j\omega L/v_0}\right]. \tag{2.5}$$

Here we encounter the *transit-time function* $[1 - \exp(-j\phi)]/j\phi$ in which $\phi = \omega L/v_0$ is called the *transit angle*. It often plays an important role in transit-time problems. We will meet it again in the discussion of semiconductor transit-time diodes in Chapter 5. The transit-time function in the complex plane is shown in Fig. 2.3. At low frequencies it is equal to unity and the anode current is a faithful image of the grid voltage. At higher frequencies, however, where $\omega L/v_0$ is of the order of unity, in other words

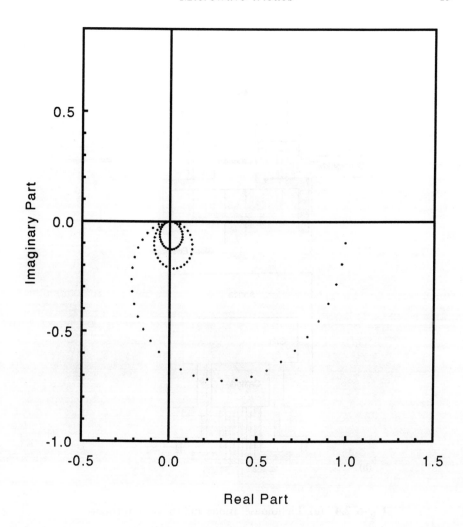

Figure 2.3 The transit-time function.

the transit time is comparable with the signal period, the amplitude decreases and a phase delay is introduced. This causes the response of the anode current, when the frequency becomes higher, to decrease rapidly. This is the second effect that causes the triode to have a finite cutoff frequency.

This analysis is of course very simplified. In a real triode there is a potential difference between the anode and the grid which accelerates the electrons. This makes the transit-time function more complicated but does not change its character. Also, a transit-time effect occurs in the cathode–grid

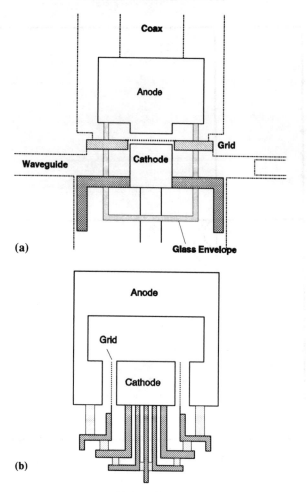

Figure 2.4 (a) 'Lighthouse' triode and (b) coaxial triode.

space. However, it is hoped that this analysis suffices to show the effect of transit time on triode performance.

The only remedy against both the parasitic reactances and the transit-time effects is to make the dimensions of the triode smaller, but there are of course limits to this. In practice the applicability of triodes does not go beyond a few gigahertz. Therefore ways have been sought to escape from the limitations imposed by transit time and instead put it to work. The first successful solution for this task was the *klystron*, to be treated in the next section. Still, in the low-frequency microwave range triodes are very much in use when high powers are required. Figure 2.4 shows two designs of microwave triodes. That shown in Fig. 2.4(a) is the so-called lighthouse triode, the name being derived from its typical shape. The design is aimed at keeping the cathode–grid and grid–anode distances small and providing a low-reactance mounting in coaxial or waveguide circuits. The tube in Fig. 2.4(b) is a coaxial design with the same objectives. This tube has a much greater power capability owing to its larger cathode area. Thanks to its coaxial construction it fits quite naturally into coaxial line circuits. Tubes of this type can deliver powers of many kilowatts and are used in TV transmitters [1].

2.3 THE KLYSTRON

2.3.1 Principle

Suppose that we have an electron gun that emits a parallel beam (Fig. 2.5(a)). We can modulate the velocity of this beam by sending it through holes in two parallel plates 1 and 2, between which an a.c. voltage exists. To provide a uniform axial electric field between the plates the holes are provided with fine grids. The a.c. field produces alternating accelerations and decelerations on the beam. After the plates the electrons enter the so-called *drift space*, where the fast electrons gradually overtake the slow ones. We can represent this graphically in an *Applegate diagram* (Fig. 2.5(b)).

Since klystrons are intended for operation at microwave frequencies it is not practical to use plates connected by wires. Instead one uses microwave resonators or *cavities*. In klystrons the cavity has a typical form that is narrow in the middle and wider at the edge (Fig. 2.6). It was invented by W. W. Hansen of Stanford University in 1937 as the high frequency equivalent of an LC circuit. Because of the likeness of its cross-section with

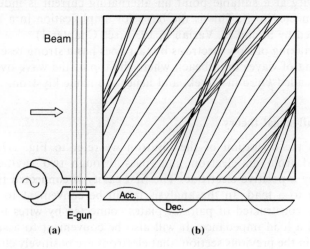

Figure 2.5 (a) Klystron and (b) Applegate diagram.

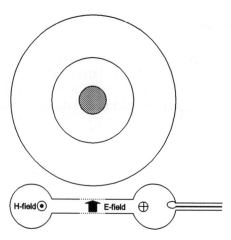

Figure 2.6 Rhumbatron top view and cross-section.

a set of rhumba balls he called it the *rhumbatron*. The microwave signal is usually coupled in and out via a small loop at the end of a coaxial line. The value of the equivalent load impedance depends on the load at the end of the coaxial line and on the strength of the coupling between the electromagnetic field and the wire loop. Since the loop couples mainly with the magnetic field the coupling will be at a minimum when the plane of the loop is parallel to the magnetic field lines and at a maximum when both are orthogonal. The velocity modulation leads to a strong density modulation as can be seen from the density of lines in the diagram and if we place a second cavity at a suitable point an alternating current is induced in the circuit connected to it. This is the basis for amplification in a klystron. It was invented by S. and R. Varian in 1937 (see Chapter 1).

The overtaking of slow electrons by fast ones has a strong resemblance to the breaking of waves on a beach where the top of the wave overtakes the base. It is called *klyzo* in Greek, and hence the name klystron.

2.3.2 Small-signal theory

To discuss the operation of the klystron we refer to Fig. 2.7. Electrons are emitted from an electron gun, pass through two cavities and are collected by a positive electrode. The first cavity is connected to a source, the second to a load. In the analysis it will be convenient to think of a klystron as constructed of pairs of plates connected by wires to a voltage source and a load impedance. It will also be convenient to assume, as we have done in the previous section, that electrons are positively charged. This saves quite a few minus signs and the results are the same.

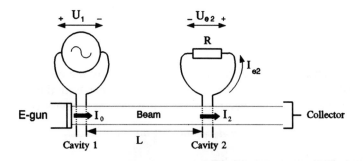

Figure 2.7 Klystron model.

Suppose that in the electron gun the electrons are accelerated by a d.c. voltage U_0. In the rest of the tube up to the second cavity there is no d.c. field. They then arrive at cavity 1 with a velocity

$$v_0 = (2qU_0/m)^{1/2}.$$

Suppose the density is n_0; then the d.c. beam current is

$$I_0 = qn_0v_0A$$

where A is the cross-sectional area of the beam. Between the first two plates an a.c. voltage

$$U_\sim = U_1 \sin \omega t \tag{2.6}$$

is applied. We assume that the distance d_1 between these plates is so small that the time d_1/v_0 the electrons spend between the plates is short compared with the period $2\pi/\omega$ of the a.c. signal. This is called the *quasi-static approximation* because for the electrons it is as if they were in a static electric field. Also, we will assume whenever it is convenient that the a.c. amplitude U_1 is small compared with the d.c. accelerating voltage U_0. This is called the *small-signal approximation*. It enables us to split the variables in d.c. and a.c. quantities and to discard products of two a.c. quantities since they are of second-order magnitude. The d.c. quantities, in the present case U_0, v_0 and n_0, are supposed to be known.

The electrons are now accelerated in the positive half-period of the a.c. signal and decelerated in the negative half-period. Introducing the *modulation index* $M = U_1/U_0$ the velocity directly behind the cavity can be written as

$$v = v_0\left(1 + M \sin \omega t\right)^{1/2} \approx v_0\left(1 + \tfrac{1}{2} M \sin \omega t\right), \quad M \ll 1, \tag{2.7}$$

where t is the time of arrival at cavity 1. Since the current is conserved the density has undergone the opposite modulation (the accelerated electrons are pulled a little further apart and the decelerated ones are pushed together):

$$n = n_0\left(1 + M \sin \omega t\right)^{-1/2} \approx n_0\left(1 - \tfrac{1}{2} M \sin \omega t\right), \qquad M \ll 1. \qquad (2.8)$$

It will be instructive to calculate the power carried by the beam. It is found by multiplying the number of electrons that pass per unit time at a given place with the energy per electron. In front of the first cavity this gives

$$P_0 = n_0 v_0 \tfrac{1}{2} m v_0^2 = \tfrac{1}{2} m n_0 v_0^3. \qquad (2.9a)$$

Emerging from the cavity we have

$$P_2 = \tfrac{1}{2} m n_0 \left(1 + \tfrac{1}{2} M \sin \omega t\right)^{-1/2} v_0^3 \left(1 + M \sin \omega t\right)^{3/2} = P_0\left(1 + M \sin \omega t\right). \qquad (2.9b)$$

The time average of this is equal to P_0 so apparently it has cost no energy to modulate the beam. This is a consequence of our quasi-static approximation where we have neglected the transit time between the plates. In practice there will always be some dissipation, first because the transit time in the modulating cavity is not zero and second because there are always losses in the circuit.

With this modulated velocity the electrons enter the space between the two cavities, which is called the *drift space*. The time that the electrons need to arrive at the second cavity is (if L is the length of the drift space)

$$\tau = t + \frac{L}{v(t)} \approx t + \tau_0\left(1 - \tfrac{1}{2} M \sin \omega t\right), \qquad \tau_0 = \frac{L}{v_0}, \qquad (2.10)$$

where again the small-signal approximation has been applied. To calculate the current at cavity 2 we have to keep in mind that conservation of charge holds, i.e. $dQ = I_1 \, dt = I_2 \, d\tau$, which yields for I_2 after differentiation of equation (2.10)

$$I_2 = \frac{I_0}{1 - \omega \tau_0 M/2 \cos \omega t} \approx I_0(1 + X \cos \omega t)$$

$$\approx I_0[1 + X \cos \omega(\tau - \tau_0)] \qquad (2.11)$$

where $\omega \tau_0$ is called the *transit angle* and $X = \omega \tau_0 M/2$ the *bunching parameter*. Note that now the current is modulated, and that the modulation depth can be greater than that of the original velocity modulation. This is because the transit angle can be made as large as we please by increasing the length of the drift space L. The velocity modulation is the same as before but the density modulation has increased enormously because the faster electrons have gained on the slower ones.

The beam current now passes through cavity 2 and induces a current in the external circuit connected to these plates. Suppose for simplicity that this circuit consists of a resistor R. The induced current can be calculated with the help of the Ramo–Shockley theorem mentioned in the previous section and discussed in Appendix 2.A. It is the spatial average of the beam current flowing between the plates. There are also currents induced by the drift space and the space between cavity 2 and the anode but these can be neglected: they are carried off to ground and do not contribute to the cavity current. (This does not hold for the direct current. Here the outer spaces do contribute induced currents and these compensate exactly the direct current induced between the plates. So there is no net direct current coming out of the plates.) Therefore it is a reasonable approximation to assume that the induced alternating current is equal to the beam current between the plates. Over the external resistor R now a voltage develops equal to

$$U_{e2} \approx I_0 R X \cos \omega(\tau - \tau_0) \tag{2.12}$$

with the polarity shown. It produces a field between the plates that decelerates the electrons when I_{e2} is positive and accelerates them when it is negative. The velocity modulation induced by this field can be calculated as

$$v_{2i} = \frac{q U_{e2}}{m d_2} \frac{d_2}{v_0} = -\frac{v_0}{2U_0} I_0 X R \cos \omega(\tau - \tau_0). \tag{2.13}$$

This can be much greater than the original velocity modulation given by equation (2.6). Equally important is that it is in antiphase with the beam current given by equation (2.11). This means that more electrons are decelerated than accelerated so that kinetic power is taken from the beam and delivered to the load resistor R. Since in our simplified analysis no power was necessary to modulate the beam we have infinite power amplification. The voltage amplification is finite and equal to

$$G_v = \frac{I_0 X R}{U_1} = \frac{\omega L}{v_0} \frac{I_0 R}{2 U_0}. \tag{2.14}$$

In reality we of course have a resonant cavity which presents a more complicated load than a simple resistor. Seen from the place where the electrons cross the plates this cavity can be represented approximately by a parallel LCR resonant circuit, whose impedance we can write as

$$Z_{res} = \frac{R}{1 + j Q(\omega/\omega_0 - \omega_0/\omega)} \tag{2.15}$$

where $Q = R/\omega_0 L$ and $\omega_0 = 1/(LC)^{1/2}$. At the resonant frequency ω_0, Z is real and has the value R, so that for this frequency the expressions derived above remain valid. Outside this frequency Z will decrease in magnitude and a phase shift is introduced so that the conversion of kinetic beam energy into electric power will be less efficient. The operating range of a klystron thus covers only a small frequency band around f_0.

If we do not want to use the small-signal approximation we have to keep the exact form of equation (2.6). In Appendix 2.B it is derived that in this case equation (2.11) is replaced by

$$I_2 = I_0\left\{1 + 2\sum_1^\infty J_n(nX)\cos[n\omega(\tau - \tau_0)]\right\} \tag{2.16}$$

For small X, $J_1(X)$ can be approximated by $\frac{1}{2}X$ and the higher-order terms are of order X^2 so that equation (2.11) is recovered. When X becomes higher the Bessel function increases less than linearly with X and eventually starts to decrease again (Fig. 2.B.2). This means that the amplification saturates when U_1 is increased to large values. Also, in this range of operation the harmonic content of the signal increases enormously, resulting in non-linear distortion. On the other hand, this points the way to another application of the klystron, namely as a frequency multiplier.

2.3.3 Multicavity and extended-interaction klystrons

The *two-cavity klystron* described above can be used as an amplifier with narrow bandwidth (ca. 2%). When more amplification or output power is

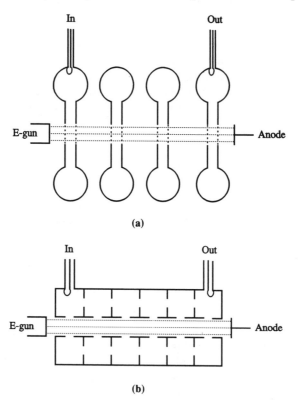

Figure 2.8 Multicavity klystron (a) and extended interaction klystron (b). Reproduced with permission from Staprans, A., McCune, E. W. and Ruetz, J. A. (1973) *High-Power Linear Beam Tubes*, Proc. IEEE, **61**. © 1973 IEEE.

desired one can put two of them in series. There is, however, a cleverer way to increase the amplification, by putting more cavities in series in the same klystron (Fig. 2.8(a)). In the intermediate cavities no power is extracted but the induced current is used to generate extra velocity modulation, thus giving the same effect as a cascade of amplifiers.

To explain this theoretically we can refer to equation (2.13). It has been remarked there already that the velocity modulation induced by the second cavity can be much greater than the original one, so extra amplification is provided. What one has to do in the present case is to take care that the cavity presents a non-resistive impedance to the gap. Then the velocity modulation is out of phase with the beam current and no power is extracted from the beam. A step further is to couple two successive cavities not only via the electron beam but also through their electromagnetic fields (Fig. 2.8 (b)). This has the advantage that the velocity-modulating voltage is divided over two gaps, so the voltage can be increased without the danger of vacuum breakdown. This is important for high-power tubes. Such a klystron is called an *extended-interaction klystron* and is an intermediate step towards the traveling wave tube, to be treated in the next section. Multicavity klystrons with up to seven cavities in total are used as high-power amplifiers. These can deliver more than 1 MW at 350 GHz. Such powers are used in particle accelerators, where the field of an electromagnetic wave is used to accelerate charged particles, and in fusion plasma machines where the r.f. energy is used to heat up the plasma. Figure 2.9 shows a high-power klystron which has a total length of about 6 m.

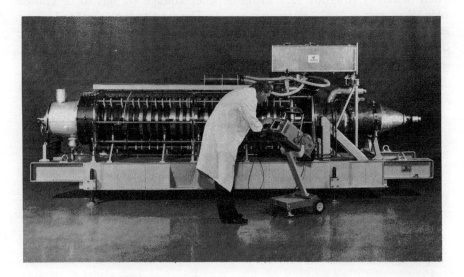

Figure 2.9 High-power multicavity klystron.

2.3.4 Reflex klystron

By feeding back the output signal of a klystron to the input one can make
an oscillator. This can however, be done in a simpler way by leaving out
the second cavity and letting the beam reflect from an electrode at a
negative potential, the so-called *repeller*. The returning beam induces a
current in the first cavity and if this happens with the right phase the whole
will oscillate. This is called a *reflex klystron* (Fig. 2.10). The oscillation
frequency can be tuned with the repeller voltage, but has of course to
remain within the bandwidth of the cavity. Tuning over a greater band can
be done by mechanically adjusting the cavity. In Appendix 2.C a theoretical
analysis of the operation of a reflex klystron is given. The result is a set of
two equations, one for the oscillation frequency ω and one for the normalized oscillation amplitude X:

$$\omega\tau_0 = \frac{3\pi}{2} + 2m\pi + \arctan[Q(\omega/\omega_0 - \omega_0/\omega)] \tag{2.17a}$$

$$\frac{2J_1(X)}{X} = \frac{2U_0}{\omega\tau_0 I_0 R}[1 + Q^2(\omega/\omega_0 - \omega_0/\omega)^2]^{1/2} \tag{2.17b}$$

where τ_0 now is the time the electrons take to travel to the repeller and
return to the cavity. Its value is given in Appendix 2.C. The left-hand side
of equation (2.17b) is at the beginning, when the oscillation starts, equal to
unity and decreases as the oscillation amplitude X increases. Therefore to
start an oscillation the right-hand side must be smaller than unity, a
condition which is most easily fulfilled when the frequency ω is close to ω_0.
Then to satisfy equation (2.17a) the repeller voltage U_R has to be chosen
such that $\omega_0\tau_0 = 3\pi/2$, assuming $m = 0$. The oscillation amplitude now
grows until equation (2.17b) is satisfied exactly. If we change U_R from the
optimum value, the right-hand side of equation (2.17b) will increase and the

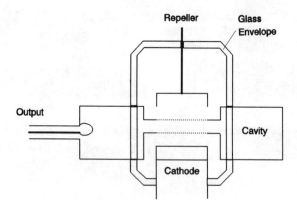

Figure 2.10   Reflex klystron.

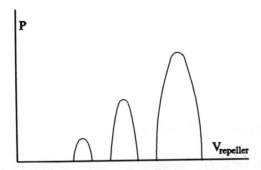

Figure 2.11 Output power against repeller voltage of a reflex klystron.

oscillation will stabilize at a lower value of X. When the frequency is too far off resonance the oscillation never starts. Therefore the output of a reflex klystron as a function of repeller voltage will look as shown in Fig. 2.11. The secondary peaks at smaller values of U_R result from m in equation (2.17a) being greater than zero. Of course this derivation is still very simplified; for instance, no account is taken of the fact that τ is also modulated when X is large. However, it is hoped that the analysis is sufficient to give some insight into the operation of a reflex klystron.

The reflex klystron was used much in the past as a local oscillator and in measurement setups. For these applications it is completely replaced by transistor or diode oscillators. Only at millimeter wave frequencies, up to 200 GHz, is the reflex klystron still in use.

Although the klystron uses transit time it is transit-time limited itself. In the foregoing it has been assumed that the distance between the plates (or the flat planes of the rhumbatron) is very small. This, however, is finite and when the transit time of the electrons in the cavity is of the order of magnitude of the signal period then the response of the beam to the field in the cavity is also attenuated. To escape from this limitation as well as the narrow bandwidth of the klystron ways to achieve a more distributed interaction between the electron beam and the electromagnetic field have been sought. This led in the end to a whole new class of microwave tubes, classified under the name *traveling wave tubes*.

2.4 TRAVELING WAVE TUBES

2.4.1 Principle

The trouble with the klystron is that the electric field which modulates the electron beam is in fact a standing wave through which the beam runs with great velocity. This produces a transit-time limitation while at the same time

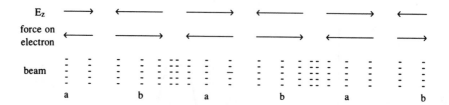

Figure 2.12 Density waves on an electron beam.

a standing-wave structure, being a resonator, is always a narrow band device. A waveguide is inherently broad band and moreover offers the possibility of continuous interaction between the field and the electron beam. To understand how a wave and an electron beam traveling together can lead to amplification we use a simple picture introduced by Brillouin [2]. Suppose we have an electric field in the *z*-direction given by

$$E_z = E_1 \cos(\omega t - kz).$$

Evidently this has a wave nature and the phase velocity is $v_{ph} = \omega/k$. Suppose now that there is a uniform beam of electrons traveling in the same direction with the same velocity (Fig. 2.12) where we as observers are traveling along, also with the velocity v_{ph}. Now, electrons at positions a will experience a retarding force while those at positions b will be accelerated. Consequently they will be driven together into bunches as indicated in the figure. Suppose now that the bunched beam moves a little faster than the field pattern. Then the bunches of high electron concentration will move slowly into the retarding field and will give up energy to the field. This energy will be used to increase the amplitude of the electric field and so we are converting kinetic energy of the beam into electromagnetic energy of the wave. This is the principle of the *traveling wave tube*. On the other hand, if initially the electrons are going slower than the phase velocity of the wave they will be caught up by the accelerating field and their speed will be increased at the cost of field energy. This is the principle of the *linear accelerator*. Of course, if the structure were infinitely long the electrons would alternatingly pass through retarding and accelerating fields and nothing would happen effectively. However, in a structure of finite length things can be arranged such that one of the two effects prevails.

2.4.2 Waves on an electron beam

Another way to look at the problem is to make a mathematical analysis of an electron beam modulated by a wave-like signal. In section 2.3 we already discussed a velocity-modulated beam which eventually became density modulated too; however, in that section we ignored the forces that the electrons exert upon each other. The density modulation produces an a.c. electric field which in turn influences the electron velocity. In the following we will take full account of this interaction. To make the analysis tractable we have to introduce some idealizations. In particular we assume the beam to be infinitely wide in all directions, and to have a uniform d.c. velocity v_0 in the positive z direction. All a.c. components are supposed to have a space–time dependence of the form

$$\exp[j(\omega t - k_z z)]$$

which describes a wave propagating in the positive z direction. Density waves with this space–time dependence will produce an electric field in the z direction. We will neglect the d.c. electric field generated by the average electron density n_0. (We can for instance assume that the electron charge is neutralized by an equal density of positive ions which are so heavy that they are not moved by the electric field. In reality this does not occur of course, but the beam has a finite cross-section and the beam charge will produce a constant electric field in the radial direction.) The a.c. electric field along the z coordinate will modulate the velocity of the electrons which will lead to density modulations as in the klystron. The latter in its turn will produce an electric field and the circle is closed.

The relevant equations describing this interaction are as follows:

- the equation of motion for the electron,

$$\frac{\partial v}{\partial t} + (v \cdot \nabla)v = \frac{q}{m} E;$$ (2.18)

- the continuity equation for the electron density,

$$\frac{\partial n}{\partial t} + \nabla \cdot nv = 0;$$ (2.19)

- Poisson's equation for the field generated by the density modulation,

$$\nabla \cdot \varepsilon E = \rho.$$ (2.20)

These equations are non-linear and in general not solvable analytically. We can make them solvable by using the small-signal approximation as we did before in the analysis of the klystron. This linearizes the equations and if we furthermore use the harmonic time–space dependence given above we can write the variables in the form

$$v = v_0 + \mathrm{Re}[v_1 \exp(j\omega t - kz)].$$

For the first-order components we then obtain

$$j\omega v_1 - jk_z v_0 v_1 \qquad = \frac{q}{m} E_1$$ (2.21)

$$j\omega n_1 - jk_z n_0 v_1 - jk_z v_0 n_1 = 0$$ (2.22)

$$-jk_z \varepsilon_0 E_1 \qquad = qn_1$$ (2.23)

where the index 0 indicates d.c. quantities and the index 1 a.c. quantities. We now have a homogeneous system of equations which only has a solution when the eigenvalue equation is satisfied:

$$\frac{q^2 n_0}{\varepsilon_0 m} - (\omega - k_z v_0)^2 = 0 \tag{2.24}$$

or

$$k_z = \frac{\omega \pm \omega_p}{v_0} \tag{2.25}$$

where the *plasma frequency* $\omega_p = (q^2 n_0 / \varepsilon_0 m)^{1/2}$ has been introduced. A further discussion of this quantity is given in Chapter 4.

The result is that two charge density waves are possible, a *fast wave* (minus sign) with a phase velocity greater than v_0 and a *slow wave* (plus sign) with a phase velocity smaller than v_0. Both are longitudinal, that is the electric field and the density modulation are in the direction of propagation. It is now interesting to look at the power transport in these waves. The kinetic energy transported by the beam per unit area and per unit time is equal to

$$P_{\text{kin}} = nv \frac{1}{2} mv^2 \tag{2.26}$$

where

$$n = n_0 + \text{Re} \{n_1 \exp[j(\omega t - k_z z)]\}, \quad v = v_0 + \text{Re} \{v_1 \exp[j(\omega t - k_z z)]\}.$$

Inserting this in equation (2.26) we obtain zero-order terms containing only d.c. quantities, first-order terms containing products of d.c. and a.c. quantities, and second-order terms containing products of two a.c. quantities. The first give the average energy transport

$$P_0 = \frac{1}{2} mv_0^3 n_0,$$

the second give terms that average to zero over time and the third category gives a correction due to the a.c. modulation of the beam. The second-order term gives

$$P_{\text{a.c.}} = \frac{1}{2} \text{Re}(\frac{3}{2} mn_0 v_0 v_1^2 + \frac{3}{2} mv_0^2 v_1 n_1)$$

which using equation (2.23) can be written as

$$P_{\text{a.c.}} = \frac{3}{4} mn_0 v_0 |v_1|^2 \frac{\omega}{\omega - k_z v_0} \tag{2.27}$$

If we now substitute k_z from equation (2.25) we find that the a.c. power is positive for the fast wave and negative for the slow wave. In other words, the kinetic energy of the beam *increases* when the beam is modulated by a wave going *faster* than the electrons (linear accelerator mode) and *decreases* with a modulating wave that is *slower* (amplifier mode). The conclusions of the mathematical analysis are thus in agreement with the Brillouin argument.

2.4.3 The traveling wave tube

The continuous interaction demands that the wave and the electrons run at approximately the same speed and this poses a difficulty. The velocity of waves in simple waveguides is usually of the order of magnitude of the velocity of light (in coax, for example, $c/\varepsilon^{1/2}$) but with reasonable beam voltages we can only attain electron velocities of about one-tenth of the velocity of light. Another thing is that obviously a wave with an electric field component in the propagation direction is needed, so a simple coax where the fundamental mode is a TEM wave will not do.

Rudolf Kompfner (Chapter 1) had the brilliant idea of rolling up the inner conductor of a coaxial line into a helix. Assuming that the velocity of the wave along the wire remains the same, the velocity in the axial direction will be reduced by a factor given by the circumference divided by the pitch (Fig. 2.13). In the beginning he sent the beam between the helix and the outer conductor and it worked, but he soon discovered that he could obtain more amplification if he sent the beam through the middle of the helix. This

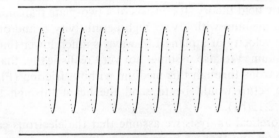

Figure 2.13 Coaxial line with inner helical conductor.

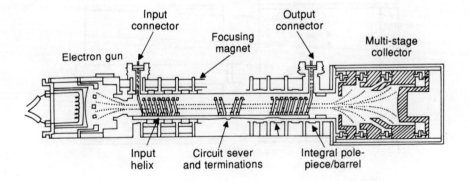

Figure 2.14 Traveling wave tube.

is because inside the helix a much stronger axial electric field arises than outside and we have seen that this is necessary to produce velocity modulation in the beam. It appeared in this case even possible to leave out the outer conductor completely and this is the form that traveling wave tubes have to the present day (Fig. 2.14). The news of Kompfner's invention quickly reached John Pierce of Bell Laboratories and he gave within a few months a theoretical analysis of the operation of the traveling wave tube (TWT).

2.4.4 Theoretical model of the traveling wave tube

In the model of Pierce the helix is substituted by a uniform transmission line. The coupling with the beam is accounted for on the one hand by assuming that the electrons experience an axial field equal to the derivative $-dV/dz$ of the voltage on the transmission line and on the other hand by assuming a distributed capacitive coupling between the beam and the line.

Since this model, although it is already strongly simplified, is still difficult to follow through, we will here use a different model that only uses electromagnetic field theory. It consists of a two-plate transmission line that is filled with a medium with a very high permittivity ε_r and/or permeability μ_r, so that the velocity of a plane EM wave is much lower than the velocity of light in vacuum. We also assume, not very realistically, that we can send a beam of electrons through this medium without friction (Fig. 2.15(a)). A more realistic setup would be to send the beam through a hole in the medium (Fig. 2.15(b)).

For the theoretical analysis we assume that the electrons can move only in the z direction and that we have a space–time dependence of the form

$$\exp[j(\omega t - k_x x - k_z z)].$$

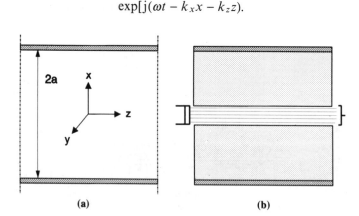

(a)   (b)

Figure 2.15 Model of TWT for theoretical analysis: (a) parallel plate waveguide used as a model; (b) more realistic configuration.

For this case the equations of Maxwell can be written as

$$jk_z E_y = -j\omega\mu_r\mu_0 H_x \tag{2.28a}$$

$$-jk_z E_x + jk_x E_z = -j\omega\mu_r\mu_0 H_y \tag{2.28b}$$

$$-jk_x E_y \qquad = -j\omega\mu_r\mu_0 H_z \tag{2.28c}$$

$$jk_z H_y = j\omega\varepsilon_r\varepsilon_0 E_x \tag{2.29a}$$

$$-jk_z H_x + jk_x H_z = j\omega\varepsilon_r\varepsilon_0 E_y \tag{2.29b}$$

$$-jk_x H_y \qquad = j\omega\varepsilon_r\varepsilon_0 E_z + J_z. \tag{2.29c}$$

Because of the restricted space dependence there are two independent systems. Interesting for us is the TM mode with field components H_y, E_x and E_z, of which E_z provides the interaction with the electrons. J_z is the electron current. We can express it in E_z with the help of the small-signal equations (2.21) and (2.22). In spite of the fact that we now also have a dependence on the x coordinate we can continue to use these equations because we have assumed that the electrons can move only in the z direction. Poisson's equation (2.23) is now replaced by Maxwell's equations above. We find for the small-signal a.c. component of the beam current

$$J_z = qn_0 v_1 + qv_0 n_1 = -j\omega\varepsilon_r\varepsilon_0 \frac{\omega_p^2}{(\omega - k_z v_0)^2} E_z \tag{2.30}$$

with

$$\omega_p^2 = \frac{q^2 n_0}{\varepsilon_r\varepsilon_0 m}.$$

$f_p = \omega_p/2\pi$ is called the *plasma frequency*. When we substitute this expression for J_z into Maxwell's equations the result is a homogeneous set of equations of which the determinant must be zero. This yields an eigenvalue equation:

$$(k_m^2 - k_z^2)\left[1 - \frac{\omega_p^2}{(\omega - k_z v_0)^2}\right] - k_x^2 = 0 \tag{2.31}$$

where $k_m^2 = \omega^2\varepsilon_r\varepsilon_0\mu_r\mu_0 = \omega^2/c_m^2$ (c_m is the velocity of light in the medium). In this equation there still appear two unknowns, k_x and k_z. We can determine k_x from the boundary conditions on $x = \pm a$. Suppose for E_z

$$E_z = A \exp(-jk_x x) + B \exp(+jk_x x).$$

At $x = \pm a$, E_z must be zero because there are metal planes in the z direction. This gives the condition

$$\sin(2k_x a) = 0 \quad \text{or} \quad k_x = \frac{m\pi}{2a}, \quad m = 1, 2, 3, \ldots .$$

In the following we consider only the fundamental mode for which $m = 1$, so we have $k_x = \pi/2a$. The eigenvalue equation now becomes, after removal of the fraction,

$$\left(k_z^2 - k_m^2 + \frac{\pi^2}{4a^2}\right)(k_z v_0 - \omega)^2 - \omega_p^2(k_z^2 - k_m^2) = 0. \tag{2.32}$$

When in this expression ω_p is allowed to approach zero we are left with two decoupled equations for the electromagnetic wave and the beam, with solutions

$$k_z^2 - k_m^2 - \frac{\omega_a^2}{c_m^2} \tag{2.33a}$$

and

$$k_z = \frac{\omega}{v_0} \tag{2.33b}$$

where

$$\omega_a = \frac{\pi v_m}{2a} \text{ and } c_m = c/(\varepsilon_r \mu_r)^{1/2} \tag{2.33c}$$

of which the first describes the fundamental electromagnetic mode of the parallel plate waveguide (with cutoff frequency ω_a) and the second the beam wave (actually the two waves described by equation (2.25) which have merged because $\omega_p \to 0$). It is customary to visualize this kind of ω–k relation by means of an ω–k diagram. For the decoupled waves this is done in Fig. 2.16(a). Here v_0 is chosen greater than c_m, the velocity of light in the dielectric that fills the waveguide. There is now an intersection at the frequency

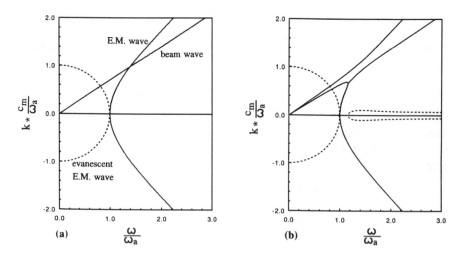

Figure 2.16 (a) EM wave and beam wave and (b) coupled waves.

$$\omega_0^2 = \frac{\omega_a^2}{1 - c_m^2/v_0^2} \tag{2.34}$$

and in this point, where the phase velocities of both waves are equal, we expect a strong interaction when ω_p has a finite value.

The complete eigenvalue equation (2.32) has four solutions: these solutions can be all four real, or two real and two conjugate complex or two sets of two conjugate complex. In Fig. 2.16(b) the ω–k diagram is sketched for $\omega_p/\omega_a = 0.2$ and $v_0/c_m = 1.4$. We see that in the region around the intersection point a strong interaction occurs resulting in two waves with complex k, one of which is growing and one is damped.

For an analytic approximation we look at the point $\omega = \omega_0$. Substituting this in the eigenvalue equation (2.32) we obtain

$$\left(k_z^2 - \frac{\omega_0^2}{c_m^2} + \frac{\omega_a^2}{c_m^2}\right)(k_z v_0 - \omega_0)^2 - \omega_p^2\left(k_z^2 - \frac{\omega_0^2}{c_m^2}\right) = 0. \tag{2.35}$$

To simplify this, define $K = (v_0/\omega_0) k_z$. Then, with equation (2.34) we obtain

$$(K^2 - 1)(K - 1)^2 - \frac{\omega_p^2}{\omega_0^2}\left(K^2 - \frac{v_0^2}{c_m^2}\right) = 0. \tag{2.36}$$

For $\omega_p = 0$ we have solutions $K = \pm 1$. Let us take ω_p very small and substitute $K = 1 + \delta$ and $K = -1 + \varepsilon$ respectively, where δ and ε are small. The first approximation yields

$$2\delta^3 = \frac{\omega_p^2}{2\omega_0^2}\left(1 - \frac{v_0^2}{c_m^2}\right)$$

or

$$\delta = \left[\frac{\omega_p^2}{2\omega_0^2}\left(\frac{v_0^2}{c_m^2} - 1\right)\right]^{1/3}\left(-1, \frac{1}{2} + \frac{j}{2}\sqrt{3}, \frac{1}{2} - \frac{j}{2}\sqrt{3}\right). \tag{2.37}$$

That is, the intersection point of the two uncoupled characteristics is split into three waves, of which one is undamped, one is damped and one is growing. The last one is the interesting one of course. The second approximation ($K = -1 + \varepsilon$) yields

$$\varepsilon = \frac{\omega_p^2}{8\omega_0^2}\left(\frac{v_0^2}{c_m^2} - 1\right). \tag{2.38}$$

This is real, which means that the negative branch of the electromagnetic wave remains undamped.

The model presented above has a big shortcoming. It replaces the inherently broad band helix by an electromagnetic structure that has a cutoff frequency. As a result amplification is obtained only in a limited frequency band. In a TWT with a dispersion-free EM circuit (the ω–k

diagram is a straight line through the origin) amplification would be obtained over an infinite frequency range. On the other hand the model is not as theoretical as it may seem. In fact the emission of radiation by a charged particle beam in a medium where the light velocity is lower than the particle velocity is known as Cerenkov radiation and is the cause of the eerie bluish light emitted by nuclear reactor cores immersed in water. Also, recently the idea of sending the electron beam through a hole in a high-permittivity medium has been realized and is known as a *Cerenkov maser* [3]. In this case the electrons have relativistic velocities which makes the analysis more complicated.

2.4.5 Applications of the traveling wave tube

Microwave traveling wave tubes are mainly used as wide-band amplifiers, e.g. as transponders in communication satellites. At the present time the helix is still much used as a waveguide because of its unique broad-band properties. A helix comes close to the ideal dispersion-free waveguide but not quite. Figure 2.17 shows the ω–k diagram of a real helix. The phase velocity is higher at the lower frequencies. The reason for this is that it is not a homogeneous structure but has a perodicity. If one uses a helix as is, the bandwidth obtained is less than one octave, enough for many purposes. When more bandwidth is desired the dispersion of the helix must be compensated, e.g. by loading the circuit with metal vanes that conduct mainly in the axial direction [4]. This way bandwidths of over two octaves can be reached.

A point of concern is the wave traveling in the backward direction, the negative root of equation (2.33a) and $K = -1$ in equation (2.36). This wave will be excited at the end of the tube and since it is undamped it can carry electromagnetic energy back to the beginning of the tube. It thus provides a feedback and, when the roundtrip gain is greater than unity, oscillations

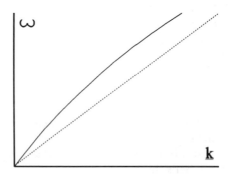

Figure 2.17 ω–k diagram of a practical helix.

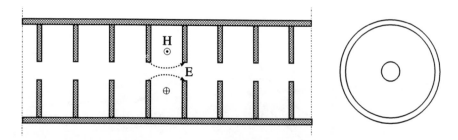

Figure 2.18 Slow wave structure in waveguide.

will result. Since the amplification of TWTs is high (100 dB is quite feasible) oscillation is almost certain to happen unless precautions are taken. The usual remedy is to introduce damping in the waveguide, e.g. by coating the supports of the helix with graphite. This attenuates both the amplifying wave and the back-traveling wave so, if the initial amplification is 100 dB, an attenuation of a little more than 50 dB is sufficient to make the roundtrip gain less than unity while still having 50 dB of useful amplification left.

In high-power TWTs coating is not possible since it leads to arcing and breakdown. Then the solution is to interrupt the helix for a short length, the two ends on both sides of the cut being terminated in matched loads. This so-called *sever* provides enough attenuation to prevent oscillation. Other types of traveling wave tubes use slow-wave structures that consist of coupled resonators (Fig. 2.18). To understand that this slows down the wave it is best to picture it this way: the electromagnetic field energy has to be passed on from one resonator to the other and this costs time so that the wave progresses more slowly.

2.4.6 Traveling wave tube variations: backward-wave oscillator, gyrotron

There are many types of tubes that make use of the traveling wave principle. The best known of these is the *magnetron* which will be discussed in the following paragraph. Another is the *backward-wave oscillator*. Here use is made of a wave circuit with a negative slope in the ω–k diagram (Fig. 2.19).

The phase velocity ω/k is positive so the wavefronts can travel along with the beam and interaction is possible. The group velocity $d\omega/dk$, however, is negative so that the electromagnetic wave transports the energy in the opposite direction. Since at the ends of the tube there are always reflections a feedback occurs. Now an amplifier with feedback will oscillate if the roundtrip gain is larger than unity. In this case that condition is easily fulfilled and the tube will oscillate on the frequency where the phase velocity of the electromagnetic wave is equal to the beam velocity, i.e. the

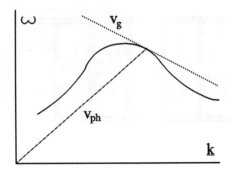

Figure 2.19 Backward wave.

frequency given by the intersection point of the two curves. The interesting feature of this construction is that the oscillation frequency can simply be changed by changing the beam voltage. So this tube can be tuned electronically over a large band. It is therefore much used in so-called sweep oscillators. In recent years it has been more and more replaced by transistor oscillators, particularly at the lower microwave frequencies.

A possible way to make a waveguide that propagates backward waves is to use a periodic structure, an example of which has already been given in Fig. 2.18. By a theorem of mathematics, known as Floquet's theorem, the solution of the wave equation in a structure that is periodic in z with period L will yield an ω–k diagram that is periodic in k_z with period $2\pi/L$. This means that there will always be parts of the diagram where $d\omega/dk_z$ is negative. Another way is to use a waveguide structure that intersects the beam only at discrete points (Fig. 2.20). Now things can be arranged such

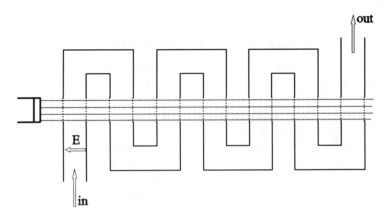

Figure 2.20 Stroboscopic backward wave.

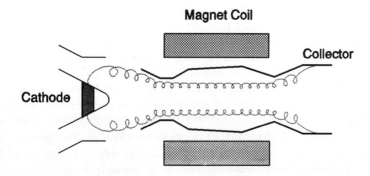

Figure 2.21 Gyrotron.

that a wave traveling the waveguide in the opposite direction to the beam is seen by the beam as a wave with forward phase velocity. This is called the *stroboscope effect* and is similar to the way in which spoked wheels in movies appear to rotate backwards.

Another interesting traveling wave tube is the *gyrotron*. Its construction is in principle simple (Fig. 2.21). An electron beam is sent at an angle to the axis into a circular cylindrical waveguide and forced by a strong axial magnetic field to go in a helical orbit. In the waveguide a circularly polarized TE_{10} wave can propagate which has an azimuthal electric field. The interaction relies on the fact that the beam velocity is so large that relativistic effects occur. A velocity modulation produces a modulation of the electron mass and this can produce bunching of the electrons which as we know is a condition for amplification of the wave. An advantage of the gyrotron is that a smooth waveguide can be used. The slow-wave structure used in other tubes has, as the frequency increases, to be made smaller in proportion to the reduction of the wavelength, which eventually leads to manufacturing problems. The gyrotron is free of these problems and can therefore be used for millimeter waves with great advantage.

Another member of the family is the *free-electron laser* [1, 3]. Here an electron beam is sent between permanent magnets with alternating polarities which create a d.c. magnetic field transverse to the beam that changes direction periodically. The periodic structure now is not in the circuit but in the d.c. velocity of the electron beam.

2.4.7 Beam–plasma amplification

The traveling wave interactions described above occur not only in devices designed for the purpose but in all systems where beams of charged particles are present and slow waves can propagate. Notorious examples of this are gaseous plasmas. A plasma is the name given to any more or less neutral

Vacuum electron devices

collection of charged particles. Neutrality implies that there are at least two charged particle species, one positive and one negative, e.g. an ionized gas or electrons and holes in a semiconductor. Gaseous plasmas occur in, for example, fluorescent light bulbs, MHD (magnetohydrodynamic) energy converters and nuclear fusion machines. Especially in the latter two cases, where the plasma is immersed in a d.c. magnetic field, many modes of wave propagation can occur, some of which have low phase velocities. Also, in these apparatuses large currents are flowing, meaning that more or less coherent streams of electrons or ions move through the plasma. If the direction and velocity of such a charged particle stream coincides with that of a propagating plasma wave interaction is possible, leading to amplification of the wave, very much as in a TWT. This goes, as we have seen, at the cost of the kinetic energy of the particle stream and the current flow in the device can be disturbed severely by these so-called beam–plasma instabilities. On the other hand the reverse effect, dissipating the energy of an electromagnetic wave in a beam–plasma interaction, has proved an effective way to heat fusion plasmas.

In the past attempts have been made to use beam–plasma amplification for microwave applications. The advantage is obvious: a plasma does not have to be machined as a metallic slow-wave structure so it could presumably be used to very high frequencies. However, there is also a big disadvantage: the electrons in gas plasmas have an electron temperature of the order of 10^4 K which means that the plasma is rather noisy. This is aggravated by the fact that there are many modes that can be amplified

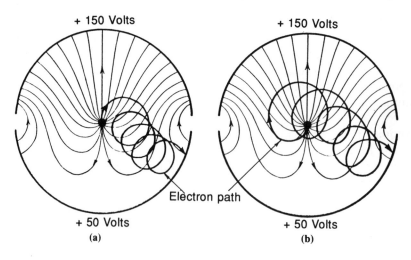

Figure 2.22 Principle of the split-anode magnetron. Relabeled from Fisk, J. B., Hagstrom, H. d. and Hartman P. L., The Magnetron as a Generator of Centimeter Waves in *The Bell System Technical Journal*, **25** (2) 167–349; published by AT&T Bell Laboratories, 1946.

besides the desired one. So, although working beam–plasma amplifiers have been demonstrated, they have found no practical application.

2.5   THE MAGNETRON

The magnetron is in principle a coaxial diode that is placed in a strong axial magnetic field. This forces the electrons into curved orbits. In the so-called *split-anode magnetron* this has been used to create a negative resistance (Fig. 2.22). Between the two halves of the anode an a.c. voltage is applied while both anodes together are on a positive d.c. potential with respect to the cathode. In one half of the period the upper anode has the highest potential and will draw the strongest current, at least as long as the frequency is low. At high frequencies, however, the transit time of the electrons starts to play a role and it can happen that the electrons arrive at the upper anode only

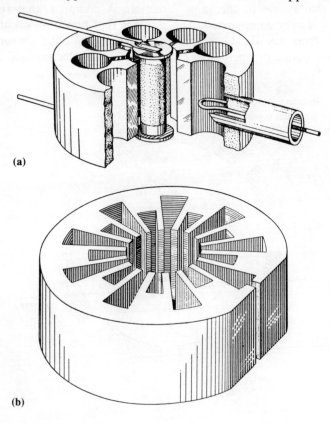

(a)

(b)

Figure 2.23   (a) Cavity magnetron and (b) rising sun magnetron. Reproduced with permission from Fisk, J. B., Hagstrom, H. d. and Hartman P. L., The Magnetron as a Generator of Centimeter Waves in *The Bell System Technical Journal*, **25** (2) 167–349; published by AT&T Bell Laboratories, 1946.

when the potential of the latter has gone down again. In this situation the anode with the highest potential draws the lowest current and vice versa. Seen from the a.c. voltage source this magnetron therefore behaves as a negative resistance.

In the microwave version, the *cavity magnetron* invented by Randall and Boot (as described in Chapter 1) in 1940, the anode is divided into a number of sections that are connected by resonators (Fig. 2.23(a)). In this system different modes of vibration are possible. The interesting one is that one in which the barriers between the cavities have alternating polarities. To force this mode one often applies *strapping*, that is connecting alternating dams by metal wires. Another way, better suited for millimeter waves, is to alternate bigger and smaller cavities, the so-called *rising sun magnetron* (Fig. 2.23(b)). We can now consider the anode as a circular delay line, which can propagate waves in both directions. One of these two can couple with the electrons that rotate in the same direction. A complete analysis of this interaction is very complicated. The result is that bunching occurs just as in the klystron, i.e. the electrons that leave the cathode uniformly distributed are bunched together in groups by the a.c. electric field that they

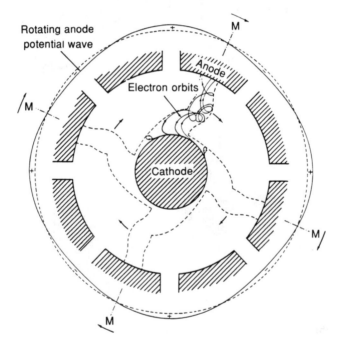

Figure 2.24 Electron cloud in the magnetron. Reproduced with permission from Fisk, J. B., Hagstrom, H. d. and Hartman P. L., The Magnetron as a Generator of Centimeter Waves in *The Bell System Technical Journal*, **25** (2) 167–349; published by AT&T Bell Laboratories, 1946.

create themselves so that the electron cloud takes the form of a wheel with four spokes (Fig. 2.24). These spokes pass the openings of the cavities just at the moment that there is a decelerating field and give energy to the electromagnetic field. Under optimal conditions they can lose 90% of their kinetic energy to the electromagnetic field so that they land on the anode with small velocity.

Magnetrons are suitable as oscillators for very great powers, especially in pulsed operation, to frequencies of 100 GHz and higher. They are therefore widely used in radar systems, and also for microwave heating in industry as well as in the kitchen (microwave ovens).

In the literature microwave tubes are often classified as O-type or M-type devices. This nomenclature was coined by French tube researchers in the 1940s and has generally been accepted. O-type tubes (from 'ordinary') are the linear-beam type devices, such as klystrons, TWTs and BWOs. M-type tubes (from 'magnetic') are the crossed-field tubes such as the magnetron.

Many other types than here described exist, for instance crossed-field (M-type) amplifiers, which are magnetron-like devices where the slow-wave structure is not closed on itself but has an interruption. This prevents oscillation, and amplification can be obtained by making an input and an output at the ends of the slow-wave structure. Another mixture of two ideas is the M-type BWO, also called carcinotron. It is used as a local oscillator at very high frequencies (300 GHz and over).

2.6 TUBE TECHNOLOGY

2.6.1 Vacuum tubes

Vacuum tubes, as the name indicates, need vacuum to operate, in the first place so that the electrons can move unhindered. However, there are other detrimental effects of residual gas molecules. They can be ionized and these ions can bombard the parts of the tubes that are at low potentials, in particular the cathode. A sustained ion bombardment will destroy a cathode in a short time. So a high vacuum is necessary, in the range 10^{-7}–10^{-9} Torr (1 Torr is 1 mm of mercury pressure and equals 1/760 atm or 133 Pa).

To maintain the high vacuum all parts of the tube must be vacuum tight. Critical points in this respect are the places where different materials are joined together. The outer envelope of a tube is made of metal, glass and/or ceramic material, ceramic especially when the tube must withstand high temperatures. So the development of tube technology is for a large part the development of metal–glass, metal–ceramic and glass–ceramic seals. These techniques started with the invention of the incandescent lamp and have reached a high degree of perfection.

The tube has to be pumped out at its fabrication, but also the liberation of gas from its inside parts during operation has to be prevented. This is done by baking the tube at a high temperature during pumping. To catch the small traces of gas that are nevertheless released in operation, *getters* are often used. These are surfaces of metals such as titanium and zirconium which adsorb most gases very strongly. In radio tubes one used to evaporate a small part of the inside of the glass envelope with a getter metal. In microwave tubes where high voltages are present this can easily lead to arcing so here solid getters are used.

2.6.2 Cathodes and electron guns

The most indispensable part of a tube is the electron emitter, the *cathode*. A good cathode should emit electrons copiously and constantly without consuming too much power. Unfortunately electrons going from a solid into vacuum always experience a potential barrier of several electronvolts. To help them overcome this barrier one has either to give them extra energy by heating the cathode material or to lower the barrier by applying a very strong electric field.

The first technique is illustrated by Fig. 2.25. Here the electron distribution is sketched at room temperature and at an elevated temperature. It is clear that at the higher temperature there are more electrons with enough energy to escape. In practice one uses two types of cathodes.

- The first type is directly heated thoriated tungsten (i.e. tungsten alloyed with 1% thorium). To obtain sufficient emission these have to be heated to about 2000 °C, so they consume a lot of power. On the other hand

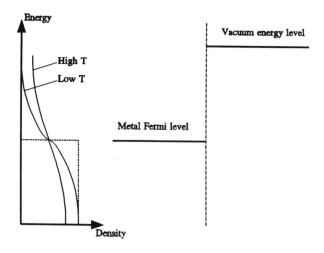

Figure 2.25 Thermionic emission.

they provide high current densities, around 3 A/cm². Also, they are resistant against ion bombardment. For these reasons their main application is in high-power tubes.

- The second type is indirectly heated oxide cathodes, called *thermionic cathodes*, nowadays consisting of a porous tungsten base impregnated with a mixture of barium and strontium oxides. These materials have low work functions and need to be heated to temperatures of about 1000 °C. Their peak current densities are lower than the previous type, about 0.2 A/cm².

Electric field emission will be treated in more detail in the next section on vacuum microelectronics.

When electrodes that focus the emitted electrons into a beam are placed in front of the cathode one speaks of an *electron gun*. An electron gun consists of a cathode and an anode with a hole in it that passes the beam. The anode is at a positive potential with respect to the cathode. The magnitude of this potential can vary from a few hundred volts in a low-power tube to hundreds of kilovolts in high-power tubes. To prevent electrons from being captured by the anode and to shape the beam a *focus electrode* (at negative potential) is often placed around the beam in the space between cathode and anode. To control or modulate the beam current a *control grid* may be used at a potential varying between negative and positive values. A grid at positive potential will intercept part of the electron beam and it will be heated by the associated dissipation. To prevent this a *shadow grid* of the same shape as the control grid is placed in front of the latter. The shadow grid is at zero potential so it does not dissipate any energy, but it screens the control grid fron the beam. A schematic cross-section of an electron gun is given in Fig. 2.26.

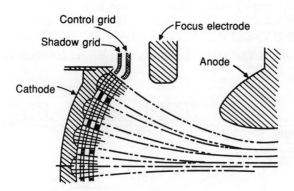

Figure 2.26 Electron gun. Redrawn from Staprans, A., McCune, E. W. and Ruetz, J. A. (1973) *High-Power Linear Beam Tubes*, Proc. IEEE, **61**. © 1973 IEEE.

2.6.3 Grids and anodes

Grids are used in triodes and klystrons. As already said in section 2.1 the grid is the most difficult part of a tube. It should define an equipotential plane and at the same time put as little material as possible in the path of the electrons while maintaining good mechanical stability. Also, the grid–cathode spacing should be small to minimize transit-time effects. However, putting a grid close to a heated cathode means that the grid will get hot as well. So materials are needed that have good electrical and thermal conductivity, are mechanically stable at high temperatures and show little secondary electron emission even at elevated temperatures. Materials that fulfill these demands are either metals with high melting points, such as molybdenum or tungsten, or pyrolitic graphite. The latter material is made by pressing a carbon compound in the form of the grid and then heating it to a very high temperature so that the compound decomposes and only the carbon remains. This is the best grid material especially for power tubes where the grid can get quite hot, not only because of the cathode but also by the grid current.

Anodes (in klystrons and TWTs they are often called collectors) have to collect the electrons which implies that the excess kinetic energy of the electrons is dissipated in the anode. Therefore the anode can become quite hot and it must consist of a good heat conductor. Like the grids they should have low secondary electron emission. To carry the heat away they are provided with cooling fins and in high-power tubes forced air or liquid cooling is applied.

In tubes such as klystrons and TWTs, where the electrons are accelerated in the electron gun, the collector can be at a lower potential than the anode of the gun. Through the interaction part of the kinetic energy of the beam is converted to r.f. energy, but the electrons never lose all their kinetic energy. So a potential that is just enough to allow the electrons to land on the collector is sufficient. For instance, if the beam is accelerated with a voltage of 1000 V, and the electrons emerging from the interaction region still have a kinetic energy equivalent to 300 V, the collector can be held at

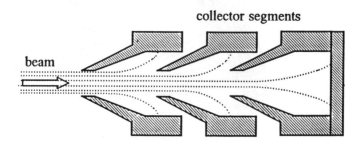

Figure 2.27 Depressed collector.

700 V. Since the collector dissipation is the product of collector voltage and beam current, a considerable amount of energy can be saved this way.

In practice the electrons come out of the interaction region with a spread in velocities. What one does then to reduce the dissipation is to make a hollow collector consisting of two or more sections (Fig. 2.27). The first section is at the highest potential and collects the slowest electrons. The faster electrons shoot past this section and are collected by the following ones which are at lower potentials. This is called a *depressed collector*. The disadvantage of it is of course that one needs a separate power supply for each collector section but in high-power tubes the extra cost of this can be compensated by the savings on energy and the cooling system.

2.6.4 Beam focusing

In multicavity klystrons, TWTs and BWOs the electron beam is very long compared with its diameter and since electrons repel each other the beam will diverge if no precautions are taken. In TWTs the beam is usually focused by an axial magnetic field. If this is strong enough the electrons will follow helical paths around the field lines. An axial magnetic field can be produced by a solenoid placed around the tube, e.g. Figs. 2.14 and 2.21, and this is the way it is done in the case of small-size lower-power tubes. For high-power tubes the necessary power consumption of the solenoid becomes excessive, and ring-shaped permanent magnets with an axial field are preferred. The direction of the field is alternated between consecutive magnets, so that one obtains a periodic magnetic field (Fig. 2.28), that focuses the diverging beam back to the centre of the tube. This is called *periodic permanent magnet* (PPM) focusing. Alloys such as Ticonal or

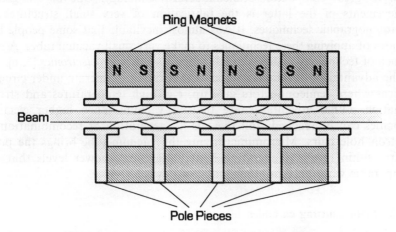

Figure 2.28 Beam focusing by permanent magnets.

Alnico can be used for the magnets but the best material from the point of view of attainable field strength and long-term stability is samarium–cobalt.

2.6.5 Noise

The main source of noise in electron tubes is the cathode. The electrons are emitted from the cathode in all directions and with a thermal velocity distribution corresponding to the cathode temperature. Both effects produce noise. Especially in thermionic cathodes which are at about 1000 °C, the velocity distribution can potentially produce a lot of noise. However, perhaps the worst noise source is the fact that the emissivity of a particular spot on an oxide cathode can vary strongly with time. This is called *flicker noise* or *burst noise* and has a spectral density inversely proportional to frequency. Finally, noise can arise from the fact that not all electrons reach the anode but some are caught by other electrodes, e.g. the grid in a triode. The partitioning of the current between the different electrodes fluctuates randomly, so this type of noise is called *partition noise*.

In section 2.2 how the electron cloud in front of the cathode provides a negative feedback on the noise in the emitted current has already been discussed. Thanks to this effect the noise is orders of magnitude lower than without it and tubes can be used as low-noise devices. Further discussion of noise will be postponed until Chapter 4.

2.7 VACUUM MICROELECTRONICS

2.7.1 Introduction

A new and interesting development in vacuum tube technology derives directly from its great competitor, semiconductor technology. One of the great achievements of the latter is the fabrication of very small structures by photolithographic techniques. It was almost inevitable that some people had the idea of applying these techniques to make very small vacuum tubes. A new branch of technology is thus started, called *vacuum microelectronics* [5, 6].

The advantage of vacuum devices is that they can operate under circumstances where semiconductors give up, e.g. high temperatures and strong radiation. Other advantages are that in vacuum much higher electron velocities can be reached, and there is no generation or recombination of electron–hole pairs. Microminiaturizing the vacuum tube brings the possibility within reach of operating it at voltage and power levels that are comparable with those of semiconductor devices.

2.7.2 Field emitting cathodes

It is obvious that one should not try to make heated cathodes on a micron scale. So the cathodes should be of the field emission type already men-

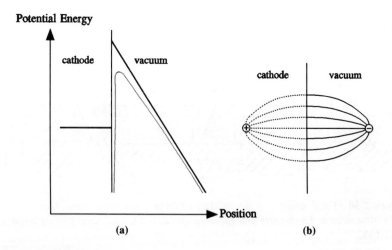

Figure 2.29   Field emission: (a) potential energy diagram; (b) image force.

tioned above. The principle is illustrated by Fig. 2.29(a). With the cathode at a negative potential with respect to its surroundings the potential outside the cathode material is increasing away from the cathode and the potential energy of the electrons is decreasing, so that the electrons are pushed away from the cathode. In the cathode the potential remains flat because cathodes are always made of metal or doped semiconductor, i.e. good conducting materials. Again, the electrons have to overcome a high barrier to be emitted from the cathode. However, now the slope of the potential makes the barrier narrower and at extremely high fields it becomes so narrow that the electrons can tunnel through it. This is called *Fowler–Nordheim tunneling*. Moreover, the peak of the barrier is lowered by the *image-force* effect. This is caused by the circumstance that an electron outside the cathode induces charges on the cathode surface, which exert an attractive force on the electron (Fig. 2.29(b)). The pattern of field lines and consequently the force is the same as if there were no interface but a positive charge at the mirror image position of the electron. The attractive force can be represented by an apparent potential that varies inversely with distance to the interface. If this potential is added to the electric potential the total potential experienced by the electron is given by the dotted line in Fig. 2.29(a). Clearly the peak of the barrier is lowered by the image force. We will encounter the same effect in metal–semiconductor junctions in Chapter 3.

The practical realization of field emitters aims at making very sharp points since electrostatics teaches us that the field lines crowd together at a point and that the maximum field strength goes up as the radius of curvature goes down. However, one point can emit only a limited stream of electrons so to obtain reasonable current densities one should have many points emitting in parallel. Here integrated circuit techniques help since they

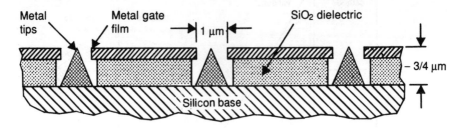

Figure 2.30 Field emitter array made in silicon. Redrawn from Hawkes, P. W. (ed.), *Advances in Electronics and Electron Physics*, **83**; published by Academic Press Inc., 1992.

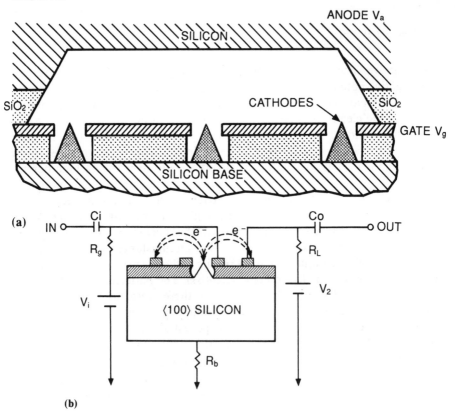

Figure 2.31 Vacuum microelectronic amplifying device. a) Redrawn from Hawkes, P. W. (ed.), *Advances in Electronics and Electron Physics*, **83**; published by Academic Press Inc., 1992. b) IEDM Digest, 89–15 © 1989 IEEE.

make it possible to fabricate large arrays of identical objects. An example of a field emitter array is shown in Fig. 2.30. Here the points are made of silicon. They are surrounded by a metal grid which is at a positive potential with respect to the points. This potential can be used to control the emitted current. Because of the geometry most of the emitted electrons will not reach the grid but go straight up.

A field emitter like this can be used as the cathode in vacuum devices (one recent application is in electron microscopes) but it can also be the basis for an amplifying device. A possibility for this is sketched in Fig. 2.31(a). This device has been called a vacuum FET [5], where FET here means field emitter triode. So far cutoff frequencies of 1 GHz have been obtained but it seems reasonable to expect that this limit will be shifted upwards. Another possibility is to place the anode at the sides so the operation looks more like that of a field effect transistor [6] (Fig. 2.31(b)).

APPENDIX 2.A RAMO'S THEOREM AND TRANSIT-TIME EFFECTS

In the treatment of components where transit-time effects occur (including the semiconductor transit-time diodes of Chapter 4) often the theorem of S. Ramo can be useful. It was introduced in 1938 by Ramo [7] and redis-covered in 1948 in a simplified form by W. Shockley, which is why it is often called the Ramo–Shockley theorem. In its simplest form it can be derived as follows: suppose we have a plane diode (Fig. 2.A.1), in which a plane layer of charge that has been emitted from the cathode moves. To avoid confusing minus signs we assume that the electron charge is positive so that the cathode becomes the positive electrode. At the point q the electric field shows a jump, whose magnitude we can find with the help of Gauss' law:

$$\varepsilon(E_2 - E_1) = Q/A \qquad (2.A.1)$$

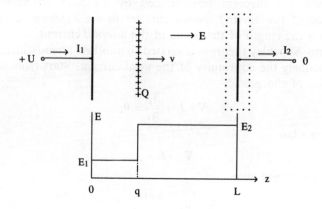

Figure 2.A.1 Ramo's theorem.

if A is the diode area. For the diode voltage we can now write

$$U = \int_0^L E \, dz = E_1 q + E_2(L - q)$$

from which we can eliminate E_1 using the previous equation:

$$U = E_2 L - \frac{Qq}{\varepsilon A}. \tag{2.A.2}$$

On the right electrode (the anode) is an amount of charge

$$Q_2 = -\varepsilon A E_2 = -\frac{\varepsilon A}{L}\left(U + \frac{Qq}{\varepsilon A}\right). \tag{2.A.3}$$

The change of this charge per unit time is the current in the connecting wire, i.e.

$$I_2 = -\frac{dQ_2}{dt} = \frac{\varepsilon A}{L}\left(\frac{dU}{dt} + \frac{Q}{\varepsilon A}\frac{dq}{dt}\right) = C\frac{dU}{dt} + \frac{Qv}{L} \tag{2.A.4}$$

where $C = \varepsilon A / L$ is the diode capacitance. We can see from this result that the total diode current is composed of the capacitive current plus a contribution from the moving charge. *The charge thus contributes to the current not only when it arrives at the anode, but during the whole time that it travels between the electrodes.* This is a consequence of the fact that the charge Q induces charges on both electrodes, which change when Q moves.

We can generalize equation (2.A.4) to a charge distribution $\rho(z)$ by integrating over z from 0 to L and dividing by L:

$$I_2 = C\frac{dU}{dt} + \frac{1}{L}\int_0^L \rho(z)v(z) \, dz. \tag{2.A.5}$$

In this form the theorem was derived by Shockley. It says that the contribution of the moving charge carriers to the *external* current is the average over the length of the diode of the *internal* current.

The Ramo–Shockley theorem is related to another, already introduced by Maxwell, namely the continuity of the total current. Start from the law of conservation of charge, i.e.

$$\nabla \cdot J_c + \frac{\partial \rho}{\partial t} = 0,$$

and Poisson's law

$$\nabla \cdot \varepsilon E = \rho.$$

These two combined give

$$\nabla \cdot \left(J_c + \frac{\partial \varepsilon E}{\partial t}\right) = 0. \tag{2.A.6}$$

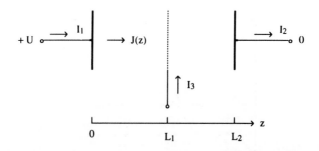

Figure 2.A.2 Ramo's theorem in a three-electrode device.

In other words, the total current is divergence free. If we integrate equation (2.A.6) over the closed surface indicated by a dotted line in Fig. 2.A.1, then we find

$$A\left(J_c + \frac{\partial \varepsilon E}{\partial t}\right) = I. \tag{2.A.7}$$

When we integrate this over z from 0 to L and divide by L, we again obtain equation (2.A.5).

In a device with more than two electrodes we have to take a little care in using Ramo's theorem. Consider Fig. 2.A.2. It follows from Ramo's theorem that charge moving in the left space induces a current in the cathode lead equal to

$$I_1 = C_1 \frac{dU_1}{dt} + \frac{1}{L_1} \int_0^{L_1} \rho(z)v(z)\, dz. \tag{2.A.8}$$

Charge moving in the right space induces a current in the anode lead of

$$I_2 = C_2 \frac{dU_2}{dt} + \frac{1}{L_2 - L_1} \int_{L_1}^{L_2} \rho(z)v(z)\, dz. \tag{2.A.9}$$

The current in the grid lead must then be found with Kirchhoff's theorem by subtracting these two: $I_3 = I_2 - I_1$.

APPENDIX 2.B LARGE-SIGNAL THEORY OF THE KLYSTRON

If we do not want to use the small-signal approximation we have to keep the exact form of equation (2.6), i.e.

$$v = v_0(1 + M \sin \omega t)^{1/2} \tag{2.B.1}$$

and instead of equation (2.8) we have

$$\tau = t + \tau_0(1 + M \sin \omega t)^{-1/2} \tag{2.B.2a}$$

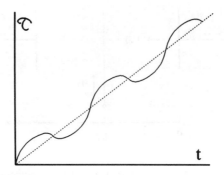

Figure 2.B.1 Double-valued arrival times.

with the derivative

$$\frac{d\tau}{dt} = 1 - X\cos(\omega t)(1 + M\sin \omega t)^{-3/2}. \qquad (2.B.2b)$$

This can be negative when $\omega t = 3\pi/2 + 2n\pi$ and $X > 1$. In this case the function $\tau(t)$ becomes double valued (Fig. 2.B.1). It means that electrons that started out at different times from cavity 1 arrive simultaneously at 2. In fact, electrons that started out earlier can arrive later. This is the "breaking" of the electron wave that gave the klystron its name. In the following we will assume that X is always smaller than unity so that $\tau(t)$ remains single valued.

Because the primary modulation is periodic in time the current modulation at cavity 2 is also periodic and can be developed in a Fourier series:

$$I_2 = I_0 + \sum_1^\infty [a_n \cos(n\omega\tau) + b_n \sin(n\omega\tau)] \qquad (2.B.3)$$

with

$$a_n = \frac{1}{\pi}\int_0^{2\pi} I_2(\tau)\cos(n\omega\tau)\, d\omega\tau, \qquad b_n = \frac{1}{\pi}\int_0^{2\pi} I_2(\tau)\sin(n\omega\tau)\, d\omega\tau.$$

Now we can use the current continuity condition $I_2\, d\tau = I_0\, dt$ and equations (2.B.2) to change from τ to t as the integration variable. The integration limits can be the same since a function periodic in τ is also periodic in t with the same period. To keep the integral manageable we use equation (2.B.2) in the small-signal approximation (2.8) assuming $M \ll 1$. The result is

$$a_n = \frac{1}{p}\int_0^{2\pi} I_0 \cos[n\omega(t + \tau_0 - \tfrac{1}{2}M\tau_0\sin \omega t)]\, d\omega t \qquad (2.B.4)$$

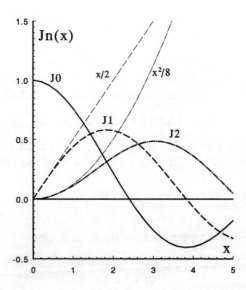

Figure 2.B.2 Bessel functions of the first kind.

and the corresponding expression for b_n. Note that in this expression we cannot use the small-signal approximation because, although M is small, the product $n\omega\tau_0 M$ can be large. Fortunately we do not have to introduce further approximations to proceed: the cosine can, using a well-known trigonometric formula, be written as

$$\cos[n\omega(t + \tau_0)]\cos[nX\sin(\omega t)] + \sin[n\omega(t + \tau_0)]\sin[nX\sin(\omega t)]$$

From the theory of Bessel functions we have the expansions

$$\cos(r\sin\theta) = \sum_{-\infty}^{\infty} J_n(r)\cos(n\theta), \qquad \sin(r\sin\theta) = \sum_{-\infty}^{\infty} J_n(r)\sin(n\theta),$$

where $J_n(r)$ is the Bessel function of the first kind and order n. These functions are shown in Fig. 2.B.2. On insertion of this expansion into equation (2.B.4) it is a matter of tedious but simple algebra to evaluate the integral. The final result is

$$I_2 = I_0\left\{1 + 2\sum_{1}^{\infty} J_n(nX)\cos[n\omega(\tau - \tau_0)]\right\}. \tag{2.B.5}$$

APPENDIX 2.C   THEORY OF THE REFLEX KLYSTRON

We can describe the reflex klystron with the same theoretical apparatus as in section 2.3, with the following changes.

- The returning electrons enter the cavity from the right so the induced current and voltage have their signs reversed.
- Since the frequency cannot now be chosen at will but is determined by the interaction of beam and circuit the complete form for the resonator impedance (2.15) has to be used instead of its value at ω_0. The induced voltage should therefore now instead of equation (2.12) be written as

$$U_{e2} = 2I_0 J_1(X)\,\text{Re}[Z_{res}\exp(j\omega\tau)] \tag{2.C.1}$$

- The transit time now is the time the electrons need to travel from the cavity to a point near the repeller and back. For the velocity in the drift space we have

$$v = v_0 - \frac{qU_R}{mL}t$$

This is zero, i.e. the electrons turn back, at a time

$$t = \frac{mv_0 L}{qU_R} = \frac{2U_0}{U_R}\frac{L}{v_0}. \tag{2.C.2}$$

The distance they travel in this time can be calculated as $d = (U_0/U_R)L$ which is greater than L if U_R is smaller than U_0. In other words, the electrons will land on the repeller when $U_R < U_0$. This is perfectly understandable since the electrons have a kinetic energy corresponding to U_0 and to destroy this energy an opposite potential of at least the same magnitude is necessary. So a reflex klystron can only operate when $U_R > U_0$.

The electrons arrive again at the cavity after a time twice that given by equation (2.C.2), i.e. the previously defined time τ_0 now is equal to

$$\tau_0 = \frac{4U_0}{U_R}\frac{L}{v_0}. \tag{2.C.3}$$

Stable oscillation of the reflex klystron is obtained when no external modulation voltage is necessary, that is when the secondary induced voltage (2.C.1) is exactly equal to the primary voltage (2.3.1):

$$U_{e2} = U_\sim = \text{Re}[-jU_1\exp(j\omega t)]$$

or

$$-jU_1 = -2I_0 J_1(X)Z_{res}\exp(j\omega\tau_0) \tag{2.C.4}$$

where equation (2.B.2a) has been used neglecting the term with M. Substituting Z from equation (2.15) we obtain two conditions:

$$\omega\tau_0 = \frac{3\pi}{2} + 2m\pi + \arctan[Q(\omega/\omega_0 - \omega_0/\omega)] \tag{2.C.5a}$$

and

$$U_1 = 2I_0J_1(X)\,\frac{R}{|\,1 + \mathrm{j}\,Q(\omega/\omega_0 - \omega_0/\omega)\,|}$$

or, after some manipulation,

$$\frac{2J_1(X)}{X} = \frac{2U_0}{\omega\tau_0 I_0 R}\,[1 + Q^2(\omega/\omega_0 - \omega_0/\omega)^2]^{1/2}. \tag{2.C.5b}$$

This implicit equation determines the oscillation amplitude X.

REFERENCES

1. Smith, B. L. and Carpentier, M. -H. (eds) (1993) *The Microwave Engineering Handbook*, Vol. I, Chapman & Hall, London.
2. Brillouin, L. (1948) Waves and electrons traveling together – a comparison between traveling wave tubes and linear accelerators. *Phys. Rev.*, **74**, 90–2.
3. Granatstein, V. L. and Alexeff, L. (eds) (1987) *High-Power Microwave Sources*, Artech, Boston, MA.
4..Gilmour, A. S. (1986) *Microwave Tubes*, Artech, Dedham, MA.
5. Brodie, I. and Spindt, C. A. (1992) Vacuum microelectronics, in *Advances in Electronics and Electron Physics*, Vol. **83**, Academic Press, Boston, MA.
6. Greene, R. F., Gray, H. and Spindt, C. (1989) *Vacuum Microelectronics*. Proc. Int. Electron Devices Meeting 1989. IEEE, New York, pp. 15–9.
7. Ramo, S. (1939) Currents induced by electron motion. *Proc. IRE*, **27**, 584–5.

FURTHER READING

Liao, S. Y. (1980) *Microwave Devices and Circuits*, Prentice-Hall, Englewood Cliffs, NJ.
Slater, J. C. (1950) *Microwave Electronics*, Van Nostrand, Princeton, NJ.
Pierce, J. R. (1950) *Traveling Wave Tubes*, Van Nostrand, Princeton, NJ.
Hamilton, D. R., Knipp, J. K. and Kuper, J. B. H. (1948) *Klystrons and Microwave Triodes*, MIT Radiation Laboratory Series, McGraw-Hill, New York.
Hutton, R. G. E. (1960) *Beam and Wave Electronics in Microwave Tubes*, Van Nostrand, New York.
Pierce, J. R. (1954) *Theory and Design of Electron Beams*, Van Nostrand, Princeton, NJ.
Ginzton, E. L. (1975) The $100 idea. *IEEE Spectrum*, **12**, 30–9.
Kompfner, R. (1976) The invention of traveling wave tubes. *IEEE Trans. Electron Devices*, **23**, 730–8.
Boot, H. A. H. and Randall, J. T. (1976) Historical notes on the cavity magnetron. *IEEE Trans. Electron Devices*, **23**, 724–9.
Flyagin, V. A., Gapunov, A. V., Petelin, M. I. and Yulpatov, V. K. (1977) The Gyrotron. *IEEE Trans. Microwave Theory Tech.*, **25**, 514–21.

PROBLEMS

2.1 Calculate, using equations (2.10)–(2.13), the power dissipated in the load resistor R. Show that it is equal to the kinetic power lost by the electron beam. Show that, using a realistic assumption about the maximum amplitude of the induced velocity modulation v_{2i}, the generated a.c. power cannot exceed the initial d.c. power, as one would expect.

2.2 A klystron amplifier should deliver 1 MW at 400 GHz. The efficiency may be assumed to be 100 %. The accelerating voltage should not exceed 100 kV. Calculate the beam velocity and the necessary beam current. What should the diameter of an oxide cathode be to produce this current (section 2.6.2)? Use the results of the previous problem to calculate the necessary value of the bunching parameter X and the minimum value of the transit time τ_0. What is the interaction length that is needed for this transit time?

2.3 A reflex klystron is designed for a frequency of 2 GHz. The anode voltage is 400 V and the distance from the cavity to the repeller is 2 mm. Calculate the repeller voltages at which the tube will oscillate.

2.4 An electron beam is accelerated by a voltage of 1000 V. Calculate the electron velocity and compare it with the velocity of light. Do you expect significant relativistic effects? If the beam current is 100 mA and its cross-section is 1 mm^2, calculate the electron density and the corresponding plasma frequency. Calculate the phase and group velocities of the fast and slow waves at frequencies of 1 and 10 GHz. Calculate the a.c. power transported by the waves when the velocity modulation depth is 1%.

2.5 A traveling wave amplifier according to Fig. 2.15(a) is characterized by the following data: $\varepsilon_r = 100$, $\mu_r = 1$; $v_0 = 4 \times 10^7$ m/s; $f_a = 2$ GHz; $I_0 = 1$ A per centimeter width in the y direction. Calculate the beam voltage necessary for the given velocity. Calculate c_m and check whether it is smaller than v_0. Calculate a, ω_0, and k_z and λ at ω_0. Calculate the electron density and the plasma frequency. Calculate the reduced growth factor δ and the non-reduced growth factor. Calculate the (electric field) amplification and the power amplification (in decibels) for a length of 20 cm.

3

Semiconductor materials

3.1 A VERY BRIEF REVIEW OF BASIC SEMICONDUCTOR PHYSICS

3.1.1 Introduction

In microwave devices we are to a greater extent than usual directly confronted with the physical properties of the semiconductor material. Dimensions are so small and time scales so short that one is working close to the mean free paths and the mean free times of the charge carriers. In the new extremely thin layer devices that are now under development even the wave nature of the electron cannot be hidden anymore. For a good understanding of microwave semiconductor devices it is therefore necessary to have some understanding of the physics involved. What follows is not intended as a crash course in semiconductor physics but as a brief memory-refreshing review for those who previously had a course in semiconductor physics.

3.1.2 Band structure

In a semiconductor crystal the electrons move in a three-dimensional periodic lattice formed by the atoms. The electron's motion obeys the laws of quantum mechanics and is described by Schrödinger's equation which is a wave equation. Its solution is called a wavefunction and from the wavefunction of an electron all its dynamical properties can be derived. In particular the squared modulus of the wavefunction (which in general is a complex function) gives the probability to find the electron at a specific point at a specific time.

The simplest wavefunction is a plane wave:

$$\psi = \exp[-i(\omega t - k_x x - k_y y - k_z z)] \tag{3.1}$$

where ω is the angular frequency and $k = (k_x, k_y, k_z)$ is called the wavevector. In wave propagation there is always a definite relation between ω and k and this is often graphically displayed in an ω–k diagram. Examples of

ω–k diagrams have already been encountered in Chapter 2. In the following we will see that such diagrams can also be used to illustrate the behavior of electrons in semiconductors.

A very basic axiom of quantum mechanics now states that the energy of an electron having the wavefunction (3.1) is given by $W = \hbar\omega$ and its momentum by $p = \hbar k$ where $\hbar$ is Planck's constant. (In physics it is customary to denote the energy of an electron by E. However, the author, having an electrical engineering background, prefers to reserve E for the electric field and to use W for the energy.) To find its velocity is a little more complicated. The wavefunction (3.1) has the same amplitude everywhere and an electron described by it could be anywhere with equal probability. Usually electrons are more localized than this and a better way to describe an electron would be a wavefunction covering a small band of frequencies which would be more localized in space and time. Schematically this is shown in Fig. 3.1. Electrical engineering students have learned in their telecommunication courses that such a wavepacket moves with a *group velocity*

$$v_g = (d\omega/dk_x,\ d\omega/dk_y,\ d\omega/dk_z) \tag{3.2a}$$

which in general differs from the *phase velocity*

$$v_f = (\omega/k_x,\ \omega/k_y,\ \omega/k_z). \tag{3.2b}$$

In this case the group velocity is the velocity with which the electron moves.

In a periodic structure the solution of a wave equation generally is of the form (3.1) multiplied by a function that has the same periodicity as the structure. The wavefunction of an electron in a crystal thus can be written as

$$\psi = P(r)\exp[-\,i(\omega t - k_x x - k_y y - k_z z)] \tag{3.3}$$

Now a theorem, known in mathematics as Floquet's theorem and in solid state physics as Bloch's, states that for waves in periodic structures the

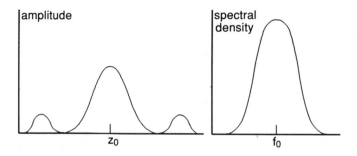

Figure 3.1 A wavepacket and its frequency spectrum.

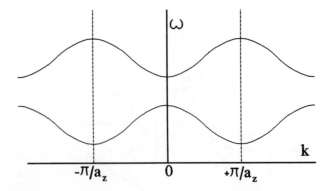

Figure 3.2   Periodic ω–k_z diagram.

ω–k relation is periodic in k, the periodicity being the inverse of that in space. In other words, if our crystal has a period a_z in the z direction, then $\omega(k_z)$ will be periodic with period $2\pi/a_z$. So the ω–k diagram of an electron will have the general appearance of Fig. 3.2. It now turns out that all wavefunctions with $k_z = k_0 + 2m\pi/a_z$ (m an integer) are indistinguishable from that with $k_z = k_0$ and then Pauli's exclusion principle tells us that they all describe one and the same electron. Also we are free to choose the origin of the k coordinates. A translation in k space gives an extra exponential that can easily be absorbed in the periodic function $P(r)$ of equation (3.3). So we can conveniently choose $k = 0$ at the extremal points. This means that we do not have to worry about k values outside the interval $-\pi/a_z \ldots + \pi/a_z$. All electrons we can ever have in our semiconductor can be described by k_z values within this interval. Corresponding considerations hold of course for the x and y directions. This restricted part of k space is called a Brillouin zone after L. Brillouin whom we already encountered in Chapter 2. For the semiconductors we are most interested in the Brillouin zone is as shown in Fig. 3.3.

Of course the ω–k_z diagram of Fig. 3.2 is only a subset of the complete four-dimensional ω–k diagram, namely the part for which $k_x = k_y = 0$. On a different scale it is also a $W(k_z)$ diagram (since $W = \hbar\omega$). In it we can distinguish regions of W where k_z has a real value and regions where it is imaginary. The former are called energy bands, and the latter forbidden bands or gaps. The highest energy band that is completely filled with electrons at zero absolute temperature is called the valence band, because it is filled with the outer shell electrons that determine the chemical valence of the material. The band directly above this, which in semiconductors is empty at zero temperature, is called the conduction band. The width along the energy axis of the forbidden band between them is called the *bandgap*. At temperatures above zero part of the electrons in the valence band can hop over to the conduction band leaving holes in the valence band behind.

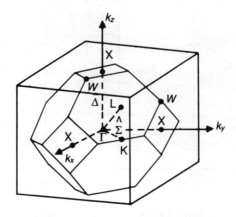

Figure 3.3 Brillouin zone for cubic semiconductors such as Si and GaAs. Redrawn from *Physics of Semiconductor Devices, second edition*; © copyright Sze, S. M. (1981), by permission of John Wiley & Sons, Inc.

The electrons in the conduction band and the holes in the valence band cause the semiconductor to be electrically conductive. In the forbidden band at least one of the three k components is imaginary, meaning that in one direction there is no running wave but a stationary exponentially decaying vibration. Aternatively, all k components are real but ω is imaginary, giving a wave amplitude that dies out in time (equation 3.1). So an electron can stay in this region only temporarily or as an evanescent Schrödinger wave. This is important for tunnel diodes (Chapter 4).

The group velocity of equation (3.2a) can, converting to energy, also be written as

$$v_{gi} = \frac{\partial \omega}{\partial k_i} = \frac{1}{\hbar} \frac{\partial W}{\partial k_i}, \quad i = x, y, z, \tag{3.4}$$

which means that in principle it is a function of k_x, k_y and k_z. The acceleration under applied fields is

$$\frac{\mathrm{d}v_{gx}}{\mathrm{d}t} = \frac{1}{\hbar} \frac{\mathrm{d}}{\mathrm{d}t} \frac{\partial W}{\partial k_x} = \frac{1}{\hbar} \left(\frac{\partial^2 W}{\partial k_x^2} \frac{\mathrm{d}k_x}{\mathrm{d}t} + \frac{\partial^2 W}{\partial k_x \partial k_y} \frac{\mathrm{d}k_y}{\mathrm{d}t} + \frac{\partial^2 W}{\partial k_x \partial k_z} \frac{\mathrm{d}k_z}{\mathrm{d}t} \right). \tag{3.5}$$

With the definition of momentum given above Newton's law is

$$F = \frac{\mathrm{d}p}{\mathrm{d}t} = \hbar \frac{\mathrm{d}k}{\mathrm{d}t}$$

which inserted in equation (3.5) gives

$$\frac{\mathrm{d}v_{gx}}{\mathrm{d}t} = \frac{1}{\hbar^2} \left(\frac{\partial^2 W}{\partial k_x^2} F_x + \frac{\partial^2 W}{\partial k_x \partial k_y} F_y + \frac{\partial^2 W}{\partial k_x \partial k_z} F_z \right) \tag{3.6a}$$

In vector notation

$$\frac{\mathrm{d}\boldsymbol{v}}{\mathrm{d}t} = \left[\frac{1}{m^*}\right]\boldsymbol{F}, \text{ with } \left[\frac{1}{m^*}\right]_{ij} = \frac{\partial^2 W}{\partial k_i \, \partial k_j}. \tag{3.6b}$$

So we can continue to use Newton's law, albeit in a modified form in which the mass has a tensor character. In this case we speak of an anisotropic band. This approach to charge carrier transport in semiconductors is known as the *effective mass approximation.*

In the valence band the energy of an electron decreases as we go away from the band extremum, so the effective mass is negative. Thus a hole, which is in every respect the negative of an electron, has a positive effective mass as well as a positive charge and the derivation above can be used for holes too. In cubic crystals, thanks to the symmetry of the cubic lattice there are no off-diagonal terms and the effective mass tensor is diagonal. Nearly all our technical semiconductors, in particular Si, Ge, GaAs and most other III–V compounds, fall in this class. When $m_x^* = m_y^* = m_z^*$ we speak of an isotropic band. Then surfaces of constant energy are spheres. This is approximately the case in the compound semiconductors. In Si and Ge the effective mass in the direction of a main axis is different from those in the orthogonal directions. Constant energy surfaces now are ellipsoids. In Fig. 3.3 these surfaces are drawn in the Brillouin zone. Note also that m_{xx} is equal to the radius of curvature of the W–k_x curve: so, the more strongly a band is curved, the lighter is its effective mass.

In a band minimum we can develop $W(\boldsymbol{k})$ in a Taylor series in which only even powers of k appear owing to the symmetry of the relation

$$W = W_0 + \frac{\hbar^2}{2}\left(\frac{k_x^2}{m_x^*} + \frac{k_y^2}{m_y^*} + \frac{k_z^2}{m_z^*}\right) + O(k^4). \tag{3.7a}$$

If the higher-order terms can be neglected even at higher k values we speak of a parabolic band: the effective masses are constant and the group velocity is simply

$$v_i = \frac{\hbar k_i}{m_i^*}, \quad i = x, y, z. \tag{3.7b}$$

In the compound semiconductors (GaAs, InP etc.) the conduction band is isotropic but not parabolic: at higher energies W increases more slowly than according to equation (3.7a), in other words the effective mass appears to increase. Often the W–k dependence is approximated by the formula of Kane:

$$W(1 + \alpha W) = \frac{\hbar^2 k^2}{2m_e^*} \tag{3.8a}$$

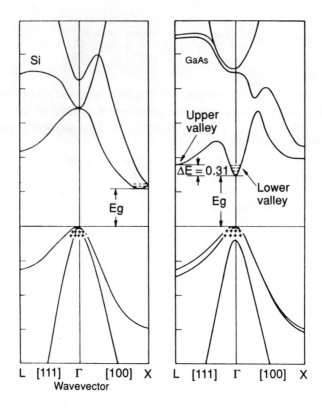

Figure 3.4 The energy bands of Si and GaAs. Relabelled from Cohen, M. L. and Chelikowsky, J. R., *Electronic Structure and optical Properties of Semiconductors*; published by Springer Verlag, 1988.

where

$$\alpha = \frac{1}{W_g}\left(1 - \frac{m^*}{m_0}\right) \tag{3.8b}$$

W_g is the bandgap and m_0 the rest mass of the electron. Note that the non-parabolicity is greater the smaller the bandgap is.

In Fig. 3.4 the band structures of silicon and gallium arsenide are given. There is a lot of detail but the most important features are as follows.

- In Si the conduction band minima are not at the center of the Brillouin zone but at six points on the main crystal axes, the so-called X points. There are higher minima but these are so elevated in energy that electrons will rarely reach them. Also, which cannot easily be seen from this figure, the effective mass in a particular minimum is greater in the directions orthogonal to the main axis on which that minimum is located than parallel to that axis. As a consequence surfaces of constant energy are are ellipsoids with their long axis parallel to the main crystal axis.

- In gallium arsenide the conduction band minimum is at the zone center (the Γ point) and the effective mass is isotropic. Constant-energy surfaces now are spheres.
- In both materials the valence band maximum is at the zone center but there are more valence bands with different effective masses.

In the foregoing discussion we have assumed a crystal with perfect periodicity and the result is an undamped electron wavefunction when the energy has the right value. This implies that the electron moves undisturbed through the whole crystal so that every semiconductor should be a super-conductor. In practice this is not observed of course. The periodicity of the lattice is spoiled by various causes, such as built-in imperfections (impurity atoms) and thermal vibrations of the lattice atoms around their equilibrium positions. Since all atoms are coupled to their neighbors by interatomic forces these vibrations are not independent but propagate as acoustic waves through the crystal lattice. The energy of these waves is quantized into quanta $\hbar\omega$ which are called *phonons*. Both the built-in imperfections and the lattice waves interrupt the smooth propagation of the wavefunction and cause the electron to deviate from its straight course and this is the reason why semiconductors are not superconductors.

3.1.3 Band bending

In the neighborhood of the conduction band minimum the W–k relation can always be approximated by the Taylor series development (3.7a) and the group velocity is given by equation (3.7b). Now equation (3.7a) can also be written as

$$W = W_0 + \tfrac{1}{2}(m_x^* v_x^2 + m_y^* v_y^2 + m_z^* v_z^2). \tag{3.9}$$

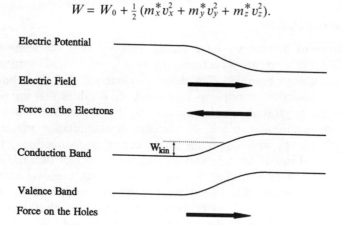

Figure 3.5 Band diagram.

We can interpret this equation as giving the total electron energy as a sum of potential energy, W_0, and kinetic energy, the rest of the right-hand side. This is important for device analysis: potential energy depends on the electric potential V, which is position dependent. So W_0 is equal to $-qV$ plus a constant, the value of which is not important. It is therefore customary to plot the conduction band minimum as a function of position, giving an upside-down picture of the electric potential (Fig. 3.5). Since the gap between conduction band and valence band is constant the latter runs parallel with the former.

The conduction band curve in this diagram gives the minimum energy an electron can have at every position. Above this it can have kinetic energy. The slope of the conduction band is a measure of the local electric field and the force on the electron as indicated in the figure. The same considerations are valid for holes: since electron energy increases upward in this diagram, hole energy will increase downward and the minimum energy of a hole is given by the valence band curve. Extra kinetic energy moves them downward in the diagram. Also, since the holes are in the same electric field, they will experience an equal force to that of the electrons but in the opposite direction.

If an electron moves from left to right in this diagram it will go against the force and kinetic energy will be converted to potential energy, at least as long as the electron does not collide with a phonon or some other lattice disturbance. The total energy will remain the same and in the diagram the electron moves on a horizontal line. It is important to have a good understanding of this type of diagram since they are used very much to explain device behavior. Examples will be given in Chapters 4 and 5.

3.1.4 Carrier statistics

In the previous section we have tacitly assumed that the semiconductor extends infinitely in all directions. In reality of course semiconductor crystals are always bounded. The wavefunctions will feel these boundaries and it is the boundary conditions that select the k values that are permitted in a particular crystal. A simple approach to this problem is to assume that we have a rectangular block of uniform semiconductor material with dimensions L_x, L_y and L_z in the three coordinate directions. It would now be natural to use for a boundary condition the fact that the electrons cannot leave the crystal, i.e. the wavefunction must go to zero at the boundaries. However, this would yield standing waves and we want the electrons to move so we prefer boundary conditions that give us running waves, such as $\psi(x + L) = \psi(x)$, so-called *periodic boundary conditions*. Substituting this into equation (3.2) we find that the allowable k values are

$$k_x = \frac{2\pi m}{L_x}, \; k_y = \frac{2\pi n}{L_y}, \; k_z = \frac{2\pi p}{L_z}, \tag{3.10}$$

where m, n and p are integer numbers. In a volume element $\Delta k_x \Delta k_y \Delta k_z$ of k space we then have a number

$$\Delta N(k) = \frac{L_x L_y L_z}{(2\pi)^3} \Delta k_x \Delta k_y \Delta k_z \tag{3.11}$$

of possible states that can be filled with two electrons each (one with spin up and one with spin down). Using the parabolic approximation (3.7a) for the case of an isotropic effective mass we can calculate the number of possible states in an energy interval ΔW, that is a spherical shell in k space with thickness ΔW, by integrating equation (3.11) over this shell. The result is

$$\Delta N(W) = \frac{L_x L_y L_z}{4\pi^2} \left(\frac{2m_c^*}{\hbar^2} \right)^{3/2} W^{1/2} \, \Delta W. \tag{3.12}$$

If we now want to convert this from number to density (number per unit volume) we have to divide it by the volume of the semiconductor, which happens to be $L_x L_y L_z$. Also dividing by ΔW we obtain the *density of states function*

$$g(W) = 2 \frac{1}{4\pi^2} \left(\frac{2m_c^*}{\hbar^2} \right)^{3/2} W^{1/2} \tag{3.13}$$

where the prefactor 2 represents the influence of spin. An important conclusion can be drawn from this formula: the larger the effective mass in a band is, the more available states there are per energy interval. This is the basis of the peculiar velocity–field characteristic of gallium arsenide discussed in section 3.2.2. The probability that an electron actually occupies a state is given by the Fermi function

$$F(W) = \frac{1}{1 + \exp[(W - W_F)/kT]} \tag{3.14}$$

where W_F is the Fermi energy, k is Boltzmann's constant and T is the absolute temperature. When the difference between W and W_F is more than a few times kT we can approximate $F(W)$ by the Boltzmann distribution:

$$F(W) \approx \exp\left(\frac{W - W_F}{kT} \right) \tag{3.15}$$

The product of the density of states function and the Fermi function gives the actual number of electrons that occupy a certain energy interval. If we integrate this from the bottom of the conduction band to the top we obtain the total number of electrons per unit volume in the conduction band, in other words the density of conduction electrons. Since in the upper part of the band there are very few electrons we may extend the upper integration limit to infinity without introducing a great error.

$$n = 2 \frac{1}{4\pi^2} \left(\frac{2m_c^*}{\hbar^2} \right)^{3/2} \int_{W_c}^{\infty} \frac{W^{1/2}\,dW}{1 + \exp[(W - W_F/kT)} = N_C \mathscr{F}_{1/2}\left(\frac{W_F - W_C}{kT} \right) \quad (3.16)$$

where

$$N_C = 2 \left[\frac{m_c^* kT}{2\pi\hbar^2} \right]^{3/2} \quad (3.17)$$

is called the effective conduction band density of states and

$$\mathscr{F}_{1/2}(x) = \frac{2}{\pi^{1/2}} \int_0^{\infty} \frac{y^{1/2}}{1 + \exp(y - x)}\,dy \quad (3.18)$$

is called the Fermi integral. This function is somewhat awkward for everyday use. At negative x it approaches asymptotically the exponential function:

$$\mathscr{F}_{1/2}(x) \simeq \exp(x), \quad x < 0. \quad (3.19)$$

At high positive x it approaches the asymptote

$$\mathscr{F}_{1/2}(x) \simeq \frac{4}{3}x \left(\frac{x}{\pi} \right)^{1/2}, \quad x \to \infty. \quad (3.20)$$

It has also been tabulated [1]. Fortunately in many cases W_F is much lower than W_C as we will see below and we can use the Boltzmann approximation (3.15) which gives

$$n \approx N_C \exp\left(\frac{W_F - W_C}{kT} \right). \quad (3.21)$$

Reasoning along the same lines we can also calculate the number of holes in the valence band. Since we are counting places that are not occupied by electrons we have to replace $F(W)$ by $1 - F(W)$ and since we are in the valence band the integration is from $-\infty$ to W_V, the maximum of the valence band. The result is

$$p = N_V \mathscr{F}_{1/2}\left(\frac{W_V - W_F}{kT} \right) \quad (3.22a)$$

and in the Boltzmann approximation

$$p \approx N_V \exp\left(\frac{W_V - W_F}{kT} \right) \quad (3.22b)$$

with

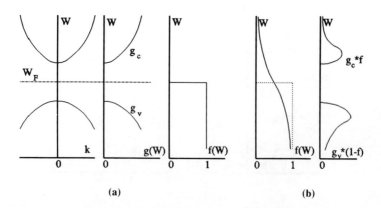

Figure 3.6 Intrinsic (undoped) semiconductor: (a) zero temperature; (b) finite temperature.

$$N_V = 2 \left[\frac{m_v^* kT}{2\pi\hbar^2} \right]^{3/2}. \tag{3.23}$$

So far we have been silent about the Fermi energy introduced by equation (3.15). It appears to govern the numbers of available electrons and holes but what determines it is unclear. In fact it is the other way around: the number of available carriers determines the position of the Fermi level. Suppose we have a semiconductor without impurities, then at zero absolute temperature all electrons are in the valence band and none in the conduction band. The Fermi level must be positioned such that in the conduction band its value is zero whereas in the valence band it must be unity. This means that it must lie in the forbidden gap: let us assume in the middle (Fig. 3.6(a)). At a slightly higher temperature some electrons will acquire enough energy to transfer to the conduction band. They leave behind an equal number of holes so that by equating equations (3.21) and (3.22) we obtain

$$N_C \exp\left(\frac{W_F - W_C}{kT} \right) = N_V \exp\left(\frac{W_V - W_F}{kT} \right)$$

which is satisfied by

$$W_F = \tfrac{1}{2}(W_C - W_V) + kT \ln\left(\frac{N_V}{N_C} \right). \tag{3.24}$$

The situation is now like that in Fig. 3.6(b). Our assumption made above that the Fermi level will be at midgap when the temperature is zero is justified. We also may conclude that it has been permitted to use the

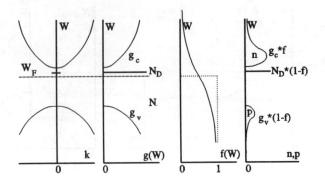

Figure 3.7 n-type doped semiconductor.

Boltzmann approximation since W_F is far away from both W_C and W_V. Note that in this case the product of n and p is a constant:

$$np = N_C N_V \exp\left(\frac{W_V - W_C}{kT}\right) = n_i^2. \tag{3.25}$$

n_i is called the *intrinsic density*.

The next step is to look at doped semiconductors. When we replace some of the atoms by foreign atoms with a different number of valence electrons we will have an excess or a deficit of electrons. For instance, if we replace silicon atoms by phosphorus or arsenic, which have valency 5, then there is one electron too many for the covalent binding with the surrounding atoms and this electron will be only loosely bound. A little energy suffices to break it loose from its parent atom and send it on the move through the crystal. These atoms are therefore called *donors*. An electron that moves through the lattice is by definition a conduction band electron. This implies that we can represent the donor atom by an energy level a little below the conduction band (Fig. 3.7). Note that we cannot assign a *k*-value to the donor energy level, because an electron bound to a donor describes a closed orbit and cannot be described by a plane wave.

Similarly, an atom from group III of the periodic system, e.g. boron or aluminum, has a deficiency of one electron and needs only a little energy to catch one from the valence band, so we can assign an energy level to it a little above the valence band. Such an atom is called an *acceptor* and the appropriate picture is shown in Fig. 3.8.

In compound semiconductors such as GaAs group VI elements (e.g. sulfur, selenium, tellurium) can act as donors because they preferably replace arsenic atoms in the lattice. Likewise, group II elements (beryllium, zinc) prefer gallium sites and become acceptors. The behavior of group IV elements is interesting. They can act as donors or acceptors depending on the atoms they

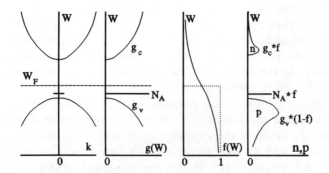

Figure 3.8 p-type doped semiconductor.

replace, which in turn depends on the specific element and on the crystal growth technique. It has been found that silicon mainly goes to gallium sites so it can be used as a donor whereas carbon goes to arsenic sites and becomes an acceptor. With these dopants there is always a certain degree of compensation, i.e. donors and acceptors are present simultaneously. A fully compensated semiconductor, i.e. having equal densities of donors and acceptors, is not conducting at all. What happens in this case is that all electrons from the donor atoms drop down to the acceptor levels instead of going to the conduction band.

In the case of doped semiconductors the position of the Fermi level is determined by the balance between electrons in the conduction band, those in the valence band and those on the donor and acceptor levels. In the case of an n-type semiconductor the conduction band electrons have come from the valence band or from the donor atoms. Some electrons have stayed behind on the donors so the balance equation is

$$n + n_D = N_D + p$$

where n_D is the number of electrons on donor atoms. This gives in the Boltzmann approximation

$$N_C \exp\left(\frac{W_F - W_C}{kT}\right) + N_D \exp\left(\frac{W_F - W_D}{kT}\right) = N_D + N_V \exp\left(\frac{W_V - W_F}{kT}\right). \quad (3.26)$$

We can write this as a quadratic equation in n:

$$n\left[1 + \frac{N_D}{N_C}\exp\left(\frac{W_C - W_D}{kT}\right)\right] - N_D - \frac{n_i}{n} = 0 \quad (3.27)$$

of which the only possible solution is

$$n = \frac{N_D + (N_D^2 + 4An_i)^{1/2}}{2A} \quad \text{with} \quad A = 1 + \frac{N_D}{N_C}\exp\left(\frac{W_C - W_D}{kT}\right). \quad (3.28)$$

This result shows that, at room temperature where $W_C - W_D$ is of the same order as kT, when N_D is larger than n_i but small compared with N_C, A is close to unity and n is equal to N_D, that is all donors are ionized. This changes when N_D approaches N_C or when the temperature is lowered substantially. In the latter case one speaks of freeze-out: the electrons are, as it were, frozen to their donors.

When we now look at the position of W_F we find that it has moved upwards towards the conduction band. At high donor densities it even moves into the band. In this case the exponential approximations cannot be used anymore, and we have to use the full Fermi function and its integral, equations (3.15a) and (3.18). In this case we speak of *degenerate* material. Similar deductions can be made for p-type material. The Fermi level moves towards the valence band as the acceptor doping increases.

There are also impurities which can release or capture an electron but which have energy levels close to the middle of the gap. These are called traps because it takes a lot of energy (about half the bandgap) to release an electron into the conduction band and it is therefore a process with a low probability. An electron that is captured by such an atom can sit there for a long time, in the range of microseconds to milliseconds. Material that contains many of these traps therefore is usually semi-insulating because it contains neither free electrons nor holes. Examples are copper in Si and chromium or oxygen in GaAs. In the past often semi-insulating GaAs substrates were made by doping them with Cr to compensate for the unintended donors or acceptors they contained. Nowadays substrates can be grown so purely that they are semi-insulating without adding Cr.

3.2 IMPORTANT BULK MATERIAL PROPERTIES

3.2.1 General considerations

Semiconductor materials for microwave applications have to fulfill special demands. They have to work at extremely high frequencies, which means among others that capacitive currents can become large unless capacitances are kept small. As a rule of thumb, the capacitive current (which is always present) should not be greater than the resistive current. A second requirement is that the time a charge carrier needs to traverse a device should be small, otherwise transit-time delay and the associated phase shifts will occur. Both demands can only be met by shrinking the device dimensions as far as possible. If one still wants to obtain reasonable powers out of these small devices the current densities and electric fields have to be high. High electric fields are also necessary for high carrier velocities which we need for short traversal times.

Limits on high power are, however, set by the maximum allowable temperature rise a device can stand. The maximum temperature a material can sustain depends for a large part on its bandgap: the thermal generation of carriers is proportional to a Boltzmann factor $\exp(-E_g/kT)$ so the maximum sustainable temperature is roughly proportional to the bandgap. Another consideration is formed by the temperature hardness of contacts. The combination of high temperature and high current density causes metal from the contact to diffuse into the semiconductor (*electromigration*), causing device properties to deteriorate. Very often this is found to be proportional to another Boltzmann factor $\exp(-E_a/kT)$ where E_a is called the activation energy. Metals that diffuse easily, such as Al or Au are not suitable for high-temperature contacts. Instead one should use refractory metals such as Ti or W.

The temperature rise a device experiences is proportional to the power dissipation and inversely proportional to the heat conductivity of the material. Evidently one wants materials with high heat conductivity. Also, the heatsinking is an important consideration in the packaging and mounting of microwave devices.

The maximum value of the electric field is determined by the field strength at which carrier multiplication by impact ionization starts to become significant. Since this process causes runaway current and excess noise one usually wants to avoid it, except in devices where it can be put to use, such as IMPATT diodes (Chapter 5). There is no hard limit for current density as such but the limits on dissipation and electric field combined give a limit on current density.

These considerations lead to the following desired material properties:

- *high velocities* of charge carriers, i.e. high mobilities and saturation velocities, for which reason the working charge carriers in microwave devices are the electrons, so one always has n-channel FETs and n–p–n bipolar transistors;
- *low relative permittivity;*
- *high coefficient of heat conduction;*
- *high bandgap.*

The most used materials in the microwave field are silicon and gallium arsenide, of which the latter is becoming increasingly important since about 1970. The use of Si is a natural extension of its abundant use at lower frequencies. The original reason to introduce GaAs was that the v–E characteristic of this material shows a negative differential mobility (section 3.2.2), which opens the possibility of negative-resistance devices. The first practical application of GaAs was therefore in Gunn diodes, which will be described in Chapter 5. Other materials from the III–V family, notably InP, show the same property. Nowadays the main reason for employing GaAs is that transistors with much higher cutoff frequencies and lower noise figures can be made. The following table gives a comparative overwiew of the merits of Si and GaAs.

Property (at room temperature)	Si	GaAs
Bandgap (eV)	1.1 (indirect)	1.4 (direct)
Electron mobility (m^2/(V s))	0.15	0.85
Negative differential μ	No	Yes
ε_r	11.4	12.9
Specific heat (J/(g K))	0.7	0.325
Thermal conductivity (W/(cm K))	1.31	0.48
Carrier lifetime	Microseconds	Nanoseconds
Technology	Well developed	In development

As one can see the difference in permittivity is not of much importance. Regarding thermal conductivity, silicon has a definite advantage which is partially offset by the higher bandgap of gallium arsenide.

Carrier lifetime is not important for microwave devices provided that it is long compared with the time the carriers spend in a device. In microwave devices this will be of the order of picoseconds so even the short lifetime of GaAs (about 10^{-9} s) poses no problem.

More serious is the difference in technological development. The technology of silicon started to be developed around 1950, that of gallium arsenide around 1970. Although GaAs could profit from some developments in Si technology, mainly photolithography, a number of other processes, in particular crystal growth but also etching and contacting, had to be newly developed. With all the effort going in the continuing development of Si technology this time lag of about 20 years remains the same. In section 3.3 we will discuss the technological processes that are relevant for microwave devices.

3.2.2 Carrier velocities

The demand of low resistance and reasonable powers, i.e. high current density, leads to high doping densities. Unfortunately this has a negative effect on the low-field mobilities because of the scattering of the electrons by the ionized dopant atoms. Hilsum [2] has given a simple relationship that describes the dependence of the low-field mobility on doping concentration reasonably well for the most popular semiconductors:

$$\mu(N_D) = \frac{\mu_0}{1 + (N_D/N_{D0})^{1/2}} \tag{3.29}$$

where $N_{D0} \approx 10^{17}$/cm^3. This shows that already at a donor density of 10^{17}/cm^3 the mobility is halved. The net result is that efforts to increase the current by increasing the carrier concentration are partly offset by the decrease of the mobility by ionized-impurity scattering. Tricks have therefore been invented to separate the charge carriers from the ionized donors. More about this will be said in section 3.3.

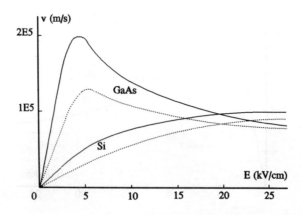

Figure 3.9 v–E characteristics of electrons in Si and GaAs: ——, undoped, ·····, donor concentration of $10^{17}/cm^3$.

In Fig. 3.9 the drift velocities of electrons in GaAs and Si are given for a range of electric field strengths that occur in microwave devices. As the figure shows, the velocity in GaAs exceeds that of Si for a large range of field strengths. This means shorter transit times and higher cutoff frequencies for comparable device dimensions. Equally important, it also means lower parasitic resistances. The influence of doping, mentioned above, can also be seen here. At high fields it is less than at low fields. This is because at high fields the electrons have higher velocities and their trajectories are less influenced, simply because they spend less time in the field of an ion.

What the figure also shows is a region of differential negative resistance in the v–E characteristic of GaAs. The cause of this is the existence of higher-lying local minima in the conduction band (Fig. 3.4). In the Brillouin zone of Fig. 3.3 the lowest minimum is in the center, the so-called Γ point, and the higher minima are off center in the L points. In the literature therefore the lower minimum is called the central valley or the Γ valley and the higher minimum the satellite valley or the L valley. The effective mass in the latter is much higher than in the central valley and the mobility therefore much lower. The energy gap between the Γ and L valleys is about 0.3 eV so at room temperature and low electric fields few electrons will be in the L valley.

At higher field strengths the electrons acquire more energy and gradually more of them will have an energy above the minimum of the L valley. These electrons can make a transfer to the satellite valley and most of them will since, because of the high effective mass, the density of states is much higher in the L valley than in the Γ valley (equation (3.13)). However, in the L valley they have a v–E relation different from that in the central valley. Again because of the higher effective mass the velocity will be lower at the

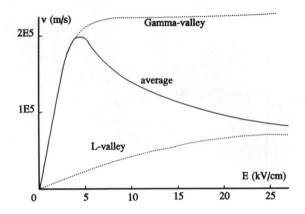

Figure 3.10 *v–E* characteristics of the Γ and L valleys in GaAs.

same field strength. The *v–E* characteristic of all electrons together will be
an average of the two, shifting gradually from the Γ valley characteristic to
that of the L valley as the field becomes higher (Fig. 3.10). In the transition
region the resultant characteristic has a negative slope. The effect was
predicted to exist in multivalley semiconductors by Ridley and Watkins in
1961. Shortly afterwards Hilsum pointed out that GaAs would be the most
suitable material and in 1963 it was experimentally discovered by I. B. Gunn,
after whom it is called the Gunn effect. Its microwave applications will be
discussed in Chapters 5, 9 and 10.

Important for transistors is another interesting phenomenon, which is
known under the name *velocity overshoot* (Fig. 3.11). When an electric field
in a semiconductor is switched on abruptly, the velocity does not immediately reach the static value that follows from the *v–E* characteristic. First

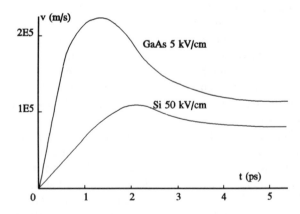

Figure 3.11 Velocity overshoot.

there is an acceleration phase in which the velocity increases linearly with time. Then collision processes start to operate, by which the acceleration decreases and after some time collisions and field are in equilibrium so that a stationary velocity is attained. In the acceleration phase an overshoot can occur which especially in III–V materials can be large. The significance of this is appreciated if we integrate the velocity over the time the overshoot phenomenon lasts. It is found that, although this time is extremely short, of the order of picoseconds, the electrons travel a distance of about 0.5 µm in GaAs and 0.1 µm in Si. In modern submicron devices this means that the electrons spend a great deal of their traversal time in a non-stationary situation where the velocity is higher than follows from the stationary v–E characteristic.

What makes GaAs even more interesting is that it is a member of a whole family of semiconductors, the so-called III–V compounds. In this family belong all binary combinations of group III elements (In, Al, Ga) with group V ones (P, As, Sb), but also ternary and quaternary alloys such as $Al_xGa_{1-x}As$ and $In_xGa_{1-x}As_yP_{1-y}$. This makes it possible to design materials with desired properties.

3.2.3 Bandgap

In addition to a higher electron velocity GaAs has another intrinsic advantage over Si: because of its large bandgap the thermally generated charge carrier density is very low so that the specific resistance of undoped material is very high, about $10^8 \, \Omega$ cm (Si never has a higher resistance than $10^4 \, \Omega$ cm). The material is then called *semi-insulating*. As has been explained above in impure material it is necessary to add Cr to compensate for the unintentional donors. The semi-insulating property of GaAs substrates makes it easy to isolate devices from each other and also offers the possibility of integrating microstrip circuits with transistors on the same substrate into so-called *MMICs* (monolithic microwave integrated circuits). This can be extended to the integration of optical and electronic components into OEICs (optoelectronic integrated circuits). MMICs will be discussed in Chapter 11.

Another peculiarity of GaAs is its surface. Here the periodic structure of the crystal ends and the concept of energy bands that follows from this periodic structure ceases to be valid. The surface atoms therefore can have energy levels in the forbidden gap. These are called surface states. In GaAs they have an energy level near the middle of the gap and this is nearly independent of the material (oxide, metal etc.) that covers the surface. These surface states capture electrons so that the surface is negatively charged and consequently the Fermi level at the surface is *pinned* at a level of about 0.7 eV below the conduction band (Fig. 3.12). Electrons in the semiconductor are now repelled from the interface and band bending as

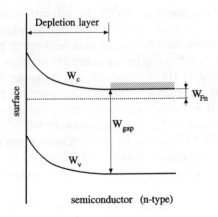

Figure 3.12 Fermi level pinning.

discussed in section 3.1.3 occurs. Note that the band bending is such that the Fermi level remains constant throughout the semiconductor. This is generally true for systems in equilibrium.

A nice consequence of the Fermi level pinning is that it is not difficult to make good Schottky barriers on GaAs. Most metals have a barrier potential of *ca.* 0.7 V. The drawback is that in both n-type and p-type material, regardless of surface treatment, there is always a depletion layer at the surface, even at free surfaces. In field effect transistors this can have undesirable consequences (section 6.2). Another drawback is that MOS transistors on GaAs will always have positive threshold voltages of about 0.5 V.

3.2.4 Heat flow and device temperature

All semiconductor devices dissipate heat when in operation. This is especially true for microwave devices whose dimensions are small and which often use effects that occur only at high electric fields (e.g. avalanche or Gunn effects). The dissipation causes a rise of the device temperature and since device operation nearly always becomes worse as the temperature goes up it is important to keep this temperature rise low. Good *heat sinking* is therefore important, that is to say the device has to be mounted on a carrier that can conduct the heat away quickly. A typical example of a mounted device is shown in Fig. 3.13(a). The active device region is in the upper layer of a chip which is soldered onto a metal base.

The heat has to be carried away through the substrate, the solder joint and the metal base. For the latter copper is usually chosen because of its good thermal conductivity. In extreme cases the heat sink is made of industrial diamond which is metal plated all around, diamond being the

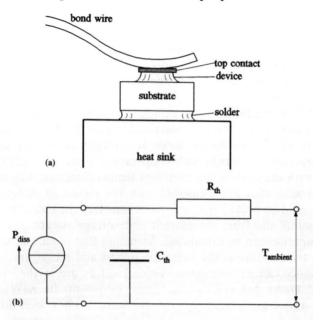

Figure 3.13 (a) Mounted device and (b) thermal circuit.

best heat conductor of all solids. Another precaution often taken is to etch the substrate thinner from the standard value of about 400 μm to about 100 μm which is the thinnest that can be reasonably well handled. Still there is a resistance to the heat flow and a temperature difference between the device and the base is set up to overcome this resistance. Since the device itself has a capacity to store heat we can describe its thermal behavior in first approximation by a simple RC circuit (Fig. 3.13(b)), where the dissipated power is represented by a current source and the temperature by an electric potential. The governing equations are

$$\frac{\mathrm{d}T_{\mathrm{dev}}}{\mathrm{d}t} = \frac{1}{C_{\mathrm{th}}}(P_{\mathrm{diss}} - Q_{\mathrm{th}}) \tag{3.30}$$

and

$$T_{\mathrm{dev}} - T_{\mathrm{amb}} = R_{\mathrm{th}}Q_{\mathrm{th}} \tag{3.31}$$

where P_{diss} is the dissipated power and Q_{th} is the heat flow, both in watts. R_{th} (K/W) is the heat flow resistance and C_{th} (J/K) the heat capacity. The latter is equal to the volume of the device multiplied by the specific heat capacity and the heat resistance is calculated in the same way as an electrical resistance from the dimensions of the structure, the thermal conductivity taking the place of the electrical conductivity. Obviously the heat resistance is composed of several parts and, since the substrate has heat capacitance also, a distributed RC network would give a better description of the whole. However, in practice very often this simple model is sufficient. From these equations it is clear that when a device is switched on the temperature rises with a time constant $R_{th}C_{th}$ to a final value $R_{th}P_{diss}$ over the ambient temperature. If one wants to operate a device under pulsed bias to keep the temperature rise low, the pulse length has to be smaller than this time constant.

The heat flow resistance can in principle be calculated but this is not easy since it is composed of different parts and not all the thermal conductivities are well known. It can be measured by making use of the fact that the device characteristics change with temperature. First a d.c. characteristic is measured with the mounting at ambient temperature and then a number of pulsed characteristics are measured with the mount at different elevated temperatures. The pulsed characteristics intersect the d.c. characteristic at different points and from the current and voltage at each intersection the power dissipation can be calculated. Assuming that at an intersection point the device temperature is the same for pulsed and d.c. operation the heat flow resistance can be calculated. Typical values are in the range 10–100 K/W. For GaAs MESFETs one finds values of 10 K/W per square millimeter device area, but only after the substrate has been thinned down to 60 µm.

3.3   HETEROJUNCTIONS, QUANTUM WELLS

Many of the different materials in the III–V family can be grown on top of each other. However, prerequisite for a usable combination is that both materials have the same type of crystal lattice with equal lattice parameters, otherwise the junction plane will be full of crystal defects which are detrimental to device performance. One then obtains a so-called *heterojunction* with interesting properties. In particular, since different materials in general have different bandgaps, there will be a discontinuity at the junction in either the conduction band or the valence band or both. The

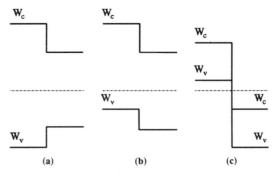

Figure 3.14 Heterojunction discontinuities: (a) normal; (b) staggered; (c) overlapping.

discontinuity can be of several types (Fig. 3.14). Modern crystal growth methods, to be discussed in section 3.4, make it possible to fabricate these junctions very sharply, even within an atomic distance. This creates new physical effects with possibilities for new devices. As an illustration, take a heterojunction of GaAs and $Al_xGa_{1-x}As$. By a happy coincidence the Al and Ga atoms are about the same size so that the lattice constant of AlAs differs very little from that of GaAs. Consequently all compositions of $Al_xGa_{1-x}As$ are well matched to GaAs and the interface contains very few misfit defects. This heterojunction is of the normal type with the bandgap difference of about 65% in the conduction band. Because $Al_xGa_{1-x}As$ has a higher bandgap (increasing with x) the electrons in the $Al_xGa_{1-x}As$ now have a higher potential energy than in the GaAs. Now if we make the AlGaAs n type and leave the GaAs undoped then the electrons will spill over from the AlGaAs into the GaAs. With the right choice of layer thickness and doping concentration we can ensure that the AlGaAs is wholly depleted and all electrons stay in the GaAs. Thus we have separated the electrons from their donors and we can have a high electron density together with the mobility of undoped material. This structure can be used as the channel for an FET, as will be discussed in Chapter 6.

We can take this idea one step further and sandwich a thin GaAs layer between two layers of AlGaAs as in Fig. 3.15(a). Now we have a potential well whose width is fully controlled by the growth process. Electrons in this well bounce back and forth between the walls and cannot escape. We have to remember, however, that the behavior of electrons is governed by Schrödinger's equation whose solution is a wavefunction. The bouncing back and forth of the electron is translated into two waves reflecting from the interfaces, resulting in a standing wave, very similar to the particle-in-a-box problem of elementary quantum mechanics. Two of these waves are

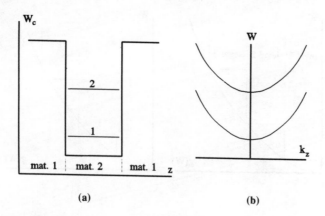

(a)   (b)

Figure 3.15 (a) Quantum well and (b) W–k relation.

sketched in the figure. This means that the component of the wavevector transverse to the interface can only take discrete values, approximately equal to π/d, $2\pi/d$ etc., if d is the width of the GaAs layer. For each of these the wavevector parallel to the interface can take any value. The energy of an electron is, to a first approximation, given by

$$W = \frac{\hbar^2(k_x^2 + k_y^2 + k_z^2)}{2m^*}. \tag{3.32}$$

Now, if the z direction is transverse to the interface, k_z can only take discrete values $n\pi/d$. The conduction band is split up into so-called *subbands*, each characterized by its number n. The minimum energy in each subband is $\hbar^2(n\pi/d)^2/2m^*$, and the W–k relationship looks like that in Fig. 3.15(b). Instead of equation (3.11) we now have, since electrons in a subband can have only one k_x value, each allowing two electrons because of spin,

$$\Delta N(k_\parallel) = \frac{L_x L_y}{(2\pi)^2} \Delta k_x \, \Delta k_y, \qquad k_\parallel = (k_x, k_y). \tag{3.33}$$

Converted to energy this gives, instead of equation (3.13),

$$g(W) = 2 \frac{L_x L_y}{4\pi^2} \frac{2m_c^*}{\hbar^2}. \tag{3.34}$$

In other words, the density per energy interval is now constant. For each additional subband the same amount is added, which gives the picture of Fig. 3.16(a). Multiplying equation (3.34) with the Fermi function and integrating one obtains the equivalent of equation (3.16) for a subband:

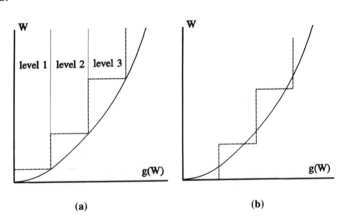

(a)   (b)

Figure 3.16 (a) Density of states for 3D and 2D electrons and (b) a more accurate comparison.

$$n_s = \frac{1}{\pi} \frac{m_c^* kT}{\hbar^2} \ln\left[1 + \exp\left(\frac{W_F - W_i}{kT}\right)\right]. \tag{3.35}$$

The density has been decorated with an index s because it is a density not in space but in the x–y plane, a so-called *sheet density*.

Because the electrons can now move freely only in two dimensions one often speaks of a *two-dimensional electron gas* (2DEG). This has some remarkable properties.

- Since the minimum energy of the electrons is $\hbar^2(\pi/d)^2/2m^*$ above the minimum of the conduction band and, by the same reasoning, the holes have their minimum energy somewhat below the valence band maximum, the apparent bandgap of the material has increased. In optical devices such as light-emitting diodes and lasers this can be seen (literally) in a shortening of the emitted wavelength.
- The density of states is lower at the same energy, because only one value of k_z is allowed per subband. However, putting it this way is misleading: we have to look at the density of states at the minimum energy the electrons can acquire, because this is where most electrons will be. If we do this we see that the density of states is actually much higher in a quantum well (Fig. 3.16(b)). This means that we can pack a lot of electrons in a small volume.
- The scattering of electrons by lattice vibrations (phonons) turns out to be weaker than in the case of three-dimensional electrons, leading to higher mobilities, especially at low temperatures.

The combination of high electron density and high electron mobility, that is made possible by quantum wells, is favorable for field effect transistors with high transconductances and high cutoff frequencies.

3.4 TECHNOLOGY OF III–V SEMICONDUCTORS

3.4.1 Crystal growth

Microwave technology on GaAs started after the discovery of the Gunn effect in 1963 and developed quickly. Epitaxial growth from the gas phase was tried, using AsH_3 and $GaCl_3$, but it soon was found that the best material quality was obtained by the use of liquid phase epitaxy (LPE). This technique relies on the fact that when a solution of a few percent of arsenic in molten gallium is cooled down, GaAs crystallizes. One often uses the fact that·a solution can be supercooled a few degrees below the temperature where liquid and solid are in equilibrium. When the supercooled solution is brought in contact with a crystal, the so-called *seed*, crystallization starts and continues until the equilibrium temperature is reached.

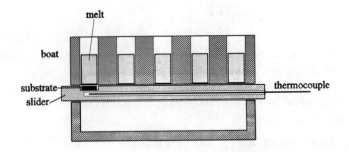

Figure 3.17 Graphite boat for LPE.

In the case of GaAs the temperature is in the range 750–800°C and the solution is kept in a graphite boat. The seed is a thin wafer of substrate material, placed on a graphite slider that can be moved under the boat (Fig. 3.17). If the wafer is monocrystalline the deposited GaAs will be monocrystalline too. By adding small amounts of other elements to the solution n- or p-type layers can be grown and by adding aluminium it is possible to grow $Al_xGa_{1-x}As$. The boat has therefore a number of compartments so that a succession of layers can be grown in one run.

Although LPE produces excellent material quality and has been quite successful (it has made possible for instance the manufacture of semiconductor lasers that operate at room temperature) it has disadvantages that made people look for other growth techniques. The main disadvantages of

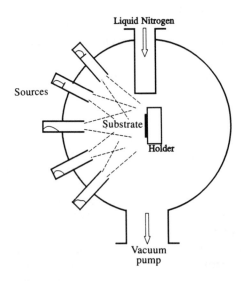

Figure 3.18 Molecular beam epitaxy.

LPE are that the layer thickness is difficult to control accurately, the interfaces between layers of different composition are not sharp and the layers cannot be made uniform over large wafer areas. Two growth techniques have emerged that do not have these disadvantages, i.e. *molecular beam epitaxy* (MBE) and *metal–organic vapor phase epitaxy* (MOVPE). Molecular beam epitaxy is nothing but a very refined high vacuum evaporation technique. It is illustrated in Fig. 3.18.

The elements making up the layer to be grown are evaporated from so-called *effusion furnaces*, i.e. tubes of an inert material such as tungsten, which are heated to a temperature at which the contents slowly evaporates. The beams of particles arrive at the substrate and combine there to form the desired material. To obtain monocrystalline layers the substrate has to be heated to a temperature which can range from 300 to 700 °C.

High demands are put on the vacuum in an MBE machine. At the pressures normally used in evaporating equipment, around 10^{-7} Torr (1 Torr is 1 mm of mercury) the amount of residual gas molecules arriving at the substrate would be of the same order as the evaporated materials, and much of this would be built into the growing crystal, making it useless as a semiconductor. One therefore tries to maintain a vacuum of 10^{-11} Torr and to this end not only big vacuum pumps are necessary but also a shroud of stainless steel which is cooled by liquid nitrogen around the substrate. All this contributes to make MBE a quite expensive affair, not only to install but also to keep running. In return one obtains good material quality, extremely sharp interfaces (to within one atomic monolayer) and excellent uniformity (a few percent over a 3 in wafer).

The growth velocity in MBE is very small, about one monoatomic layer per second, or 1 μm per hour. Evidently this is not a technique for thick layers. One the other hand, since the process can be automated it is easy to grow very complicated structures, e.g. a *superlattice* of a hundred layers of GaAs and AlAs, each only a few nanometers thick. This is something LPE can never do.

The second technique, metal–organic vapor phase epitaxy (Fig. 3.19), is a further development of the ordinary vapor phase epitaxy. In the latter the gases AsH_3 and $GaCl_3$ (plus dopants such as SiH_4) are led over a heated

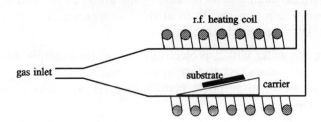

Figure 3.19 Metal–organic vapor phase epitaxy.

substrate and react there to form GaAs. The disadvantage of this process is that HCl, which is formed as a byproduct, can react again with GaAs. This makes the process difficult to control. The way out is to use gaseous compounds that are not easily formed back once they are dissociated. These are the metal–organic compounds such as trimethylgallium $Ga(CH_3)_3$ (usually abbreviated to TMG) and triethylgallium $Ga(C_2H_5)_3$ (TEG). With these compounds the reaction goes only one way and is much easier to control.

With MOVPE (which is also called metal–organic chemical vapor deposition or MOCVD) results comparable with those of MBE can be obtained, although it is more difficult to achieve the same sharpness of interfaces and to make stacks of a great number of layers with good control from beginning to end. Another disadvantage is that some of the gases used are quite toxic so elaborate safety precautions are necessary.

On the positive side the growth rate can be faster and it is easier to grow on a number of wafers at the same time. Add to this the absence of high vacuum problems and many people believe that MOVPE is better suited for production than MBE.

3.4.2 Doping

Doping is usually done during crystal growth. Silicon can also be doped after the growth from gaseous or solid sources at temperatures around 1000 °C. In the III–V technology this is not a workable method since most group V elements, especially phosphorus and arsenic, evaporate easily at higher temperatures. This makes it impossible to use temperatures higher than about 800 °C in processing III–V materials because at these temperatures the group V component would evaporate, thereby destroying the surface layer of the material.

A technique that can be used for all materials is *ion implantation*. In this technique dopant atoms are ionized, accelerated by a high electric voltage (from 50 kV to 1 MV) and shot into the semiconductor. In this way a thin surface layer can be doped. Since the crystal lattice is severely damaged by the high-energy ions, a high-temperature annealing step is necessary afterwards. In Si this is done at 900 °C; in the III–V compounds the temperature cannot be higher than 800 °C because of the above-mentioned evaporation of the group V components. Even at this temperature precautions have to be taken. There are mainly three methods:

- capping the wafer with a protective layer of, for example, Si_3N_4;
- putting two wafers face to face in the furnace;
- annealing in an atmosphere containing large amounts of the volatile component.

Especially for Si bipolar transistors and GaAs MESFETs (Chapter 5) ion implantation is a much used technique.

3.4.3 Etching

In the past, etching of III–V compounds was done by wet chemical techniques. A great variation of etchants is known but they all contain some oxidizing agent, e.g. H_2O_2 or H_2SO_4. Recently plasma etching techniques have come in use, in which an ionized gas or gas mixture is used as the etchant. When the gas is inert, such as argon, it is the kinetic energy of the ions that knocks off atoms from the surface. More commonly a mixture of an inert gas with one that is chemically reactive, e.g. CH_2Cl_2, CH_2F_2 etc., is used. Now the etching is done by a reaction of F or Cl with the semiconductor, which produces volatile compounds. The process is enhanced by the energy of the ions.

Plasma etching has the advantage of much better control of the etched thickness. A disadvantage is that the ions tend to damage the crystal structure in a thin layer near the surface.

3.4.4 Metallization

A device is useless unless it is electrically connected to the external world and these connections are always made by metal. One can distinguish three types of metal contacts to a semiconductor: ohmic, Schottky and metal–insulator–semiconductor. Metals are usually deposited by evaporation or by sputtering. Evaporation is done by heating a crucible containing the metal with, for example, resistance wire. The crucible is made of a material with a high melting point, e.g. tungsten or graphite. Very often the heating current is run through the crucible itself. A disadvantage of this so-called resistance heating is that material of the crucible can evaporate too so that the deposited layer contains many impurities. A cleaner method is to heat the metal by hitting it with an intense electron beam. By concentrating the beam on the centre of the metal only this part is evaporated and the crucible remains relatively cold.

Sputtering is a technique in which the semiconductor wafer and a plate of the metal to be evaporated, the target, are placed opposite each other in a gas discharge. The polarity of the target is negative with respect to the semiconductor so that the heavy gas ions are attracted to it. These ions, which have energies of many kilovolts, impinge on the target and break atoms loose from it. These atoms then fly through the relatively thin gas atmosphere (10^{-3} Torr) and attach to the semiconductor.

Sputtering is used for metals that have low vapor pressures such as tungsten and platinum and also for insulators. A disadvantage is that the target atoms land on the semiconductor with considerable energy so that a thin surface layer of the semiconductor may be damaged. Nowadays, when devices are made of thinner and thinner layers, this becomes an important consideration.

(a) Schottky contacts

Schottky contacts are used either as rectifying junctions for detector and mixer applications or as gate contacts in MESFETs. Many metals, including Au, form good Schottky barriers on GaAs. However, pure Au alloys very easily with GaAs so that any subsequent process step in which the temperature is raised could deteriorate the Schottky barrier. Therefore titanium is preferred followed by platinum and gold. The Au again provides low resistance and good bonding and the Pt is necessary as a barrier against diffusion of the Au into the GaAs.

(b) Ohmic contacts

The ideal ohmic contact provides resistanceless access of current to a semiconductor in both directions. In practice this ideal can only be realized approximately. What one tries to achieve is in fact a Schottky barrier with low height and narrow width so that electrons can easily tunnel through it. Narrow width is obtained by doping the layer under the contact as highly as possible and low height by selecting the right metal. In GaAs a search for the latter is a disappointing affair because, as already said above, as a result of Fermi level pinning the Schottky barrier height is always about 0.7 V independent of the metal. Therefore one selects a metal that has otherwise good properties, in particular gold. The properties that make Au a good contact metal are low resistivity, resistance against oxidation and its softness which makes it easy to bond wires on.

The high n doping under the contact is achieved by alloying the Au with some 12% of germanium. A heat treatment after deposition at about 400 °C diffuses the Ge atoms into the GaAs, thus forming the desired highly doped n-type layer. Unfortunately during this process the Au, by its surface tension, has a tendency to contract into balls, destroying the uniformity of the contact. To prevent this usually nickel is added as a *wetting agent* either by alloying it into the Au before deposition or by evaporating it separately in a sequence Ni–AuGe–Ni.

The heat treatment is traditionally done in a furnace under an inert gas flow such as argon for a time of 10–30 min. This produces an alloying of the metal with the upper layer of the semiconductor. The interface of the alloyed layer can be quite irregular and it is difficult to control the alloy depth. Therefore, again in consideration of the thin device layers used nowadays, this method is more and more being replaced by so-called *rapid thermal annealing* (RTA) where the material is brought very quickly to the desired temperature, kept there for 10–90 s and then cooled down rapidly again. In this way the alloy depth can be kept very shallow while still producing a good contact.

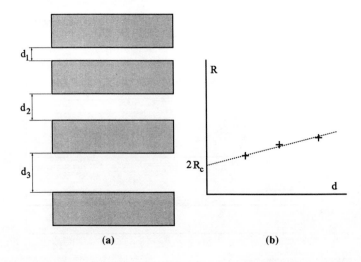

Figure 3.20 The transmission line method for measuring contact resistivity: (a) test structure; (b) results.

(c) Contact resistance

Ohmic contacts have a small, but non-negligible, resistance. Particularly in modern small-size devices the contacts can contribute significantly to the parasitic resistances. The contact resistance is defined as the resistance between the top surface of the contact metal and the first undisturbed semiconductor layer under the metal. For a uniform metallization the resistance will be inversely proportional to the area of the contact, which means that the resistance multiplied by the area, the *specific contact resistivity*, with units of ohm meter squared is the quantity characterizing the contact.

In every semiconductor technology test structures are included with which the contact resistivity can be measured. This makes it possible to monitor the quality of the contact metallization. A popular test method is the transmission line method (TLM) (Fig. 3.20(a)). In this structure a row of contact pads, all of the same size, but at different distances, are made and the resistances between adjacent contacts are measured. When these are plotted against the distance a straight line is obtained (Fig. 3.20(b)). The extrapolation of this line to zero distance is then the part of the resistance ascribed to the two contact pads. When this is divided by 2 and multiplied by the contact area, the specific resistance is obtained. Two comments should be made about this method:

- The rest resistance obtained at zero distance is not entirely due to the contact but also partly to the semiconductor material as a look at the

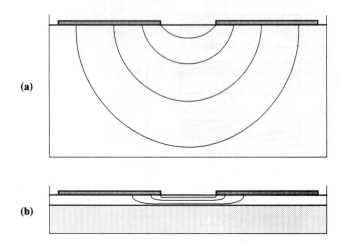

Figure 3.21 Current flows in contact resistance measurements.

current lines in Fig. 3.21(a) will show. The contact resistivity will therefore always be overestimated by this method.

- An extreme instance of this occurs when the contacts are made on a thin conducting layer on a substrate that is semi-insulating or of the opposite doping type. Then the edge of the contact will carry most of the current since the parts farther away have considerable extra series resistance between them and the uncontacted part of the semiconductor (Fig. 3.21(b)). In this case completely wrong results for the specific contact resistivity will be obtained.

On the other hand, in practical devices the current flow will often look like those in Fig. 3.21(b), e.g. near the base contact of a bipolar transistor or the source and drain contacts of a MESFET (Chapter 5). For these cases the TLM can give reliable predictions of contact resistance but one should specify the resistance per unit width (the dimension along the surface and transverse to the current path), not per unit area.

3.4.5 Insulating layers

Insulating layers can have various functions: they can be used as masking layers for etching or ion implantation, or as gate insulators in metal–insulator–semiconductor (MIS) structures. In monolithic integrated circuits they can be employed as the dielectric in capacitors or to separate first and second metallization layers. Finally they can be used to protect the bare surface of the semiconductor against the environment.

One of the most popular insulating materials in Si technology, but perhaps a bit surprisingly also in GaAs technology, is silicon dioxide SiO_2. This is of course due to its excellent electrical and mechanical properties and its chemical inertness. In Si technology it is deposited by oxidation of the silicon or by deposition from the gas phase by oxidation of Silane SiH_4. The latter technique is known as *chemical vapor deposition* (CVD). In GaAs technology it is the only applicable technique.

Attempts have been made for many years to grow a so-called *native oxide* on GaAs by oxidizing both components. This cannot be done as with Si by heating it in an O_2 atmosphere because at the high temperatures necessary the As would evaporate from the material leaving a badly damaged surface covered with Ga droplets. The only feasible way is to use an electrolytic bath with the GaAs wafer as the positive electrode. At this electrode atomic oxygen develops which reacts with both the Ga and the As. This method, which is called *anodic oxidation*, has never produced an insulator of a quality comparable with that of SiO_2 so it has never come into use.

Another good insulator is silicon nitride Si_3N_4 which is also made by CVD from a mixture of SiH_4 and nitrogen N_2 or ammonia NH_3. It is preferred for passivation of surfaces. On GaAs it has the advantage that there is no danger of oxygen diffusing into the underlying material.

Besides the oxides another class of insulating materials is formed by the organic polymers of which especially polyimide is becoming increasingly popular. It forms a hard durable layer with good electrical properties and can be used as a passivation or as a dielectric. These materials are deposited the same way as with photoresist, by spinning on a solution and letting the solvent evaporate at a mildly elevated temperature.

REFERENCES

1. Blakemore, J. S. (1962) *Semiconductor Statistics*, Pergamon, Oxford.
2. Hilsum, C. (1974) Simple empirical relationship between mobility and carrier concentration. *Electron. Lett.*, **10**, 259–60.

FURTHER READING

1. Kittel, C. (1976) *Introduction to Solid State Physics*, Wiley, New York.
2. De Cogan, D. (1987) *Solid State Devices: A Quantum Physics Approach*, Springer, New York.
3. Neamen, D. A. (1992) *Semiconductor Physics and Devices: Basic Principles*, Irwin, Homewood, IL.
4. Sze, S. M. (1981) *Physics of Semiconductor Devices*, 2nd edn, Wiley, New York.
5. Shur, M. S. (1990) *Physics of Semiconductor Devices*, Prentice-Hall, Englewood Cliffs, NJ.
6. Hess, K. (1988) *Advanced Theory of Semiconductor Devices*, Prentice-Hall, Englewood Cliffs, NJ.

PROBLEMS

3.1 If one assumes for the lowest valley in the conduction band of GaAs a Kane-type bandstructure, i.e. $W(1 + \alpha W) = \hbar^2 k^2/2m^*$, what is the maximum velocity (dW/dk) that electrons can reach, disregarding the finite size of the Brillouin zone? Take $\alpha = 0.56/\text{eV}$, $m^* = 0.067m_0$ and $\hbar = 1.05 \times 10^{-34}$ J s.

3.2 What is the energy of an electron in GaAs (Problem 3.1) at the edge of the Brillouin zone ($k = \pi/a$), if $a = 0.288$ nm? What is the corresponding velocity?

3.3 A certain device can be made in Si or in GaAs. In Si the device temperature cannot be raised above 200 °C; in GaAs the allowable temperature is 350 °C. If the chip has a cross-section of $200 \times 200\,\mu\text{m}^2$ and the substrate thickness is $100\,\mu\text{m}$, calculate the heat flow resistance for both materials. If the solder with which the chip is mounted and the heat sink contribute a resistance of 5 K/W and the ambient temperature is 20 °C, calculate the maximum power that each device can dissipate.

3.4 The heat flow resistance of a Baritt diode has been measured at 50 K/W. The diode has a diameter of $50\,\mu\text{m}$ and is made on a chip of size $600 \times 600\,\mu\text{m}^2$. The substrate thickness is $450\,\mu\text{m}$. Calculate approximately the heat flow resistance and compare it with the measured value. Calculate also the thermal capacitance. If one wants to measure this diode under pulsed bias so that its temperature does not rise significantly, give safe but reasonable values for the maximum pulse width and the minimum time between pulses.

3.5 GaAs is often made semi-insulating by doping with Cr. The supposed mechanism is that the unintentional doping is mainly n type and that the Cr atoms form acceptors with energy levels near the middle of the bandgap which capture the electrons. Show that this can work by calculating the free electron density under the following circumstances: bandgap of GaAs, 1.4 eV; Cr density, $10^{18}\,\text{cm}^{-3}$; energy level 0.7 eV below the conduction band; n-type doping at 10^{16}, 10^{17}, $10^{18}/\text{cm}^3$; level 0.05 eV below conduction band.

4

Carrier transport and noise in devices

4.1 DEVICE MODELS

To understand microwave devices or to design a circuit incorporating microwave devices we must possess some knowledge about them. A coherent body of knowledge is called a model. A device model tells us approximately how a device reacts to an electromagnetic field in which it is placed. For semiconductor devices, which in general are small compared with a wavelength, this simplifies to the response of the terminal currents to the terminal voltages. Depending on what the purpose is we distinguish two kinds of models: physical models and equivalent circuits.

The first are based on equations that describe the device physics, in other words the motion of electrons and holes in semiconductors. By solving these equations in the interior of a device with the applied voltages as boundary conditions we can calculate the (d.c. or time-dependent) current–voltage relations. This type of model is especially used in the design of new devices and the optimization of existing ones.

Equivalent networks are used in the design of circuits containing several devices where a full physical model of each device would be too time consuming. They try to represent the current–voltage relations by an equivalent network with a limited number of elements, each of which is described by an analytic and possibly simple equation. One distinguishes small-signal or linear models, in which the elements of the network are dependent on frequency and d.c. bias but not on signal amplitude, and large-signal or non-linear models in which every element is in principle described by a non-linear time-dependent differential equation. Some examples of circuit models for transistors will be given in Chapter 6.

4.2 PHYSICAL MODELING

4.2.1 Introduction

There are many types of physical models. They all have in common that a set of field equations has to be solved. In principle this should be the full set of Maxwell's equations (Chapter 2) but fortunately semiconductor devices are nearly always small compared with the electromagnetic wavelength, which means that the magnetic fields produced by the device currents have little influence. This allows us to restrict the field equations to Poisson's equation which relates the electric field E in the device to the space charge density made up of electrons (n), holes (p) and ionized donors and acceptors (N_D and N_A respectively):

$$\nabla \cdot \varepsilon E = q(N_D - N_A + p - n) \tag{4.1}$$

where ε is the dielectric permittivity of the semiconductor. This is complemented by calculations of the charge and current density distributions. For these a number of techniques are available ranging from comparatively simple ones where the charge and current densities are calculated approximately to extremely detailed ones where almost every electron is followed on its way through the device. Below they are treated in the historical order in which they were introduced.

4.2.2 The drift–diffusion model

In this model the average motion of the carriers is described by two simple equations giving the conservation of particle number and of average momentum:

$$\frac{\partial n}{\partial t} + \nabla \cdot n v_n = G - R \tag{4.2}$$

for the electron density and a similar equation for holes. The right-hand side consists of a generation term G and a recombination term R, and

$$v_n = - \mu_n E - \frac{D_n}{n} \nabla n \tag{4.3a}$$

where the electron mobility μ_n and the diffusivity D_n are introduced. The index n indicates that these are electron quantities. A similar equation exists for holes:

$$v_p = \mu_p E - \frac{D_p}{p} \nabla p \tag{4.3b}$$

Note that in the case of electrons, since the electron charge is negative and μ_n is defined as a positive number, a minus sign has to be introduced. These equations can be derived by simple statistical arguments and are moreover supported by a wealth of experimental evidence. These equations work fine for devices with large dimensions that are used at low frequencies and where the electric fields are not very high. In particular the latter condition is seldom fulfilled: in most devices the fields normally rise so high that velocity saturation (Chapter 3) sets in. As Fig. 3.9 shows this occurs in Si for fields around 10 kV/cm and in GaAs at 3 kV/cm. To accommodate this phenomenon the equations have been generalized by assuming that μ and D are instantaneous functions of the local electric field. This works well in most cases. Notable exceptions are p–n and Schottky junctions where even at zero bias high built-in electric fields exist. However, at zero bias there is an equilibrium situation so that μ and D should have their low-field values.

The drift–diffusion method with field-dependent mobility and diffusivity has been the standard method for semiconductor device simulation for many years and has been implemented in a number of device simulators, e.g. MINIMOS and PISCES. It has worked well as long as the device dimensions and the operating time scales were not too small, the transitions being roughly at $0.5\,\mu m$ and 10 ps, respectively. However, in the analysis of high-frequency Gunn diodes it soon became clear that the drift–diffusion method was not able to predict the frequency limits of the Gunn effect. The reason is that the Gunn effect is based on intervalley transfer of electrons, a process that takes a certain time, and this is not accounted for by using a field-dependent mobility. Even in the absence of intervalley transfer the average electron velocities do not follow the electric field variations instantaneously. As can be seen from Fig. 3.11 it takes a few picoseconds for the velocity overshoot to pass, so this will be noticed at frequencies over 10 GHz and device dimensions under $0.5\,\mu m$. The reason is that electrons have a thermal velocity distribution which at zero electric field has the form of a Maxwell–Boltzmann distribution. The electric field first imparts the same acceleration to all electrons but after some time collisions start to randomize the velocity again. The time constant with which this happens is called the *momentum relaxation time* and is around 1 ps at room temperature. The extra energy acquired is given off to the crystal lattice (by the emission of phonons) with a time constant called the *energy relaxation time*. Typically this is an order of magnitude larger than the momentum relaxation time. Thus, it takes something of the order of the energy relaxation time to establish a stationary velocity distribution and this phenomenon which is of great influence on the high-frequency behavior of devices is not accounted for in the drift–diffusion method. To remedy this shortcoming the drift-diffusion model is extended into what is usually called the *hydrodynamic model*, sometimes the energy relaxation model, and which is gaining increasing popularity among device modelers. In this model the above-mentioned processes of momentum and energy relaxation are taken care of by introducing the electron temperature as the variable on which μ and D depend, for which then an extra differential equation is necessary. To understand this system of equations it is necessary to go a little deeper into the statistical mechanics of electron gases.

4.2.3 The Boltzmann equation

In any semiconductor device a large number of electrons and/or holes are moving around. For instance, in a MESFET with a gate length of $1\,\mu m$, a gate width of 100 μm, a channel thickness of 0.1 μm and a doping density of $10^{17}/cm^3$ about 10^6 electrons determine the channel current. If one includes the higher-doped contact regions this number can be an order of magnitude higher. The description of the behavior of such a many-particle ensemble is a task for kinetic gas theory which was developed in the middle of the last century to analyze and explain the behavior of gases. In the early 20th century it was expanded with the inclusion of electric and magnetic fields as the controlling forces so that also ionized gases could be treated [1].

With some adaptations this theory is eminently suited to the treatment of electrons and holes in solids. The heart of it is the assumption that, since it is impossible to know the velocity of every particle at every moment, the question is turned around and velocity is treated as a coordinate, like spatial position. The ensemble is then characterized by specifying the number of electrons at a certain position having a certain velocity. Mathematically this is expressed by the *distribution function* $f(\boldsymbol{u}, \boldsymbol{r}, t)$ where $\boldsymbol{u} = (u_x, u_y, u_z)$ is the set of *velocity coordinates* and $\boldsymbol{r} = (x, y, z)$ is the set of *space coordinates*. The combination of real space and velocity space is called *phase space*. The meaning of this function, henceforth also called the DF, is that $f(\boldsymbol{u}, \boldsymbol{r}, t) \Delta u_x \Delta u_y \Delta u_z \Delta x \Delta y \Delta z$ gives the number of particles in the infinitesimal phase space volume element $\Delta u_x \Delta u_y \Delta u_z \Delta x \Delta y \Delta z$ at a time t. From the DF all desired information about the ensemble can be calculated.

In the case of semiconductors the above needs some modification. Strictly speaking the coordinate of velocity space should be not velocity but momentum. In dilute gases this means only a scale change since momentum is velocity multiplied by mass. However, as has been discussed in Chapter 3, in a semiconductor electron momentum is given by $\hbar \boldsymbol{k}$ where $\boldsymbol{k}$ is the wavenumber of the Schrödinger wave of the electron. The electron velocity $\boldsymbol{u}$ on the other hand is not simply equal to $\hbar \boldsymbol{k}/m$ but has to be derived from the energy $W(\boldsymbol{k})$ which in a non-parabolic, anisotropic energy band can be a complicated function of $\boldsymbol{k}$. In semiconductors therefore one prefers to use $\boldsymbol{k}$ instead of $\boldsymbol{u}$ as a set of coordinates.

Macroscopic quantities can be obtained as averages over the DF. By integrating the DF itself over $\boldsymbol{k}$ space one obtains the number of particles in $\Delta x \Delta y \Delta z$, that is the density in real space. Similarly, by first multiplying f with $\boldsymbol{u}(\boldsymbol{k})$ and performing the same integration one obtains the average velocity $\boldsymbol{v}$, and multiplying with $W(\boldsymbol{k})$ one obtains the average energy W. In equation form we have the following results:

● the *density* in real space is

$$n(\boldsymbol{r}, t) = \int \int \int f(\boldsymbol{k}, \boldsymbol{r}, t) \, dk_x \, dk_y \, dk_z, \qquad (4.4)$$

- the *average velocity* is

$$v(r, t) = \frac{1}{n} \int \int \int u(k) f(k, r, t) \, dk_x \, dk_y \, dk_z, \tag{4.5}$$

- the *average energy* is

$$W_e(r, t) = \frac{1}{n} \int \int \int W(k) f(k, r, t) \, dk_x \, dk_y \, dk_z, \tag{4.6}$$

- the *heat flow vector Q* is

$$Q(r, t) = \frac{1}{n} \int \int \int u(k) W(k) f(k, r, t) \, dk_x \, dk_y \, dk_z, \tag{4.7}$$

and so on. The integrations are all from $-\infty$ to $+\infty$.

The quantities n, v, W and Q are called the *higher moments* of the distribution function. They form a series that can be continued indefinitely although their physical meaning becomes less evident as we go further down in the series. If one has a Maxwell–Boltzmann distribution

$$f(u) = n \left(\frac{m}{2\pi kT} \right)^{3/2} \exp\left(-\frac{m|u - v|^2}{2kT} \right)$$

then evidently v gives the center of the distribution and T is a measure of the spread around this center. The vector Q is a measure for the deviation from a Maxwellian distribution and represents that part of the flow of energy carried by the electrons that is connected with an anisotropy in the distribution function.

The evolution of the distribution function in space and time is given by the Vlasov–Boltzmann equation, also called the Boltzmann transport equation (BTE), which is

$$\frac{\partial f}{\partial t} + \frac{\partial k}{\partial t} \cdot \frac{\partial f}{\partial k} + \frac{\partial r}{\partial t} \cdot \frac{\partial f}{\partial r} = \left(\frac{\partial f}{\partial t} \right)_c. \tag{4.8}$$

In this equation $\partial k/\partial t$ is a measure of the acceleration by electric and magnetic fields given by Newton's law,

$$\hbar \frac{\partial k}{\partial t} = q(E + u \times B), \tag{4.9}$$

and $\partial r/\partial t$ of course is the velocity u. The operators $\partial f/\partial k$ and $\partial f/\partial r$ stand symbolically for $(\partial/\partial k_x, \partial/\partial k_y, \partial/\partial k_z)$ and $(\partial/\partial x, \partial/\partial y, \partial/\partial z)$, respectively. The right-hand side is a symbolic notation for the change of the distribution function caused by collisions. It can further be specified as

$$\left(\frac{\partial f}{\partial t} \right)_c = \sum_i \int \int \int [- S_i(k, k') f(k) + S_i(k', k) f(k')] \, dk' \tag{4.10}$$

where dk' stands symbolically for dk'_x dk'_y dk'_z and $S(k, k')$ is the probability that a particle is scattered by a collision from a point k to a point k' in velocity space. The first term of the right-hand side of equation (4.10) represents processes that scatter an electron out of the volume element dk and the second represents scattering into dk. The summation is over different collision processes. The functions S_i can be calculated from the physics of the collision processes. Note that the first term in the right-hand side of equation (4.10) can be integrated directly:

$$-\sum_i \int \int \int S_i(k, k') \, dk' \, f(k) \equiv - \nu(k) f(k) \qquad (4.11)$$

where $\nu(k)$ has the dimension of a frequency and accordingly is called the *scattering frequency* or *scattering rate*. It is the sum of the scattering frequencies of the different scattering processes.

Since the BTE is a set of partial integrodifferential equations its solution is a formidable task. In the past various methods have been worked out to perform this task. The best known of these is the method of Rees [2], where the time- and space-independent BTE is solved iteratively. First the collision term (4.10) is calculated using an assumed DF. Then equation (4.8) is solved, omitting the time- and space-dependent terms, and the process is repeated using the updated DF. It can be shown that in this method each iteration step corresponds to a time step for the time-dependent BTE. In another method the dependence of the DF (again independent of time and space) on the angle between k vector and electric field is developed in Legendre polynomials and the dependence on the magnitude $|k|$ then is solved numerically [2]. In a multivalley semiconductor such as GaAs this has to be done for each valley separately and equation (4.10) has to contain

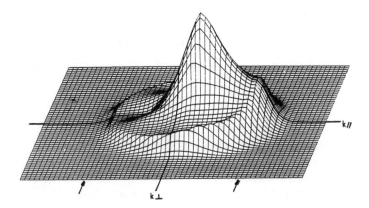

Figure 4.1 Electron distribution function in the Γ valley of GaAs at room temperature and an electric field of 10 kV/cm.

the contributions of intervalley scattering. A DF calculated with the last-mentioned method [2] is shown in Fig. 4.1. This clearly shows the peculiar form a DF can take in a semiconductor with a complicated band structure. The 'crater rim' is caused by electrons scattered back from the L valley into the Γ valley. They all have about the same energy, and thus the same magnitude of wavevector, but the angle of k is randomly distributed.

The only method that has survived to the present day is the so-called Monte Carlo method (section 4.2.4). Before going into this, however, we shall show how the BTE can be used to provide a theoretical basis for the drift–diffusion model and its extension into a hydrodynamic model.

The same procedure of multiplication and averaging that has been applied above to the distribution function to obtain the higher moments of the DF can also be applied to obtain higher moments of the BTE itself, that is differential equations for the macroscopic quantities. For instance, when equation (4.8) is integrated over k space one obtains equation (4.2), the continuity equation. When equation (4.8) is first multiplied with k and then integrated we obtain a transport equation for the average momentum which in the case of an isotropic parabolic band, where $\hbar k = m^* v$, can be simplified to

$$\frac{\partial v}{\partial t} + (v \cdot \nabla)v + \frac{q}{m} E + \frac{q}{mn} \nabla nkT_e = \left(\frac{\partial v}{\partial t}\right)_c \tag{4.12}$$

In this equation we recognize the driving force of the electric field and also a diffusion term, the one with ∇nkT. This will be discussed in more detail below. Multiplying equation (4.8) with $W(k)$ and integrating yields, again simplifying to the case of an isotropic parabolic band,

$$\frac{\partial W_e}{\partial t} + v \cdot \nabla W_e - qv \cdot E + \frac{1}{n} \nabla \cdot (vnkT_e + Q) = \left(\frac{\partial W_e}{\partial t}\right)_c \tag{4.13}$$

Here W_e is the *average* electron energy and T_e the electron temperature. They are connected by the equation

$$W_e = \tfrac{3}{2} kT_e + \tfrac{1}{2} mv^2 \tag{4.14}$$

In equation (4.13) we also recognize a driving force, namely the term $qv \cdot E$, which gives the energy input by the electric field. It is balanced by the loss term in the right-hand side which gives the energy transferred to the lattice by collisions. We also see that in each equation the next moment of the DF appears so they form a never-ending series. Unless we find a method to truncate this series we have not gained anything by going from the BTE to its higher moments. The simplest method is to truncate the set after the first two, so that only equations for the most wanted quantities n and v are retained. This leads to the *drift–diffusion model* already discussed. In this model the assumption is made that the electron temperature T_e is equal to the lattice temperature T_L. Another assumption in this model is that the collision term in the equation for v can be represented by the *relaxation time approximation*:

$$\left(\frac{\partial v}{\partial t}\right)_c = -\frac{v}{\tau_m} \tag{4.15}$$

where the *momentum relaxation time* is introduced. Finally one assumes that the relevant time scales are long compared with the momentum relaxation time (which is of the order of 1 ps) and the distance scales are long compared with the momentum relaxation time multiplied by the maximum value that v can reach. This enables us to discard the first two terms of equation (4.12) so we are left with

$$\frac{q}{m} E + \frac{q}{mn} \nabla nk T_L = -\frac{v}{\tau_m}. \tag{4.16}$$

If we now multiply this with τ_m and rearrange it a little we obtain the drift–diffusion equation in the form presented above (equation (4.3) in which form it is generally used for device modeling [3, 4]. Thus, by starting from the BTE it is possible to show clearly what are the underlying assumptions and simplifications of the drift–diffusion equation. Note also that μ and D are connected by the *Einstein relation*:

$$D_i = \mu_i \frac{kT_L}{q}. \tag{4.17}$$

It will be clear from the foregoing that the drift–diffusion technique is bound to fail at high frequencies. The first thing one can do is to restore the first two terms of equation (4.12) which have been discarded to arrive at equation (4.16). Because these terms represent the influence of the finite mass of the electrons they take care of the initial linear increase of the velocity in Fig. 3.11 but they do not lead to velocity overshoot. To account for this phenomenon one has to drop the assumption that $T_e = T_L$. This makes it necessary to extend the system of equations with equation (4.13). Also, it has been found essential to make m^* and τ_m functions of the electron energy W_e instead of the electric field. It can be shown that this leads to velocity overshoot (Problems 4.3 and 4.4).

Including an equation for electron energy one needs an expression for the collision term in this equation. Following the same line of thought as before we do this by introducing an energy relaxation time τ_e:

$$\left(\frac{\partial W_e}{\partial t}\right)_c = -\frac{W_e - W_0}{\tau_e}. \tag{4.18a}$$

W_0 is assumed to be equal to the equilibrium energy:

$$W_0 = \tfrac{3}{2} k T_L. \tag{4.18b}$$

Finally one has to make an assumption about the heat flow vector Q. The simplest assumption is to put it equal to zero, meaning that one neglects effects of the anisotropy of the distribution function. In this way one arrives at the following equation:

$$\frac{\partial W_e}{\partial t} + v \cdot \nabla W_e - q v \cdot E + \frac{1}{n} \nabla \cdot (v n k T_e) = - \frac{W_e - \tfrac{3}{2} k T_L}{\tau_e} \tag{4.19}$$

It is evident that this is still a rather crude approximation for, say, electrons in GaAs where in fact we have two populations, the light central valley electrons and the heavy satellite valley electrons. To be more correct one should have a separate set of equations for each valley and an extra term in the continuity equation to represent intervalley transfer. It has been found, however, that this does not give more accurate results than the simple approach where all valleys are lumped together. This is probably due to the fact that the DF in the central valley of GaAs has a very peculiar shape at high fields.

4.2.4 The Monte Carlo method

In this method an electron is followed through its life history which consists of a series of free flights under influence of electric and magnetic fields, interrupted by collisions. The length of each free flight, the scattering process that terminates it and the outcome of the scattering process are chosen by drawing random numbers in the following way.

First a fictitious scattering process called *self-scattering* introduced by Rees [2] is added, the outcome of which is $k' = k$ so nothing happens. Since this process changes nothing its scattering rate can be chosen freely and it is adjusted such that the total scattering rate becomes a constant:

$$\sum_i \nu_i(k) + \nu_{ss}(k) = \Gamma. \tag{4.20}$$

The probability of a free time τ between scatterings now is

$$P(\tau) = \exp(-\Gamma\tau). \tag{4.21}$$

So, if we draw a random number r with equal probability between 0 and 1, then the corresponding free time is

$$\tau = -\frac{1}{\Gamma} \ln(r). \tag{4.22}$$

Note that this inversion is much more complicated without self-scattering since then k and with it $\nu_i(k)$ are functions of time and the argument of the exponential is an integral.

The next step is to decide which scattering process will occur at the end of the free time. For a particular electron with wavevector k we can divide all $v_i(k)$ by Γ and plot them on a line $(0 \ldots 1)$. If we now again draw a random number r between 0 and 1 then, if $0 \leqslant \Gamma r \leqslant v_1$, we have process 1, if $v_1 < \Gamma r \leqslant v_1 + v_2$, process 2 etc. and if $\sum_i v_i < \Gamma r \leqslant \Gamma$ self-scattering so that the electron continues its free flight.

In the same way the outcome of the scattering process is decided by drawing random numbers. This usually is a complicated procedure since the probability of an end state k' is for most processes a complicated function of the angle between k and k'. Details can be found in the literature [5, 6].

Originally the Monte Carlo method was applied to one electron which was followed through a large number of scatterings, at least 10^4, in a constant electric field. After this the average velocity could be calculated, giving one point of the velocity–field characteristic. Nowadays, with more powerful computers, it is possible to simulate in this way many electrons simultaneously and to obtain ensemble averages at every instant of time. This is called the ensemble Monte Carlo (EMC) method. An example is shown in Fig. 4.2 where the electric field (uniform in space) is switched on at $t = 0$ [7]. The phenomenon of velocity overshoot, discussed in Chapter 3, is clearly visible.

It can be shown that the outcome of the Monte Carlo method approaches the solution of the Vlasov–Boltzmann equation when either one electron is followed over a sufficiently large number of scatterings or when an ensemble of sufficient size is used. In practice something in the order of 10^4 scatterings or 10^4 electrons is necessary.

It is also possible to calculate diffusion coefficients with the EMC method. Initially all electrons are placed at $r = 0$. During the calculation they will spread out in space and their positions are calculated by integrat-

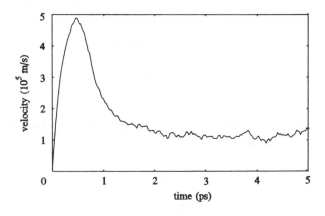

Figure 4.2 Velocity overshoot in GaAs calculated with the EMC method.

ing the velocities. It can now be shown that the spread of the electrons increases linearly with time, after the overshoot phenomenon has passed, in the form of the equation

$$\langle z^2 \rangle - \langle z \rangle^2 = 2Dt \tag{4.23}$$

where D is the diffusion coefficient discussed above.

The Monte Carlo method allows one to take account of the fine details of semiconductor band structure and scattering process and is therefore a powerful tool to analyze the behavior of charge carriers in semiconductors. Its biggest disadvantage is that it uses a large amount of computer time, especially in the EMC mode. Another problem is that one needs detailed information about band structure, electron–phonon interactions etc. which is not always available. In this case one has initially to guess certain parameters and then to deduce their most probable values by comparing calculated and measured v–E characteristics. Finally, as Fig. 4.2 shows, the results are always and inevitably noisy owing to the statistical nature of the simulation.

The EMC method can also be used for device simulation. In this case the actual number of electrons in the device is represented by a much smaller

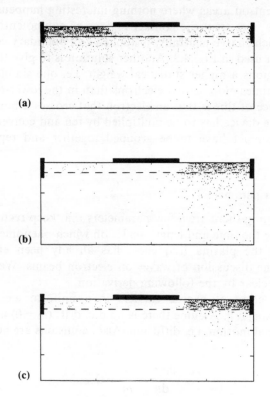

Figure 4.3 Electron distribution in an HEMT: (a) Γ, (b) L and (c) X valleys.

number in the calculations with a correspondingly greater charge. At regular time intervals Poisson's equation is solved using the calculated positions of the electrons to determine the space charge. In the next time step the electrons drift under the influence of the updated electric field. Figure 4.3 shows the electron distribution in an HEMT (Chapter 6) obtained this way [8]. Each point in this figure gives the position of a simulated electron which itself represents a few hundred real electrons.

One clearly sees the depletion region in the AlGaAs layer under the gate, and also the narrow channel at the AlGaAs–GaAs heterojunction (cf. section 5.5). The electrons in the Γ, L and X valleys are shown. At the source side and under the gate, where the electrons have not taken up much energy from the electric field, nearly all of them are in the Γ valley; however, near the drain where the highest electric fields exist most electrons are hot and have transferred to the L and X valleys. This is an example of the detailed information that can be obtained from a Monte Carlo calculation.

One problem with device simulation by the EMC method is that most electrons reside in the highly doped contact regions so that much calculation time is spent on areas where nothing interesting happens. One way to get around this is to treat the contact regions as equipotential regions, on the edge of which charge neutrality is applied as a boundary condition. This is the approach used in Fig. 4.3. Another solution is to give the electrons in the contact regions a higher statistical weight, i.e. one simulation electron represents ten times more actual electrons than in the lower-doped regions. The consequence of this is that an electron that crosses a boundary from a contact into the device has to be multiplied by ten and conversely electrons that enter a contact have to be grouped together and replaced by one electron.

4.2.5 Important parameters

In device modeling there are a few parameters that keep recurring and that often determine the time and length scales on which phenomena take place. One of these, the plasma frequency, has already been encountered in Chapter 2 in the discussion of waves on electron beams. We can make its meaning more clear by the following derivation.

Take a collection of charge carriers in a uniform a.c. electric field $E(t) = \mathrm{Re}[E_1 \exp(j\omega t)]$. Assume there is no d.c. drift ($v_0 = 0$) and neglect the effects of thermal motion, i.e. diffusion. Also, collisions are neglected. Then Newton's equation gives

$$\frac{\mathrm{d}v(t)}{\mathrm{d}t} = \frac{q}{m} E(t)$$

and with a uniform density n we obtain for the current density

$$\frac{\mathrm{d}J(t)}{\mathrm{d}t} = \frac{q^2 n}{m} E(t).$$ (4.24)

Now look at a small cylindrical space, the cylinder axis parallel to the electric field, with length d and cross-sectional area A (Fig. 4.4(a)). For the current through and the voltage across this element we can now write

$$\frac{\mathrm{d}I_\mathrm{c}(t)}{\mathrm{d}t} = \frac{1}{L} U(t), \text{ with } L = \frac{md}{q^2 nA}.$$ (4.25a)

Clearly this element of space behaves inductively owing to the inertia of the carriers. To the carrier current we now have to add the dielectric current

$$I_\mathrm{d} = C \frac{\mathrm{d}U(t)}{\mathrm{d}t}, \text{ with } C = \frac{\varepsilon A}{d}.$$ (4.25b)

Since the total current is the sum of these two we can represent the element of space by a parallel circuit consisting of the inductance L and the capacitance C (Fig. 4.4(b)). This is a resonant circuit and its resonant frequency is given by

$$\omega_\mathrm{p}^2 = \frac{1}{LC} = \frac{q^2 n}{\varepsilon m},$$ (4.26)

which is the formula for the plasma frequency already encountered in Chapter 2. Evidently the plasma frequency is the natural resonant frequency of a collection of charge carriers.

A second parameter occurring regularly, especially in semiconductor problems, is the dielectric relaxation time. In the above collisions were neglected, but here we will assume that they are dominant and that inertia can be neglected. A relaxation time model as discussed in section 4.2.2 then gives for the velocity

$$v = \frac{q \tau_\mathrm{m}}{m} E.$$ (4.27)

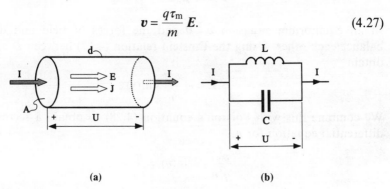

(a) (b)

Figure 4.4 The a.c. behavior of a plasma: (a) an element of space; (b) its electric network representation.

Suppose now that normally the charge of the mobile carriers is balanced by an equal opposite charge of ions, but locally there is an excess charge. This produces an electric field according to Poisson's equation:

$$\nabla \cdot E = \frac{q(n_0 - n)}{\varepsilon}. \tag{4.28}$$

Also, the equation for charge continuity has to be satisfied:

$$\frac{\partial n}{\partial t} + \nabla \cdot n\boldsymbol{v} = 0. \tag{4.29}$$

The last equation is non-linear so to solve the system we have to linearize it. To this end we assume that $n - n_0$ and $\boldsymbol{v}$ remain small. Then we can approximate $\nabla \cdot n\boldsymbol{v}$ by $n_0 \nabla \cdot \boldsymbol{v}$ and by substituting equations (4.27) and (4.28) in (4.29) we obtain

$$\frac{\partial n}{\partial t} = -\frac{n - n_0}{\tau_d}, \tag{4.30}$$

where

$$\tau_d = \frac{\varepsilon m}{q^2 n_0 \tau_m} \tag{4.31}$$

is called the *dielectric relaxation time*. Note that $q^2 n_0 \tau_m / m = q\mu n_0$ is the *conductivity* σ of the material so we can write τ_d also as $\tau_d = \varepsilon/\sigma$. Equation (4.30) shows that any small charge disturbance in a conductor will die away with a time constant equal to the dielectric relaxation time. Especially in metals with their large conductivities this time constant is extremely short, so this explains why one does not find space charge in the inside of a metal.

A third parameter often occurring in conductive media is the Debye length. It plays a role wherever diffusion is important. Take the drift–diffusion equation (4.3):

$$\boldsymbol{v}_n = -\mu_n E - \frac{D_n}{n} \nabla n.$$

In an equilibrium situation $\boldsymbol{v} = 0$ and the forces of field and diffusion balance each other. Using the Einstein relation (4.17) between D and μ we obtain

$$E = -\frac{kT}{q} \nabla \ln(n).$$

We combine this with Poisson's equation (4.28) to obtain a second-order differential equation for n:

$$\frac{\varepsilon kT}{q^2 n_0} \nabla^2 \ln(n) = \frac{n}{n_0} - 1. \tag{4.32}$$

The first fraction has the dimension of a length squared and its square root is called the Debye length:

$$\lambda_D = \left(\frac{\varepsilon k T}{q^2 n_0} \right)^{1/2} \tag{4.33}$$

Since ∇^2 can be written as $\partial^2/\partial x^2 + \partial^2/\partial y^2 + \partial^2/\partial z^2$ we can make equation (4.32) dimensionless by introducing reduced coordinates x/λ_D etc. This makes it clear that the length scale for problems described by equation (4.32) is the Debye length. As is clear from the starting equation these are problems where field and diffusion (approximately) balance each other. In semiconductors there are many instances of this, e.g. a p–n junction under small bias and the transition region between an n^+ contact and an n region. In all these cases there is outdiffusion of carriers creating a space charge, which in turn produces an electric field that opposes the outdiffusion. In the next two chapters, where we will discuss depletion layers, we will often assume for simplicity that the edges of these layers were infinitely sharp. It will now be clear, however, that this is not true and that the depletion edges are smeared out over a distance in the order of a Debye length (Problem 4.7).

4.3 EQUIVALENT NETWORK MODELS

In applications a semiconductor device is always part of a circuit. This means that a model of a device is of little use on its own. What one would like to know is how a device behaves in interaction with a circuit. The complexity of circuits can vary greatly, from a diode in a resonator to a digital integrated circuit containing thousands of transistors. In the former case it is feasible to couple a physical model of the diode as described above with a few differential equations describing the passive part of the circuit; in the latter case this is out of the question. Thus there is a need for compact device models, particularly for transistors, that make a circuit analysis feasible.

Besides the well-known circuit simulators such as SPICE which are able to handle discrete-element circuits, programs have been developed which aim at microwave circuits, which consist mostly of transmission line sections and other distributed components. Examples are COMPACT®, TOUCHSTONE® and MDS®.

In all these programs compact device models are necessary. Examples will be given in Chapter 6. These models are based on a division of the device into different regions and the derivation of an approximate analytical model for each region. At first sight one would expect that the parameters in these models can be found by physical modeling of the device as explained above. However, this is not that simple. In a physical model based on partial differential equations and/or a Monte Carlo simulation it is not so easy to draw the boundaries between different regions since the transitions are always gradual. For instance, it is very difficult to decide from a physical model which part of the charge in the depletion region of a MESFET should be allocated to the gate–source capacitance and which to the gate–drain capacitance. Another problem is that the assumptions made in a compact model to be able to obtain an analytical solution are not always compatible with the carrier transport equations in the physical model. So in practice the parameters of compact models are determined mostly by comparison with experiments.

4.4 NOISE

4.4.1 Introduction

The analysis of noise is based on the theory of stochastic signals which is treated in many eminent handbooks [9–11]. A very brief summary is given below.

A function of time $x(t)$ is called stochastic if its behavior cannot be predicted. It is caused by some process and if we had the possibility to look at a great number of similar processes we could form the *ensemble average* or *expectation value* $E[x(t)]$ as the average of all the functions $x(t)$ at a certain time. If this is independent of time we have what is called a *stationary process.*

Usually we are not that fortunate and we have only one signal to look at, e.g. the noise produced by a specific device. We can observe this signal for a long time and form the *time average*

$$\langle x(t) \rangle = \frac{1}{2T} \int_{t-T}^{t+T} x(t)\, \mathrm{d}t. \tag{4.34a}$$

If this becomes independent of time as T goes to infinity and equal to the ensemble average we speak of an *ergodic process.* When we talk of noise we usually mean the fluctuations of a quantity around its average. A measure for these fluctuations is the *variance*

$$\mathrm{var}[x(t)] = \langle [x(t) - \langle x(t) \rangle]^2 \rangle = \langle [x(t)]^2 \rangle - \langle x(t) \rangle^2. \tag{4.34b}$$

In the following we assume that we are dealing with signals whose average is zero (or from which the average has been subtracted).

Closely related to the variance is the *autocorrelation function*

$$R_{xx}(t) = E[x(\tau)x(\tau - t)] \tag{4.35}$$

where we have assumed that $x(t)$ is a real function. For a stationary process $R_{xx}(t)$ is an even function of t: $R_{xx}(-t) = R_{xx}(t)$. For an ergodic process we may approximate R by

$$R_{xx}(t) = \lim_{T \to \infty} \frac{1}{2T} \int_{-T}^{T} x(\tau)x(\tau - t)\, \mathrm{d}\tau. \tag{4.36}$$

Instead of the variance it is better to use the expectation value of the square of $x(t)$. Obviously this is a positive number and equal to

$$E\left[|x(t)|^2\right] = R_{xx}(0). \tag{4.37}$$

The importance of the autocorrelation function is that it can be Fourier transformed:

$$S_{xx}(\omega) = \int_{-\infty}^{\infty} R_{xx}(t) \exp(-j\omega t)\, dt \tag{4.38}$$

with the inverse transform

$$R_{xx}(t) = \frac{1}{2\pi} \int_{-\infty}^{\infty} S_{xx}(\omega) \exp(j\omega t)\, d\omega \tag{4.39}$$

from which it follows that

$$E\left[|x(t)|^2\right] = \frac{1}{2\pi} \int_{-\infty}^{\infty} S_{xx}(\omega)\, d\omega. \tag{4.40}$$

In other words, $S_{xx}(\omega)$ is the *spectral density* of $|x(t)|^2$. When $R_{xx}(t)$ is an even function of t, $S_{xx}(\omega)$ is an even function of ω.

The usual method to measure noise is to filter it through a narrow band filter and to square and average (Fig. 4.5). The response of the filter in the time domain is

$$y(t) = \int_{-\infty}^{\infty} x(t - \tau)h(\tau)\, d\tau \tag{4.41}$$

where $h(\tau)$ is the response to a delta function at $t = 0$. Via the cross-correlation between $y(t)$ and $x(t)$ we can calculate the autocorrelation of $y(t)$ whose Fourier transform is

$$S_{yy}(\omega) = |H(\omega)|^2 S_{xx}(\omega) \tag{4.42}$$

where $H(\omega)$ is the Fourier transform of $h(t)$. Now, from equation (4.40) we have

$$E\left[|y(t)|^2\right] = \frac{1}{2\pi} \int_{-\infty}^{\infty} S_{yy}(\omega)\, d\omega = \frac{1}{2\pi} \int_{-\infty}^{\infty} |H(\omega)|^2 S_{xx}(\omega)\, d\omega. \tag{4.43}$$

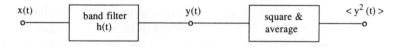

Figure 4.5 Noise measurement.

If the band filter in Fig. 4.5 has a band-pass characteristic

$$H(\omega) = 1, \quad \omega_1 < |\omega| < \omega_2$$
$$0, \text{ outside this range,}$$

then equation (4.43) can be written as

$$E\big[|y(t)|^2\big] = \frac{1}{2\pi} \int_{\omega_1}^{\omega_2} S_{xx}(\omega)d\omega + \frac{1}{2\pi} \int_{\omega_2}^{\omega_1} S_{xx}(\omega)\, d\omega$$

$$\approx \frac{\omega_2 - \omega_1}{2\pi} [S_{xx}(\omega_0) + S_{xx}(-\omega_0)] \tag{4.44}$$

where $\omega_0 = (\omega_1 + \omega_2)/2$. Evidently with a band filter we can measure the spectral density. When $S_{xx}(\omega)$ is an even function, which is the case for stationary noise, we can simplify this to

$$E\big[|y(t)|^2\big] = 2 S_{xx}(\omega_0) \Delta f. \tag{4.45}$$

So what we measure is actually twice the spectral density. In practice $x(t)$ is either a current or a voltage.

Note that in the literature on device noise mostly reference is made to the measured spectral intensity $2S_{xx}(\omega)$ for which the symbol $S_x(\omega)$ is used. One also finds symbols such as $\overline{\delta i^2}$ or $\overline{i_n^2}$ instead of $S_1 \Delta f$. Confusion can arise when one mixes the different symbols without keeping their definitions in mind. To avoid this we will use S with a double subscript for the double-sided spectral density defined by equation (4.38). If the measured spectral intensity is meant we will denote it by a single subscript. In the following we will express everything in the measured spectral densities.

Stochastic signals in networks do not behave differently from deterministic signals. They obey Ohm's law and the power dissipated in a resistor is

$$P(t) = i^2(t)R = v^2(t)/R. \tag{4.46}$$

As a consequence the spectral density of dissipated power in an impedance Z (admittance Y) is according to equation (4.45) given by

$$S_P(\omega) = \tfrac{1}{4} S_1(\omega) \, \mathrm{Re}(Z) = \tfrac{1}{4} S_V(\omega) \, \mathrm{Re}(Y). \tag{4.47}$$

4.4.2 Diffusion noise, noise temperature

A resistor produces noise, even without bias, by the thermal agitation (Brownian motion) of the charge carriers. A famous theorem of Nyquist states that a resistor in thermal equilibrium with its surroundings at an absolute temperature T can deliver an amount of power into a matched load approximately equal to kT per hertz bandwidth, where k is Boltzmann's constant. It is to be understood here that we are talking about a practical measurement situation where the contributions of positive and negative frequencies are added, as in equation (4.45). So the power spectral density (as defined by equation (4.38)) is equal to

$$S_{PP}(\omega) = \tfrac{1}{2} kTf(\omega) \tag{4.48}$$

where $f(\omega)$ is a quantum correction factor that becomes different from unity only at infrared frequencies:

$$f(\omega) = \frac{\hbar\omega/kT}{\exp(\hbar\omega/kT) - 1}. \tag{4.49}$$

This quantum factor reflects the fact that we are talking about the energy of electromagnetic fields which is quantized in *photons* of magnitude $\hbar\omega$, where $\hbar$ is Planck's modified constant, equal to 1.05×10^{-34}. Equation (4.49) corresponds to the Bose–Einstein distribution law for photon density. The correction factor is negligible, i.e. equal to unity, for frequencies where $\hbar\omega \ll kT$, or $f \ll 2.09 \times 10^{10} T$. It can easily be verified that at room temperature this condition is fulfilled for the whole microwave frequency range, but for cryogenic systems at millimeter wave frequencies the correction becomes important and leads to a reduction of the diffusion noise.

Nyquist's theorem can be generalized to a complex impedance in which case 'matched load' means the complex conjugate of the impedance. We can replace such a noisy impedance by a noiseless impedance in series with a voltage noise source, or alternatively by a noiseless admittance parallel to a current noise source (Fig. 4.6). It is easy to see from Fig. 4.6 that the measured spectral density of the sources is given by

$$S_V(\omega) = 4kT\,\mathrm{Re}(Z), \qquad S_I(\omega) = 4kT\,\mathrm{Re}(Y). \tag{4.50}$$

In devices under bias we cannot assume thermal equilibrium and these formulas cannot be used to predict the noise. Indeed, as discussed in Chapter 3, at high electric fields the average electron energy, in other words the electron temperature, rises considerably above the lattice temperature and we should replace the ambient temperature by the electron temperature. In practice the situation can even be more complicated since the electric field in the device need not be uniform and the electron temperature will be non-uniform also. Further there may be other noise sources, as described below. In this case, assuming the noise can either be calculated by some

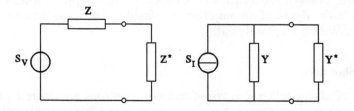

Figure 4.6 Equivalent circuits for noisy impedances.

method or be measured, it is customary to reverse the above formulas and to define an equivalent noise temperature:

$$kT_n = \frac{S_V}{4\,\mathrm{Re}(Z)} = \frac{S_I}{4\,\mathrm{Re}(Y)}. \qquad (4.51)$$

In the case of non-linear devices Z and Y are understood as the small-signal impedance and admittance, respectively. Of course this definition is meaningful only when T_n is constant over the frequency band of interest.

4.4.3 Noise measure and noise figure

If one divides the noise temperature by a reference temperature one obtains a dimensionless quantity, called the *noise measure* [12]:

$$M = \frac{T_n}{T_0}. \qquad (4.52)$$

It is natural (but not necessary) to take as the reference temperature the ambient temperature. This means that for a device in thermal equilibrium the noise measure is equal to unity. In microwave applications the reference temperature is fixed at 290 K.

The noise measure is a physical parameter characterising a device and is often not directly measurable, although in principle it can always be calculated from the device physics. A measurable quantity in microwave systems is the *noise figure F*, defined as the deterioration of the signal-to-noise ratio between input and output:

$$F = \frac{P_{s,\,in}/P_{n,\,in}}{P_{s,\,out}/P_{n,\,out}}$$

where P_s and P_n refer to signal and noise power, respectively, and $P_{n,\,in}$ is the power delivered by an input impedance having a noise temperature equal to a standard temperature. In microwave technology the latter is taken as 290 K. It is implied in the definition that the same noise bandwidth is used at input and output, which means that one can replace the noise powers by their spectral intensities. The concept of noise figure applies to mixers (Chapter 8) as well as amplifiers (Chapter 9) and is also used much as a figure of merit for transistors (Chapter 6). In Chapter 9 will be shown how the noise figure of an amplifier is related to the noise measure of the device providing the amplification.

4.4.4 Shot noise

In spite of all the talk about wavefunctions, electrons are discrete particles and an electron current passing a certain plane consists actually of a random sequence of delta functions. This gives an inevitable noisiness to the current and this noise is called *shot noise*. It can fairly easily be derived [9] that the spectral density of shot noise is equal to

$$S_I(\omega) = 2qI_0 \Delta f \tag{4.53}$$

when I_0 is the average current. This formula is valid for current that enters a device through a contact. One must bear in mind, however, that feedback effects can occur because the modulation of the injected current changes the charge distribution in the rest of the device which changes the electric field distribution. The latter can in its turn influence the injected current. A striking example of this is a vacuum diode where the electrons emitted from the cathode form a cloud in front of the cathode which creates an electric field pushing the electrons back to the cathode. An increase of the emitted current increases this field so that a negative feedback occurs. Vacuum devices therefore show full shot noise only in the saturation regime, that is the bias range where the anode voltage is so strong that there is an accelerating field all the way from the cathode to the anode.

In semiconductor devices one has in principle always a combination of shot noise, at the place where carriers enter the device, and diffusion noise throughout the device. In practice often one of the two prevails or they can be ascribed to different regions of the device. For instance, in Schottky diodes one can assume that the barrier plus depletion layer is the seat of shot noise and the undepleted layer contributes diffusion noise. These can then be calculated separately and added together. In devices where the electrons enter from a highly doped n^+ region a potential barrier is formed at the junction by the injected electrons and this produces a negative feedback as in the vacuum diode. As a result the contribution of shot noise is negligible compared with diffusion noise. The Baritt diode (Sections 5.5 and 5.8) is an example of a device which is in between the previous two. There is a potential barrier as in a vacuum diode but its negative feedback is not strong enough to make shot noise insignificant. On the other hand, because the electrons become hot there is considerable diffusion noise too.

Another remark that has to be made here is that the above formula suggests that the shot noise goes to zero as the direct current goes to zero. One should realize, however, that in semiconductor devices at zero bias there are normally two currents in opposite directions which compensate each other. So the resulting direct current is zero but the noise is not, since the noise contributions of the two opposing currents are uncorrelated and have to be added. An example of this will be given in Section 8.1.3, where noise in Schottky diode detectors is discussed.

4.4.5 Multiplication noise

Carrier multiplication (also discussed in Chapter 5) can be a strong source of noise. Not only is the noise in the injected current multiplied by the same factor that multiplies the current but the multiplication process itself is of a stochastic nature so it adds extra noise. Devices that operate on the basis of multiplication, e.g. IMPATT and photoavalanche diodes are therefore very noisy. Yet devices where the multiplication plays only a secondary role can show increased noise even when the current shows no effect of the extra generated carriers.

It will be shown in Chapter 5 that a current injected into an avalanche region from the outside is multiplied by a factor $M_A = 1/(1 - N_A)$ which can become very large. The same can be shown to hold for currents injected internally by the multiplication process [13]. The mean squared current fluctuations then will be multiplied by M_A^2. These fluctuations can be assumed to have a shot noise character and so they are related to the direct current by equation (4.64). The total noise including multiplication effects should then be $qM_A^2 I_0$. In ref. 15 it is shown that when one uses a simple model where the hole ionization coefficient is a constant factor multiplied by that of the electrons ($\alpha_p = k\alpha_n$) and M_A is large, the noise of an avalanche diode is given by the expression

$$S_{II} = qM_A^3 \left(kI_{sn} + \frac{1}{k} I_{sp} \right) \tag{4.54}$$

where I_{sn} and I_{sp} are the injected electron and hole saturation current densities defined in section 5.2.5 multiplied by the diode cross-sectional area. When $k = 1$ this reduces again to $qM_A^2 I_0$.

Suppose now that k is much smaller than unity, meaning that the electrons ionize much more than the holes; then the avalanche noise is mainly coming from the injected hole current. It follows that when I_{sn} and I_{sp} are of equal magnitude the noise can be much larger than for equal ionization coefficients. Only when the injected hole current is much smaller than the electron current can the noise be smaller. In microwave avalanche diodes (IMPATT diodes; section 5.8) the injected electron and hole currents are of comparable magnitude and materials such as silicon having widely different ionization coefficients will be noisier than materials such as gallium arsenide where the ionization coefficients are almost equal. In avalanche photodiodes (APDs) the converse is true. Here the avalanche region is made thick to provide good absorption of the light. This means that the photogeneration of electron–hole pairs is much larger at one side than at the other. So if one takes care that the light falls in on the side where the electrons are injected (the negative pole) the noise can be very much reduced in the case of Si, but not in the case of GaAs.

4.4.6 Other noise sources

In Chapter 3 it was mentioned that many impurities can have energy levels near the middle of the bandgap. At surfaces and interfaces between two different materials these states can also occur because here the regular periodicity of the crystal lattice is broken and free valences can occur. Electrons can be captured by these states from both the conduction band and the valence band. The collision term of equation (4.12) can for these processes be written as

$$\left(\frac{\partial n}{\partial t}\right)_c = -\frac{n - n_0}{\tau} \tag{4.55}$$

and the time constants connected with the capture and release processes are long, in the range of micro- to milliseconds. Since release and capture from a trap are stochastic processes they produce noise. The spectrum of this noise is Lorentzian in shape, i.e.

$$S_I(\omega) \propto \frac{1}{1 + (\omega\tau)^2}. \tag{4.56}$$

This *generation–recombination noise* is therefore concentrated at the low-frequency end of the spectrum and one might ask why we pay attention to it in a book about microwave devices. The reason is that this noise can make its influence felt at microwave frequencies: in non-linear circuits, especially oscillators, it can be mixed with microwave signals and produce sum and difference frequencies in the microwave range. This is called upconversion. The sideband noise of oscillators is for a large part determined by the low-frequency noise of the active device.

When the low-frequency noise of a device is measured one seldom sees a spectrum of the form (4.56). This is because there are always different kinds of traps with different time constants so their spectra overlap. In fact, at the lowest frequencies one usually finds a spectrum that is roughly proportional to $f^{-\gamma}$ where γ is between 0.5 and 2. This is called 1/f *noise* or *flicker noise*. The latter name comes from the vacuum tube field where the heated cathode seems to emit electrons not uniformly but in isolated spots that wander across the cathode surface. In semiconductors it is often assumed that 1/f noise can be explained as generation–recombination noise from a large number of traps with a wide range of time constants. The overlap of all these spectra produces a 1/f spectrum in a limited frequency range. However, Hooge *et al.* [14] have from a very large number of experiments concluded that there is an intrinsic 1/f noise which originates in the bulk, occurs over a virtually unlimited frequency range and can be described for a uniform device by the equation

$$\frac{S_I}{I^2} = \frac{\alpha_H}{fN} \tag{4.57}$$

where Hooge's 'constant' α_H is remarkably similar for widely different materials, such as metals and semiconductors. It usually lies in the range 10^{-5}–10^{-3}. The factor N is the total number of carriers in the device. Later work has led to the conclusion that this noise is caused by fluctuations in the carrier mobility, the cause of which is as yet unknown.

In practice we have to assume that the low-frequency noise in devices is made up of a mix of trap-generated noise (including that from surface states) and 'true' $1/f$ noise.

4.4.7 The impedance–field method

A general method to calculate the noise of a two-terminal device has been devised by Shockley and coworkers [15]. It is called the *impedance–field method* and is illustrated by Fig. 4.7.

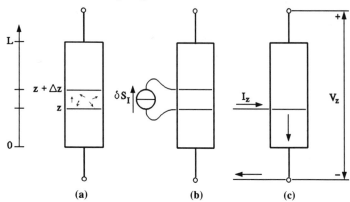

Figure 4.7 The impedance–field method.

In Fig. 4.7(a) it is indicated how the random motion of charge carriers in a section Δz of the device produces a noise current. In Fig. 4.7(b) this noise current is replaced by an external noise current source placed parallel to the device section, as we did in Fig. 4.6 for a whole device. Finally, in Fig. 4.7(c) it is illustrated how a complex current injected at a position z produces a complex voltage between the device terminals. This can be described by a transfer impedance defined as

$$Z_{zT} = \frac{V_T}{I_z}.$$

A noise current with spectral density $\delta S_{11}(\omega, z)$ injected at $z + \Delta z$ and extracted at z, as in Fig. 4.7(b), will give a voltage response at the terminals of

$$\Delta S_{VV}(\omega) = \delta S_{11}(\omega, z)\left|\nabla Z_{zT}\,\Delta z\right|^2 \tag{4.58}$$

with

$$\nabla Z_{zT} = \frac{Z_{z+\Delta z,\,T} - Z_{zT}}{\Delta z}. \tag{4.59}$$

The latter is called the *impedance field*.

We now have to evaluate the current source. For a single electron having a random velocity $v(t)$ the current induced in the segment Δz is by Ramo's theorem (Appendix 2.A) equal to

$$i_{1e}(t) = -\frac{qv(t)}{\Delta z}. \tag{4.60}$$

Since we assume that the movements of all electrons in Δz are uncorrelated the autocorrelation function of the current is the autocorrelation function of one electron multiplied by the number of electrons in Δz, which gives together with equation (4.60)

$$R_{ii}(t, z) = \left(\frac{q}{\Delta z}\right)^2 R_{vv}(t, z)n(z)A\,\Delta z \tag{4.61}$$

where $n(z)$ is the electron density and A the cross-sectional area of the device. Taking the Fourier transform we obtain

$$\delta S_{11}(\omega, z) = \left(\frac{q}{\Delta z}\right)^2 S_{vv}(\omega, z)n(z)A\,\Delta z \tag{4.62}$$

which inserted in equation (4.53) gives

$$\Delta S_{vv}(z) = q^2 An(z)\,|\nabla Z_{zT}|^2 S_w(\omega, z)\,\Delta z. \tag{4.63}$$

Assuming that the contributions of the different segments of the device are uncorrelated the total noise voltage spectral density becomes

$$S_{vv} = q^2 A \int_0^L n(z)\,|\nabla Z_{zT}|^2 S_w(\omega, z)\,dz. \tag{4.64}$$

With this formula the noise of a device can be calculated. To obtain a feeling for what it means let us apply it to the simplest possible example, a resistor in thermal equilibrium. We assume that the electron density is equal to the donor density N_D which is uniform. In this case the impedance field is simply

$$\nabla Z_{zT} = \frac{R}{L} = \frac{1}{q\mu N_D A} \tag{4.65}$$

where μ is the low-field mobility. In thermal equilibrium the average velocity in the z direction is given by

$$\langle \tfrac{1}{2} mv_z^2 \rangle = \tfrac{1}{2} kT$$

assuming the thermal energy is divided equally among the degrees of freedom. Now the autocorrelation of the individual electron velocity has an exponential dependence on time with a time constant τ_m called the *momentum relaxation time*, which also appears in the mobility (see the derivation leading to equation (4.16)), so we have

$$R_w(t) = \frac{kT}{m} \exp\left(\frac{-|t|}{\tau_m}\right) \tag{4.66}$$

from which follows

$$S_w(\omega) = \frac{kT}{m} \frac{2\tau_m}{1 + (\omega\tau_m)^2}. \qquad (4.67)$$

Inserting equations (4.64) and (4.66) into (4.63) and carrying out the integration we arrive at the result

$$S_{vv}(\omega) = \frac{2kTR}{1 + (\omega\tau_m)^2}. \qquad (4.68)$$

So we have derived Nyquist's theorem (*cf.* equation (4.50)). Note that the quantum correction factor does not appear here because we have nowhere in the derivation taken account of the quantized nature of the electromagnetic field. However, at the same time we have shown that it is valid only for frequencies well below the inverse of the momentum relaxation time. Since the latter is of the order of 10^{-13}s at room temperature, the theorem is valid throughout the microwave frequency range.

Note that in equation (4.67) the product $\tau_m kT/m$ occurs. This is in fact the low-field diffusion coefficient which has been defined earlier in this chapter (equation (4.18)). The fact that thermal noise is determined not by the electron temperature alone but by the diffusion coefficient is the reason why it is also called *diffusion noise*. Evidently the impedance–field technique becomes much more complicated in realistic models of microwave devices in which inhomogeneous and high fields can occur. Nevertheless the method has successfully been applied to among others Gunn and Baritt diodes (Chapter 5).

4.4.8 Noise in two-port devices

In two-port devices, particularly transistors, the same mechanisms as described above can cause noise. The modeling of these effects will be much more complicated than in plane diodes owing to the two-dimensional character of the electric field. Whereas in one-port devices it is fairly obvious that one can describe their noise properties by a single noise current or voltage source and by one noise parameter, this is not immediately clear for two-port devices. One suspects that there should be at least two sources and two parameters and the question is whether these can be defined in such a way that they are independent of the way the device is embedded in a circuit.

It has been found that two noise sources are necessary as well as sufficient, but one has to include the possibility that they are partially correlated. Where one places the sources is a matter of choice and depends mainly on the way one chooses to describe the device and its immediate surroundings. One may place a voltage source in series at each port as in Fig. 4.8(a) or replace these by parallel-connected current sources, Fig.

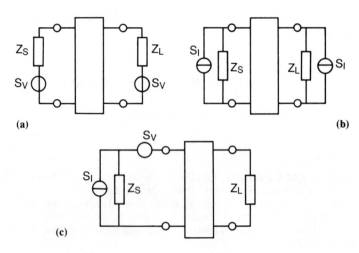

Figure 4.8 Three equivalent ways of representing the noise of a two-port device.

4.8(b), or one can represent the noise behavior by a current source and a voltage source, both at the input port (Fig. 4.8(c)).

Lange [16] has shown that the noise behavior of a two-port device can be expressed in two embedding-independent parameters, namely the optimal noise figure F_{opt} and a noise measure N which is analogous to the M of a one-port device. With reference to Fig. 4.8(c) they can be written as

$$F_{opt} = \frac{(S_e S_i)^{1/2}}{2kT_0}\left[(1 - \gamma_i^2)^{1/2} + \gamma_r\right] \tag{4.69}$$

where

$$\gamma_r = \mathrm{Re}\,\frac{S_{ei}}{(S_e S_i)^{1/2}}, \quad \gamma_i = \mathrm{Im}\,\frac{S_{ei}}{(S_e S_i)^{1/2}} \tag{4.70}$$

and

$$N = \frac{[S_e S_i(1 - \gamma^2)]^{1/2}}{4kT_0}. \tag{4.71}$$

In these expressions S_e and S_i are the spectral intensities of the voltage and current source, respectively, and S_{ei} is the corresponding quantity derived from the cross-correlation between the two. Thus, the complex coefficient γ is a measure of the degree to which the two noise sources are correlated. In Chapter 9 these definitions will be used to derive constant noise figure circles for amplifiers.

It is obvious that making the connection between device physics and the noise parameters defined above is a much more complicated matter than in a one-port device. An attempt to do this for a MESFET will be described in section 6.7. From equations (6.7.2) we can derive the noise sources of Fig. 4.8(c) as

$$S_i = 4kT_d g_{ds}(f/f_T)^2$$

$$S_e = 4kT_g r_{gs} + 4kT_d \frac{g_{ds}}{g_m^2}(1 + \omega^2 C_{gs}^2 r_{gs}^2)$$

$$S_{ei} = 4kT_d g_{ds}(\omega^2 C_{gs}^2 r_{gs} + j\omega C_{gs}).$$

REFERENCES

1. Chapman, S. and Cowling, T. G. (1960) *The Mathematical Theory of Non-Uniform Gases*, Cambridge University Press, Cambridge.
2. Rees, H. D. (1969) Calculation of distribution functions by exploiting the stability of the steady state. *J. Phys. Chem. Solids*, **30**, 643–55.
3. Hess, K. (1988) *Advanced Theory of Semiconductor Devices*, Prentice-Hall, Englewood Cliffs, NJ.
4. Shur, M. S., (1990) *Physics of Semiconductor Devices*, Prentice-Hall, Englewood Cliffs, NJ.
5. Jacoboni, C. and Lugli, P. (1989) *The Monte Carlo Method for Semiconductor Device Simulation*, Springer Series on Computational Electronics, Vienna, New York.
6. Moglestue, C. (1993) *Monte Carlo Simulation of Semiconductor Devices*, Chapman & Hall, London.
7. Widdershoven, F. P. (1984) Eindhoven University of Technology M.Sc. Thesis. *Report ET-4-84*.
8. Nederveen, K. (1989) Eindhoven University of Technology Dr. Thesis, 27 October 1989, (ISBN 90-9003048-4).
9. Van der Ziel, A. (1981) *Noise*, Prentice-Hall, Englewood Cliffs, NJ.
10. Papoulis, A. (1981) *Probability, Random Variables, and Stochastic Processes*, McGraw-Hill, New York.
11. Carlson, A. B., (1981) *Communication Systems*, McGraw-Hill, New York.
12. Haus, H. A. and Adler, R. B. (1959) *Circuit Theory of Linear Noisy Networks*, Wiley, New York.
13. McIntyre, R. J. (1966) Multiplication noise in uniform avalanche diodes. *IEEE Trans. Electron Devices*, **13**, 164–8.
14. Hooge, F. N., Kleinpenning, T. G. M. and Vandamme, L. K. J. (1981) Experimental studies on 1/f noise. *Rep. Prog. Phys.*, **44**, 479–532.
15. Shockley, W. Copeland, J. A. and James, R. P. (1966) The impedance–field method of noise calculation in active semiconductor devices, in *Atoms, Molecules and the Solid State* (ed. P. O. Löwdin), Academic Press, New York.
16. Lange, J. (1967) Noise characterization of linear twoports in terms of invariant parameters. *IEEE J. Solid-State Circuits*, **2**, 37–40.

FURTHER READING

1. Van Someren Greve, S. C. (1989) Eindhoven University of Technology Dr. Thesis, 9 November 1989 (ISBN 90-9000778-4).
2. Pospieszalski, M. W. (1989) Modeling of noise parameters of MESFETs and MODFETs and their frequency and temperature dependence. *IEEE Trans. Microwave Theory Tech.* **37**, 1340–50.

PROBLEMS

4.1 How much time does it take to accelerate an electron in the lowest valley of GaAs to an energy of 0.32 eV (this is the energy where the next higher valley starts)? Compare this with the momentum relaxation time which can be calculated from $\mu = q\tau_m/m^*$. What do you conclude from this comparison?

4.2 Derive an equation for the threshold field in GaAs (i.e. where $dv/dE = 0$) under the following assumptions:

- the conduction band has two valleys, a central (lower) valley and a satellite (higher) valley, separated by an energy Δ;
- the effective mass in both valleys is constant and the satellite valley has the highest mass;
- the energy distribution in the central valley follows the Boltzmann law $\exp(- W/kT_c)$ and all electrons reaching the energy Δ transfer immediately to the satellite valley;
- the increase of drift velocity and energy in each valley is described by equations (4.14), (4.16) and (4.18), in which the time- and space-dependent terms may be neglected;
- the momentum and energy relaxation times are constant and the same for both valleys and the energy relaxation time is much greater than the momentum relaxation time.

The result is an implicit equation which is not analytically solvable. Show, however, that it is governed by two parameters: kT_L/Δ and m_c/m_s. Show that no solution is possible when $m_c > 0.5m_s$. Estimate the threshold field using the following values:
$m_c^* = 0.067m_0$, $m_s^* = 0.22m_0$, $\tau_m = 10^{-13}$ s, $\tau_e = 10^{-12}$ s, $T_L = 300$ K, Boltzmann's constant $k = 1.38 \times 10^{-23}$ J/K, $q = 1.602 \times 10^{-19}$ C.

4.3 Carrier transport in homogeneous material and a uniform electric field can be described by the following set of equations which includes the effect of energy relaxation:

$$\frac{dv}{dt} = \frac{q}{m^*} E - \frac{v}{\tau_m(W)} \tag{4.p.1}$$

$$\frac{dW}{dt} = qvE - \frac{W - W_L}{\tau_e}. \tag{4.p.2}$$

To model high-field effects τ_m is assumed to be a decreasing function of W approximated by

$$\tau_m = \frac{\tau_{m0}}{(1 + aW/W_L)^2}. \tag{4.p.3}$$

Write a computer program to solve this set of equations by explicit time integration:

$$v^{k+1} = v^k + \Delta t \times \text{(right-hand side equation (4.p.1))}^k$$

$$W^{k+1} = W^k + \Delta t \text{ (right-hand side equation (4.p.2))}^k$$

(this is simple enough to be done on a programmable pocket calculator). Do this for electric fields of 0.5 and 5 kV/cm. Make an estimate of the necessary time step and the number of time steps required. Take the following set of data:
$\tau_m = 10^{-13}$ s, $m^* = 0.07m_0$, $a = 0.25$, $\tau_e = 10^{-12}$ s, $W_0 = 0.038$ eV. Is velocity overshoot possible in this model? If yes, at which field strength?

4.4 In the previous problem, add an energy-dependent mass according to $m^* = m_0^*[1 + b(W/W_L)^2]$ with $b = 0.02$ and see what happens.

4.5 Derive an equation for the small-signal frequency–dependent conductivity σ_1 of a semiconductor that obeys equations (4.p.1) and (4.p.2) of Problem 4.3. Linearize the equations by writing all variables as the sum of a d.c. part and a (small) a.c. part with harmonic time dependence. Develop $\tau_m(W)$ in a Taylor series and retain the first two terms. To keep the formulas legible, introduce the following parameters:

$$\sigma_0 = \frac{q\tau_m(W_0)}{m^*}, \qquad \alpha = \frac{q^2\tau_e}{m^*}\left|\frac{d\tau_m}{dW}\right| \qquad \left(\frac{d\tau_m}{dW} < 0\right)$$

Plot $\mathrm{Re}(\sigma_1)$ and $\mathrm{Im}(\sigma_1)$ against frequency taking $\tau_m(W)$ according to equation (4.p.3) of Problem 4.3 and the numerical data also as in that problem. Consider two cases: one where the d.c. electric field is such that $\sigma_1(f = 0)$ is positive and one where it is negative.

4.6 Split the distribution function $f(k, r, t)$ into a symmetric part f_s and an antisymmetric part f_a, with $f_s(-k) = f_s(k)$ and $f_a(-k) = -f_a(k)$. Suppose the collision term can be written as

$$\left(\frac{\partial f}{\partial t}\right)_c = -\frac{f_s - f_0}{\tau_s} - \frac{f_a}{\alpha_a}$$

where f_0 is the equilibrium distribution. Show that now in the transport equations for v and W the momentum and energy relaxation times are identical with τ_a and τ_s, respectively.

4.7 Consider the n^+–n junction depicted in Fig. 4.P.1. Write down the equations and boundary conditions for the electron density on both sides. To make the equations solvable assume that $|n_i - n_{Di}| \ll n_{Di}$, $i = 1, 2$, and linearize (hint: take $n_1 - n_{D1}$ and $n_2 - n_{D2}$ as the unknowns). Calculate the interface density and the approximate distance over which the space charge layer stretches on both sides.

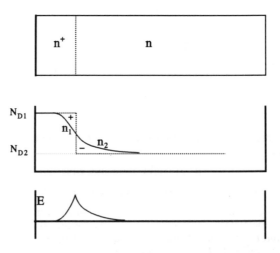

Figure 4.P.1 Electron outdiffusion at an n^+–n junction

4.8 What is the noise measure of an ideal resistor at a temperature *T*?
4.9 Calculate the noise measure of an ideal Schottky diode as a function of current. Take for the diode admittance the differential value d*I*/d*V*. After you have done this check with section 8.1.3.

5

Diodes

5.1 INTRODUCTION

The generic term diode will in this chapter be used for all non-linear two-terminal devices, i.e. not only asymmetric structures such as p–n junctions but also symmetric ones such as Gunn (n^+–n–n^+ or Baritt (p^+–n–p^+) diodes. Many different types of diodes exist nowadays which have all kinds of applications in microwave technology. They can be classified in various ways. According to application we can divide them into three main groups: frequency conversion, frequency or amplitude control and signal generation. According to their microwave impedance we can distinguish between passive and active devices, referring to the real part of the impedance being positive and negative, respectively. It so happens that both classifications almost coincide in the sense that passive diodes are used for frequency conversion and control and active diodes for signal generation. In the following we will discuss the many physical diode structures that exist nowadays. Applications will be discussed in Chapters 8, 9 and 10.

5.2 P–N JUNCTIONS

The theory of p–n junctions is treated in many excellent basic texts on semiconductor devices [1–3] and here we will highlight only the aspects relevant for microwave applications.

5.2.1 Characteristics

In a typical p–n junction (Fig 5.1), depletion layers exist on both n and p sides owing to outdiffusion of carriers. Although the carrier concentrations decrease gradually, it is often assumed that the depletion layer has sharp edges. If also the transition from p to n doping is assumed to be abrupt, it is easy to calculate the electric field in the junction. The peak field is given by

$$E_p = \frac{qN_D w_N}{\varepsilon_0 \varepsilon_r} = \frac{qN_A w_P}{\varepsilon_0 \varepsilon_r} \tag{5.1}$$

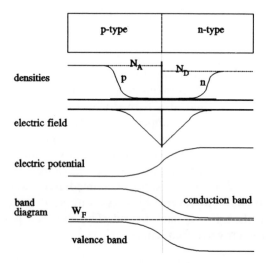

Figure 5.1 p–n junction: composition, space charge, electric field, potential and band diagram are shown from top to bottom.

where w_N and w_P are the depletion layer widths on the n and p sides, as shown in the figure. Integration of the field gives the electric potential and the latter divided by the electron charge $-q$ gives the potential energy of the electrons in the conduction and valence bands. We encounter here an example of band bending, already discussed in section 3.1.3.

The condition that the Fermi level must be flat in equilibrium, i.e. without an external voltage applied, yields the potential difference between the n and p sides, called the *diffusion voltage* or *built-in voltage*:

$$qV_D = W_{gap} - (W_c - W_F)_n - (W_F - W_v)_p. \tag{5.2}$$

The built-in voltage V_D can be calculated if one assumes that far from the junction the electron and hole densities are equal to the donor, respectively acceptor densities, and that in equilibrium they are related to the Fermi level by equations (3.16) and (3.22a). When the doping concentrations are not too high, the Boltzmann approximations equations (3.21) and (3.22b) may be used and we can write

$$qV_D = W_{gap} - kT \ln(N_c/N_D) - kT \ln(N_v/N_A). \tag{5.3}$$

When the concentrations are high, for instance in tunnel diodes, one has to replace the logarithms by the inverse of the Fermi function defined in section 3.1.

Since, on the other hand V_D is the integral of the electric field in the depletion layer it can also be expressed in the width of this layer:

$$V_D = \frac{qN_D w_N^2}{2\varepsilon_0\varepsilon_r} + \frac{qN_A w_P^2}{2\varepsilon_0\varepsilon_r} = \frac{qN_A N_D (w_P + w_N)^2}{2\varepsilon_0\varepsilon_r(N_A + N_D)}. \tag{5.4}$$

To arrive at the last equality equation (5.1) has been used.

When bias is applied V_D has to be replaced by $V_D - U$ where the externally applied voltage U is taken positive in the forward direction. Then equation (5.4) indicates that at a forward bias of about V_D the depletion layer should disappear. What happens in reality is that before this point is reached injection of electrons into the p-doped region and holes into the n-doped region occurs. So there is always negative charge in the p region and positive charge in the n region. This *stored charge* can be a source of trouble because it does not disappear immediately when the diode is switched from forward to reverse bias. As long as it is there the diode is conducting even though it is back-biased. This makes a p–n junction unfit as a detector at microwave frequencies.

5.2.2 Current and capacitance

The current–voltage characteristic of p–n junctions is extensively discussed in many handbooks on semiconductor devices [1–3]. In principle it has the well-known exponential form

$$I = I_s\left[\exp\left(\frac{qU}{nkT}\right) - 1\right] \tag{5.5}$$

where I_s is called the reverse saturation current. The ideality factor n is equal to unity in the ideal case but can be slightly higher in practice. At high forward bias the current rises more slowly than this formula predicts owing to space charge effects of the injected electrons and holes and to the non-negligible voltage drop over the series resistance of the bulk p and n regions.

The capacitance follows from the change with bias voltage of the charges on both sides of the junction and is for the simple structure of Fig. 5.1 equal to

$$C = \frac{\varepsilon A}{w_N + w_P} = A\left[\frac{\varepsilon q N_D N_A}{2(N_D + N_A)(V_D - U)}\right]^{1/2}. \tag{5.6}$$

In the reverse direction the diode behaves as a voltage-dependent capacitance with low losses since the reverse current is small. The voltage dependence of the capacitance is used in *varactor diodes* (section 5.3.2). Of course this expression is only approximate and like equation (5.4) it goes completely wrong in forward bias when the bias voltage approaches V_D. It is nevertheless used much in circuit simulations to model reverse-biased diodes.

In diodes with non-uniform doping profiles large deviations from the formula above can occur even at reverse bias. The gate capacitance of ion-implanted MESFETs (Chapter 6) is an example of this. On the other hand, using a more detailed analysis, the capacitance–voltage characteristic can be used to determine the doping profile (section 5.3.3).

5.2.3 p–i–n diodes and step recovery diodes

For some applications an intrinsic or low-doped p-type layer is inserted between the (highly doped) p and n layers. This is the case with p–i–n *diodes* and *step recovery diodes*.

The structure, charge density and electric field under forward bias of a p–i–n diode are sketched in Fig 5.2. The d.c behavior is the same as that of an ordinary p–n junction; it shows the familiar exponential characteristic. At microwave frequencies there is a difference. Since the electric field in the intrinsic layer is low, carrier transport is mainly by diffusion which is a slow process. The transit time through this region is approximately given by $t_b = L_b^2/2D_{el}$, where L_b is the base thickness and D_{el} the diffusion coefficient of the electrons. For a base thickness of $10\,\mu m$ and a diffusivity of $0.001\,m^2/(V\,s)$ a transit time of about $5\,\mu s$ is found. The only other process that can influence the charge density is recombination which is also slow. So it takes much longer than the period of a microwave signal to change the charge distribution. Only the carrier velocity is modulated by the microwave signal. The microwave conductance can therefore be written as

$$G_\sim = q(\mu_n n + \mu_p p)\frac{A}{L} \tag{5.7}$$

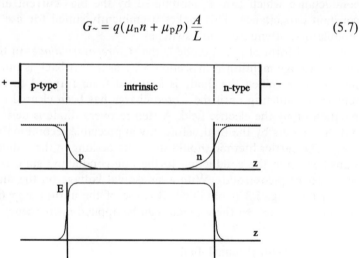

Figure 5.2 p–i–n diode.

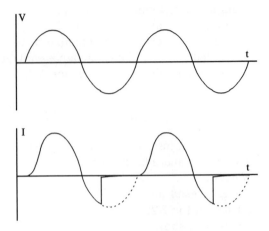

Figure 5.3   Waveforms of a step recovery diode.

where A is the diode area and L the thickness of the i layer. Now since the carrier densities are proportional to the direct current this opens the possibility of using this diode as a frequency-independent conductance which is controlled by the direct current. The capacitance is determined by the width of the region with low carrier concentration which is made up mostly by the intrinsic layer. It is only weakly influenced by the d.c. bias so only the real part of the impedance is varied by the control current. For microwave signals the p–i–n diode thus behaves as a parallel circuit of a conductance which can be modulated by the bias current and an almost constant capacitance. This makes it eminently suited for use in amplitude modulators, attenuators and switches.

A special form of p–i–n diode is the *step recovery diode*. In this device the charge storage mentioned in section 5.2.1 as a drawback is put to work in a positive way. When the diode is switched from forward to reverse bias it remains conducting until the stored charge has been cleared out by recombination or by the electric field. A step recovery diode is designed to sweep out the carriers by the field before any appreciable recombination has taken place. The carrier lifetime should therefore be long in these diodes. Then the transition from the conducting to the non-conducting state is very fast, of the order of picoseconds. With a sinusoidal voltage on the diode a current like that in Fig. 5.3 is obtained. Because of the abrupt step this current is rich in harmonics, so these diodes can be applied in frequency multipliers.

5.2.4   The classical tunnel diode

A tunnel diode is a p–n junction where both sides are so heavily doped that the Fermi level is in the conduction band on the n side, and in the valence

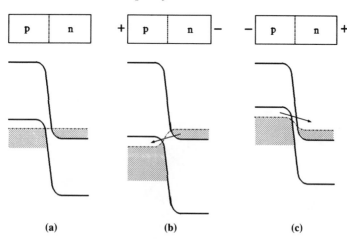

(a)   (b)   (c)

Figure 5.4 Tunnel diode.

band on the p side. It was discovered in 1958 by L. Esaki that these diodes have an N-shaped *I–V* characteristic in the forward direction. This can be explained as follows.

A look at the band diagram (Fig. 5.4), shows that the valence band at the p side is higher than the conduction band at the n side. Moreover, the spatial separation between the two bands is very small in the transition region as can be inferred from equation (5.4). This makes it possible for electrons to *tunnel* through the forbidden gap from the valence band to the conduction band and vice versa. At zero bias both tunnel currents will balance exactly and no net current results. If a small bias is applied in the reverse direction the Fermi level at the n side will be raised somewhat and more electrons are available for tunneling at the n side than at the p side, so that a net reverse current will result. Increasing the reverse bias will increase the Fermi level difference and at the same time reduce the distance the electrons have to cross so that the current keeps on increasing strongly. With a small bias in the forward direction initially the same happens as in the reverse direction: the current increases owing to the difference of the Fermi levels. In this case, however, the bands are pushed farther apart and eventually the tunnel current starts to decrease again, going to zero for high bias. This does not mean that the diode current goes to zero, however: at sufficiently high forward bias electrons and holes can cross the junction while remaining in their own bands, the same way as in normal p–n junctions. The total current is the sum of tunnel current and normal diode current and has the form shown in Fig. 5.5. In the region of minimum current we have a third contribution, called the *excess current*. It arises from other causes, mainly tunneling via impurity levels near the middle of the gap [1].

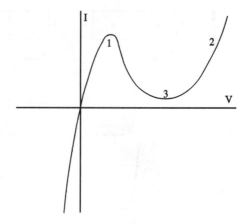

Figure 5.5   Tunnel diode characteristic.

The theory of tunneling in p–n junctions involves quite a bit of quantum mechanics [4] and will not be given here. The results can very well be approximated by a rather simple expression:

$$I = I_\mathrm{p} \frac{U}{V_\mathrm{p}} \exp\!\left(1 - \frac{U}{V_\mathrm{p}}\right). \tag{5.8}$$

Here I_p is the peak current and V_p the voltage at which the peak appears. The latter is equal to

$$V_\mathrm{p} = (W_\mathrm{n} + W_\mathrm{p})/2q \tag{5.9}$$

where $W_\mathrm{n} = W_\mathrm{Fn} - W_\mathrm{c}$ and $W_\mathrm{p} = W_\mathrm{v} - W_\mathrm{Fp}$ are the doping degenerations on the n and p side, respectively. In this case the full equations (3.16) and

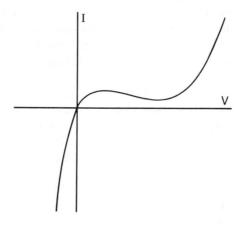

Figure 5.6   Backward diode *I–V* characteristic.

(3.22a) have to be used to calculate them: because of the degenerate dopings the Boltzmann approximation is not applicable. The calculation of the peak current is too involved to be given here. It is found that it increases with increasing doping densities and with decreasing bandgap. The latter property makes germanium a somewhat better material than silicon although both are used in practice. Gallium arsenide is difficult to dope degenerately (p type at least) and is not used normally for tunnel diodes.

If one reduces the doping densities of a tunnel diode a situation is reached where the reverse current is still large but the peak in the forward direction has disappeared. Around the origin the characteristic now resembles the inverse of a normal diode characteristic (Fig. 5.6) and can be used for detection. Because of their upside-down characteristic these diodes are called *backward diodes*.

5.2.5 Avalanche diodes

If in a moderately doped p–n junction the bias is increased in the reverse direction, the electric field in the depletion region increases to very high values as can readily be seen from Fig. 5.1. The few electrons and holes that are present in the depletion region are accelerated in this field and although they lose energy by collisions their average energy can become quite high. Such an energetic electron (or hole) can collide with another electron in the valence band and knock the latter up into the conduction band so that an electron–hole pair is created. This is called *impact ionization*. The newly created electrons and holes are also accelerated in the field and create in turn extra e–h pairs so that a chain reaction occurs. The reverse diode current now increases rapidly. This is called *avalanche breakdown*.

The electric fields necessary to produce an avalanche are very high, in the range of hundreds of kilovolts per centimeter. Combined with the high current densities that can arise, a lot of heat will be generated and great care has to be taken in the design and mounting of these diodes or they will burn out in the process.

A simplified mathematical analysis of the avalanche process can be made as follows. Suppose we have the situation depicted in Fig. 5.7. Since the holes are traveling from left to right the hole current will be increased to the right by the ionizations. Similarly, the electron current increases to the left. The continuity equations for electrons and holes both have the same generation term in the right-hand side:

$$\frac{\partial n}{\partial t} + \nabla \cdot n\boldsymbol{v}_n = \alpha_n n \left| \boldsymbol{v}_n \right| + \alpha_p p \left| \boldsymbol{v}_p \right| \tag{5.10a}$$

$$\frac{\partial p}{\partial t} + \nabla \cdot p\boldsymbol{v}_p = \alpha_n n \left| \boldsymbol{v}_n \right| + \alpha_p p \left| \boldsymbol{v}_p \right|. \tag{5.10b}$$

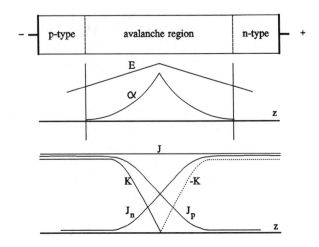

Figure 5.7 Electron and hole currents in an avalanche diode.

Here v_n and v_p are the electron and hole velocities and α_n, α_p are called the *ionization coefficients* of electrons and holes, respectively. The interpretation of the right-hand side is that the number of ionizations by electrons is proportional to the number of electrons, which is obvious, and to their average velocity, which means that on the average an electron has to travel

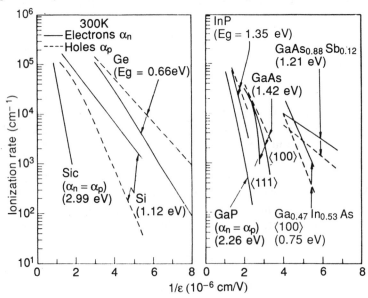

Figure 5.8 Ionization coefficients. Relabeled from *Physics of Semiconductor Devices, Second edition*; © copyright Sze, S. M. (1981), by permission of John Wiley & Sons, Inc.

a certain distance before it has gained enough energy to produce an ionization. Of course this distance depends on the electric field and this is accounted for by making α_n and α_p functions of the electric field. Since the energy necessary for the creation of an electron–hole pair is equal to the bandgap, the ionization coefficients will also depend on bandgap, being higher for low-bandgap materials such as Ge. Figure 5.8 shows the ionization coefficients for some materials.

Before going on let us introduce some more approximations.

● The ionization coefficients for electrons and holes are equal:

$$\alpha_n = \alpha_p = \alpha.$$

● Keeping in mind that the holes are traveling from left to right and the electrons in the opposite direction, we assume that their velocities are constant and equal to the saturation value:

$$v_p = - v_n = v_s.$$

The first assumption is seldom fulfilled (Fig. 5.8); the second is reasonable since at the high fields involved the charge carriers will travel at their saturated velocities which are not much different for electrons and holes.

With these assumptions the currents are given by

$$J_p = qpv_p = qpv_s, \quad J_n = - qnv_n = qnv_s.$$

Looking for a d.c. solution we replace J_n and J_p by the total direct current

$$J_0 = J_p + J_n$$

and the difference current

$$K_0 = J_p - J_n.$$

Substituting these into equation (5.10) with $\partial/\partial t = 0$ and $\nabla = (0, 0, \partial/\partial z)$ we obtain

$$\frac{\partial}{\partial z} J_0 = 0 \qquad (5.11a)$$

$$\frac{\partial}{\partial z} K_0 = 2\alpha(z) J_0. \qquad (5.11b)$$

To solve this system we also need boundary conditions. To obtain these we assume that at the left side where the holes enter the depletion layer the hole current is equal to the value it would have when no multiplication takes place, i.e. $J_p(0) = J_{ps}$, the hole contribution to the reverse saturation current. A similar boundary condition applies to the electron current at the right side: $J_n(L) = J_{ns}$. Since the total current is J_0 everywhere we have

$$J_n(0) = J_0 - J_{ps}, \qquad J_p(L) = J_0 - J_{ns} \qquad (5.12a)$$

and

$$K_0(0) = -J_0 + 2J_{ps}, \qquad K_0(L) = J_0 - 2J_{ns}. \qquad (5.12b)$$

Using these boundary conditions we find

$$K_0(z) = 2J_0 \int_0^z \alpha(z') \, dz' - J_0 + 2J_{ps} \qquad (5.13a)$$

$$K_0(L) = 2J_0 \int_0^L \alpha(z) \, dz - J_0 + 2J_{ps}. \qquad (5.13b)$$

From equation (5.12b) the latter is equal to $J_0 - 2J_{ns}$ and we can write

$$J_0 = \frac{J_s}{1 - N_A} \quad \text{with} \quad J_s = J_{ps} + J_{ns}, \quad N_A = \int_0^L \alpha(z) \, dz. \qquad (5.14)$$

$M_A = 1/(1 - N_A)$ is called the *multiplication factor*. N_A gives the number of new electron–hole pairs that on the average are produced by an electron or hole while it is in transit through the avalanche region. Because α increases with electric field, N_A and M_A will increase as the reverse bias is raised and when N_A approaches unity the reverse current will rise very rapidly. When it becomes equal to unity, the avalanche keeps itself going independent of the current coming in from the outside.

Avalanche diodes are used in microwave technology as negative-resistance devices by combining the effects of avalanche multiplication and transit-time delay (section 5.8). They are also used as photodetectors. In this case J_s is raised by the photocurrent, i.e. the current due to electron–hole pairs generated by the absorption of photons. Clearly the photocurrent is multiplied by a factor M_A which can be very high when N_A is close to unity. This makes *avalanche photodiodes* (APDs) very sensitive detectors. Also their bandwidth is large because the avalanche zone is narrow so that the transit times of electrons and holes are short.

A problem for both microwave avalanche diodes and APDs is the noise generated by the multiplication process. It has been found that for microwave avalanche diodes the noise is lowest when the ionization coefficients of both carrier species are the same. For APDs the opposite is true: here the lowest noise is achieved when one ionization coefficient is much larger than the other.

5.3 SCHOTTKY DIODES

5.3.1 Characteristics

Schottky diodes are metal–semiconductor junctions. The structure and band diagram of a Schottky contact (named after Walter Schottky, a researcher of Siemens in Germany who in the 1930s founded the theory of

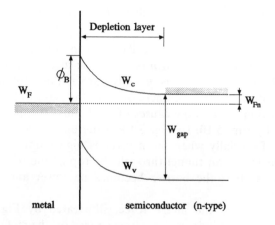

Figure 5.9 Schottky contact.

rectifying metal–semiconductor junctions) are shown in Fig. 5.9. Usually there is a difference in electron potential energy of about 0.7 eV between the metal and the conduction band of the semiconductor, the latter being on the high-energy side. This difference, expressed in volts, is called the *barrier potential* ϕ_B. Therefore, when a metal is joined to an n-type semiconductor, electrons will diffuse from the semiconductor into the metal where their energy is lower. The metal surface will be negatively charged while in the semiconductor positively charged donor atoms remain. An electric field is thus set up that pulls the electrons back into the semiconductor.

The equilibrium situation that eventually develops is sketched in the figure. A depletion layer is formed and a certain amount of band bending occurs, depending on the barrier potential ϕ_B and the doping concentration of the semiconductor. For the purpose of calculating the depletion layer width we can treat the Schottky diode as a p–n junction with an infinitely high p doping. Then it can easily be derived that the depletion layer width is equal to

$$w = \left[\frac{2\varepsilon}{qN_D} (V_D - U) \right]^{1/2} \tag{5.15}$$

with $V_D = \phi_B - W_F$, when ϕ_B and W_F are expressed in volts. Correspondingly, the differential capacitance is

$$C = \frac{\varepsilon A}{w} = A \left[\frac{\varepsilon q N_D}{2(V_D - U)} \right]^{1/2}. \tag{5.16}$$

For the current–voltage characteristic an expression similar to that of the junction diode can be derived [1–3]:

$$I = I_s\left[\exp\left(\frac{qU}{nkT}\right) - 1\right] \quad \text{with} \quad I_s = AA^*T^2\exp\left(-\frac{q\phi_B}{kT}\right) \quad (5.17)$$

where A is the diode area, A^* is the Richardson constant and I_s is called the saturation current. The factor n is called the ideality factor.

In reverse bias the current is usually larger than follows from this equation. Responsible for the increase are the same effects that were discussed in section 2.7.1 as the causes of field emission into vacuum. They are illustrated by Fig. 5.10. In Fig. 5.10(a) the barrier potential at reverse bias is shown. Especially when the n layer is highly doped, the barrier is narrow and electrons can tunnel through the top of the triangular barrier. At higher reverse bias the barrier becomes narrower and the tunneling current increases.

The second effect is the image force, illustrated by Fig. 5.10(b). An electron in the n region induces a positive charge on the metal surface. The electric field lines are the same as if there were a positive charge at the mirror position of the electron (Fig. 5.10(b)). This apparent charge is called the *image charge* and exerts an attractive force on the electron, called the *image force*, with the result that the potential energy of the electron decreases as it comes closer to the metal. The extra potential energy (which can be represented by a positive potential) subtracts from the conduction band energy giving the dotted line in Fig. 5.10(a) for the total potential energy. The net effect is that the effective barrier becomes lower. The amount of barrier lowering can be expressed in terms of the electric field at the barrier [1]:

$$\Delta\phi_B = \left(\frac{qE}{4\pi\varepsilon}\right)^{1/2} \quad (5.18)$$

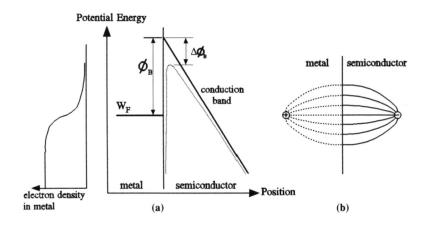

Figure 5.10 Schottky barrier in reverse bias.

where E is the electric field and ε is the appropriate value for the dielectric permittivity, equal to or lower than that of the semiconductor material. Again the reverse current is increased, now by the increase of $\Delta\phi_B$.

5.3.2 Varactor diodes

Varactor diodes make use of the fact that the capacitance of a backbiased p–n or Schottky junction is strongly voltage dependent (section 5.2.2). They can therefore be used as reactive elements to control the frequency of a signal. The equivalent circuit consists of a capacitance in series with parasitic series resistances due to the substrate the diode is made on and the contacts. In principle there is also a parallel conductance due to the reverse diode current but this is so small that it can be neglected. Hence one may conclude that the diode must have a cutoff frequency $f_c = (1/2\pi) R_S C_J$, which gives an indication of the frequency range in which it can be used. In practice one tries to stay well off the cutoff frequency because the losses introduced by the varactor increase rapidly when this frequency is approached. This is expressed by the *quality factor* $Q = f_c/f$ which should be large, i.e. at least a few hundred.

From the foregoing it is evident that R_S must be kept small. For this reason GaAs is preferred as a material for Schottky varactors because of its high electron mobility which gives a lower substrate resistance. Cutoff frequencies up to about 1000 GHz have been obtained.

Applications of varactor diodes are

- tuning of oscillators (Chapter 10),
- frequency multiplication (Chapter 8),
- mixing to a higher frequency, so-called *parametric upconversion*, and
- parametric amplifiers (Chapter 9).

5.3.3 *C–V* measurements

From equation (5.16) it can be seen that by measuring the capacitance of a Schottky diode information can be obtained about the doping concentration. If one differentiates equation (5.16) with respect to the applied voltage U one obtains, after a little rearrangement to eliminate the unknown V_D,

$$\frac{dC}{dU} = \frac{C^3}{A^2 \varepsilon q N_D}. \tag{5.19}$$

This can be extended to situations where N_D is a function of position. In this case E as a function of x is not a straight line. However, when U is changed by a small amount, the whole $E(x)$ curve moves up or down an amount dE without changing its shape. Consequently we can write

$$dU = x_d \, dE$$

where x_d is the edge of the depletion layer, measured from the metal–semiconductor junction, and dE is the same everywhere. At the depletion edge we have by virtue of Poisson's equation

$$dE = \frac{qN_D(x_d)}{\varepsilon} \, dx_d,$$

giving

$$dU = \frac{qN_D(x_d)}{\varepsilon} \, x_d \, dx_d.$$

Since the incremental charge is $dQ = AqN_D(x_d) \, dx_d$ we find as before that the differential capacitance is inversely proportional to x_d. Moreover, equation (5.19) remains valid but now with N_D equal to $N_D(x_d)$. So with x_d determined from the capacitance and $N_D(x_d)$ from its derivative we can, by varying U, probe the doping concentration as a function of depth. This is an often-used method to determine the doping profile in devices where Schottky barriers are present already, e.g. MESFETs (Chapter 6) or where a p–n junction with a much higher doping on one side is present, e.g. avalanche diodes. In device fabrication it is good practice to include a Schottky diode on the wafer for C–V measurements.

Note that C–V measurements do not give information about the doping close to the metal–semiconductor interface. To bring the edge of the depletion layer close to the metal one has to bias the diode in the forward direction and then it soon starts to draw resistive current which makes the capacitance measurement inaccurate. Also the method does not work when there are p–n junctions below the surface, as in the layer structure of a bipolar transistor. For these cases a method has been developed where the Schottky metal is replaced by an electrolyte. When this diode is reverse biased one can do a C–V measurement; when it is forward biased it etches the semiconductor. So by alternating forward and reverse bias one can work one's way through the different layers and profile each of them along the way.

5.4 TRANSFERRED ELECTRON DIODES

Transferred electron diodes (TEDs) are made of materials such as GaAs or InP and derive their interesting properties from the velocity–field characteristics of these materials which have already been discussed in Chapter 3. The device structure is very simple: a moderately doped n layer with n^+ contact layers on both sides (Fig. 5.11). If one thinks of a diode as a two-terminal device with a rectifying I–V characteristic this is not a diode. Indeed, at low voltages it behaves as a resistor. Still, it is customary to classify them as diodes and here we will follow that convention.

When the voltage exceeds a threshold value, roughly equal to the thickness of the low-doped region multiplied by the threshold field in the

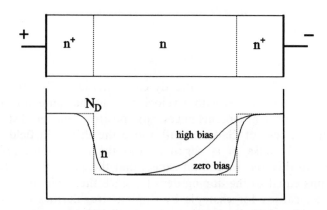

Figure 5.11 Doping profile and charge distribution of a transferred electrion diode.

v–E characteristic, deviations from the ohmic behavior start to occur. At first sight we would, assuming that the electron density remains equal to the donor density, expect that the I–U characteristic follows the v–E characteristic. This is not the case: in a region where dv/dE is negative, the charge distribution is not stable. This can be demonstrated by the following derivation.

Suppose we have a spatially uniform electron density n_0 (equal to the density of the positively charged donor atoms), a field strength E_0 and a velocity v_0 (which is a function of E_0 given by Fig. 3.10), and somewhere a small disturbance occurs so that we can write $n = n_0 + n_1$. This will give rise to a disturbance of the field, E_1, and of the velocity, v_1. Assuming that these disturbances are small we can apply small-signal theory to calculate their mutual dependence. From the continuity equation we have

$$\frac{\partial n_1}{\partial t} + v_0 \frac{\partial n_1}{\partial z} + n_0 \frac{\partial v_1}{\partial z} = 0. \tag{5.20}$$

The space charge produces an electric field according to Poisson's equation:

$$\frac{\partial E_1}{\partial z} = \frac{q}{\varepsilon} n_1. \tag{5.21}$$

The velocity is, neglecting diffusion, a function of the electric field only, and its small-signal component can be obtained from a Taylor expansion around v_0:

$$v_1 = \mu_1 E_1 \quad \text{with } \mu_1 = \left(\frac{dv}{dE}\right)_{E = E_0}. \tag{5.22}$$

This set of equations has solutions of the form

$$n_1 = N \exp(-t/\tau_d) f(t - z/v_0), \tag{5.23}$$

where

$$\tau_d = \frac{q n_0 \mu_1}{\varepsilon}$$

is the *dielectric relaxation time* already encountered in Chapter 4. We see that the disturbance moves with a velocity v_0 and has an exponential time dependence. In normal circumstances τ_d is positive and the disturbance is damped. In the case of GaAs or InP above the threshold field, however, τ_d is negative because μ_1 is negative, and the disturbance grows as it propagates. In this case we cannot expect that the electron density in the n layer remains equal to the doping density, since there will always be small disturbances, caused for example by doping irregularities or by thermal fluctuations. The *I–U* characteristic then will differ from the *v–E* characteristic.

The charge imbalance that occurs can assume different forms, dependent on the doping concentration and the length of the diode (i.e. the width of the n layer). Shockley has already shown [5] that in short diodes a static charge redistribution occurs with the effect that the *I–V* characteristic has a positive slope everywhere. He pointed out, however, that for a high-frequency signal superimposed on the d.c. bias the differential conductance of such a device can be negative. It has been found that low-doped diodes follow Shockley's model and the *I–V* curve is continuous as the voltage increases to values beyond the threshold, as shown in Fig. 5.12, curve a. In higher-doped devices the diode can switch abruptly from a state with a uniform electron distribution to a highly non-uniform state accompanied by a drop in the current, giving the bistable characteristic of Fig. 5.12, curve b. The continuous parts of this characteristic normally have positive slopes everywhere. If a negative slope is seen it may be due to a temperature rise at higher bias.

As Shockley has already remarked, the impedance of the device can have a negative real part at certain frequencies which can be used to generate

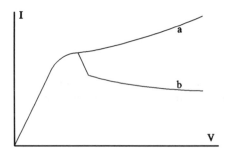

Figure 5.12   *I–V* characteristics of transferred electrode diodes.

microwave energy. Further discussion of this phenomenon is postponed until section 5.8 on negative-resistance devices at the end of this chapter.

5.5 PUNCH-THROUGH AND BULK BARRIER DIODES

Punch-through is a well-known phenomenon in bipolar transistors. A punch-through diode is essentially a transistor without a base connection (Fig. 5.13). For microwave applications p–n–p or n–p–n structures with a base width of 5–10 μm are used. Punch-through occurs in this device when the voltage across the structure is increased so much that the collector depletion layer extends to the emitter. Further increase of the voltage causes the diode to conduct current because the barrier between the emitter and the collector is lowered (Fig. 5.13(b)).

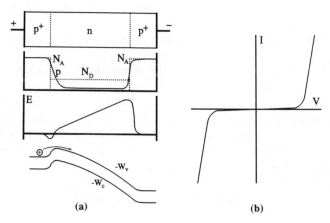

Figure 5.13 Punch-through diode: (a) physical structure, carrier density, electric field and band diagram; (b) *I–V* characteristic.

The behavior of the forward-biased junction can be modeled reasonably well with a diode characteristic:

$$J_0 = J_s\left\{\exp\left[\frac{q(V_{bi} - U_B)}{kT}\right] - 1\right\} \tag{5.24}$$

where V_{bi} is the built-in voltage of the emitter–base junction, also giving the barrier height at zero bias. At low bias the depletion layers do not touch and U_B remains close to V_{bi}, so the current is low. When the applied voltage U has reached the punch-through value, the depletion layers touch and U_B starts to decrease.

Assuming abrupt doping profiles and negligible carrier concentration in the base (compared with the ionized donor density) the electric field profile is a straight line as drawn in the figure and it is easy to calculate U_B and the voltage U:

$$U_B = \tfrac{1}{2} E_0 w_B = \frac{\varepsilon E_0^2}{2q N_D} \tag{5.25}$$

$$U = \frac{q N_D (L - w_B)^2}{2\varepsilon} - U_B = \frac{q N_D L^2}{2\varepsilon} - L\left(\frac{2q N_D U_B}{\varepsilon}\right)^{1/2} \tag{5.26}$$

from which follows

$$U_B = \frac{\varepsilon}{2q N_D L^2}\left(\frac{q N_D L^2}{2\varepsilon} - U\right)^2. \tag{5.27}$$

The bias voltage where conduction starts is given by equation (5.26) with $U_B = V_{bi}$. After this point the current increases very fast as can be seen by combining equations (5.27) and (5.24), and an *I–V* characteristic as in Fig. 5.13(c) is obtained.

   This structure can be used as a transit-time negative resistance device (section 5.8.3). The *I–V* curve also suggests applications as a detector but with standard fabrication techniques the punch-through voltage will always be at least a few volts, so the detector would have to be biased. Also, the curvature of the characteristic is less than that of a Schottky diode. However, thin layer growth techniques such as MBE make it possible to fabricate a variant consisting of a thin n–i–n structure with a very thin (even monoatomic) highly doped p layer in the middle (Fig. 5.14). This is called a *bulk barrier diode* and can be designed to have a low punch-through voltage. Also, by placing the p layer off-center the characteristic can be made asymmetric which is necessary for use as an unbiased detector.

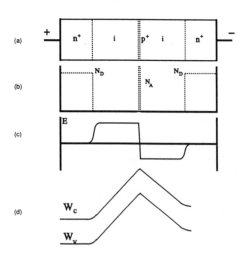

Figure 5.14 Bulk barrier diode: (a) structure; (b) doping profile; (c) electric field; (d) band diagram.

5.6 HETEROJUNCTION AND QUANTUM WELL DIODES

5.6.1 Heterojunctions

Heterojunctions have already been briefly described in Chapter 3. Usually band bending will occur but the details will strongly depend on the discontinuities in the bands, and the doping on both sides. Take, for example, the AlGaAs–GaAs heterojunction, which is used in many devices, among others various transistors (Chapter 6) and tunneling devices (Section 5.6.2).

As already mentioned in Chapter 3 the lattice constant of $Al_xGa_{1-x}As$ differs very little from that of GaAs so that the interface contains very few defects. This heterojunction is of the normal type (Chapter 3) with the bandgap difference of about 65% in the conduction band. Because $Al_xGa_{1-x}As$ has a higher bandgap (increasing with x) the electrons in the $Al_xGa_{1-x}As$ now have a higher potential energy than in the GaAs.

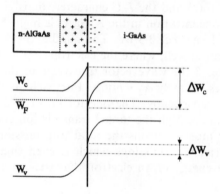

Figure 5.15 Band profile of an AlGaAs–GaAs heterojunction.

If we make the AlGaAs n type and leave the GaAs undoped then the electrons will spill over from the AlGaAs into the GaAs (Fig. 5.15). Depending on doping concentration a more or less extended depletion layer develops in the AlGaAs. The positive charge in this layer pulls the electrons in the GaAs back towards the junction. In this way a potential well is formed as shown in the figure, which can be used as the channel for a so-called high electron mobility transistor, as will be discussed in Chapter 6. The current–voltage characteristic of this structure is governed by the fact that the electrons can easily go from the AlGaAs to the GaAs but in the opposite direction find a potential barrier on their way. So this junction will have a rectifying characteristic.

If on the contrary we dope the AlGaAs p type, the holes will spill over into the GaAs, since they too have a lower energy there, and a potential well for holes will be formed. This structure too will have a rectifying characteristic but with reversed polarity compared with the other one.

5.6.2 Quantum well and superlattice diodes

In recent years it has become possible with new growth techniques such as MBE, to grow very thin layers with sharp interfaces from materials with different bandgaps. When the layer widths are of the same order of magnitude as the Schrödinger wavelength of the electron, then we can influence the behavior of the electron directly through its wavefunction. We can make the wave reflect, resonate etc. A typical example is the *double-barrier resonant tunneling* (DBRT) diode. It consists of a layer of GaAs between two layers of AlGaAs. All three are very thin, *ca*. 5 nm. At the outside n-type doped GaAs contact layers are made (Fig. 5.16). The physical structure and the conduction band profile at three bias values are sketched, in Fig. 5.16(a) and the *I–U* characteristic is given in Fig. 5.16(b).

Because the Al concentration in the AlGaAs is high the conduction band discontinuities are large, *ca*. 0.3 eV. The middle GaAs layer now has the character of a potential well, where the electrons are locked in. As we have discussed in Chapter 3 the wavevector transverse to the layers can assume only discrete values, namely $k_z \approx m\pi/a$. For each m there is a subband which the electron can occupy. The one for $m = 1$ is drawn in the figure. Because the barriers are high the tunnel transmission coefficient of each of them is small. For most electrons the total transmission is the product of the two single-barrier transmissions which is even smaller. This changes, however, when the energy of an electron that tries to cross the barriers is

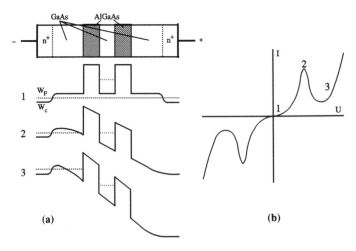

Figure 5.16 Double-barrier resonant tunneling diode: (a) structure and band diagram for three bias voltages; (b) *I–U* characteristic.

equal to that of the discrete level in the well. Then the wave character of the electron manifests itself and constructive interference occurs between the waves reflected from both barriers. The wavefunction now fits between the barriers and the wave amplitude can rise substantially. This is comparable with the situation in an optical or microwave resonator. At resonance the transmission through the resonator is large.

It can be shown that, if the current transmission coefficients of the separate barriers are T_1 and T_2, respectively, the total transmission is given by

$$T = \frac{T_1 T_2}{1 + R_1 R_2 - 2(R_1 R_2)^{1/2} \cos \phi} \tag{5.28}$$

where R_1 and R_2 are the reflections of the two barriers. These satisfy

$$R_{1,2} = 1 - T_{1,2}.$$

The angle ϕ is approximately equal to $k_z a - \pi$, where k_z is the z component of the wavevector in the well. If k_z has a value that belongs to the discrete energy level in the well, then $\phi \approx 0$ and a resonance in the transmission occurs. When T_1 and T_2 are small, R_1 and R_2 are approximately equal to 1 and the numerator can, at resonance, become very small so the total transmission is much greater than the product $T_1 T_2$. In fact, when T_1 and T_2 are equal, the resonant transmission is exactly unity.

At zero or low bias, the upper band diagram of Fig. 5.16(a), most electrons are below the resonance level and the current that gets through is small. In the middle band diagram the Fermi level coincides with the subband level in the well and there are many electrons that can tunnel through resonantly, so the current is large. If we increase the bias further the discrete level in the well sinks below the conduction band on the left and there are no electrons left that fulfill the resonance condition. The result is the *I–U* characteristic of Fig. 5.16(b). The increase of the current at still higher voltages is caused by electrons that go over the top of both barriers.

Thus, the *I–V* characteristic of a DBRT diode is quite similar to that of the ordinary tunnel diode, showing a negative-resistance region. However, because the structure is symmetric the characteristic is the same in both directions. One can make it asymmetric by giving the barriers different widths or heights. Indeed, it is an advantage of these diodes over the p–n junction tunnel diode that we have more parameters to adjust so they can be better optimized. We also have the possibility to use other material combinations from the III–V system such as InGaAs/InAlAs and InAs/GaSb.

When we alternate a great number of GaAs and AlGaAs layers a so-called *superlattice* results. The name is derived from the fact that we have superimposed a new periodicity upon that of the crystal lattice. Because of the coupling between the wells the discrete energy levels of Fig. 5.16 now

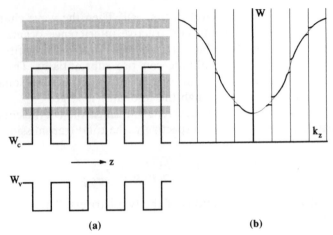

Figure 5.17 A superlattice: (a) band diagram and minibands; (b) Brillouin zone.

widen into *minibands*. We can also describe it by saying that that the Brillouin zone is split up in smaller zones (Fig. 5.17).

When an electron is moved by an electric field to the top of a miniband its effective mass goes to infinity and its mobility becomes very low. This could lead to a decrease of velocity with field as in the transferred-electron effect of section 5.4. *I–V* characteristics with negative slopes have been measured for superlattice devices.

5.7 SUPERCONDUCTING DIODES

The theory of superconductivity is far too complicated to be explained in a few words, so the reader is referred to the existing literature [6]. The superconducting device that is most interesting for microwave applications is the SIS (*superconductor–insulator–superconductor*) diode. It consists of two superconductors separated by a very thin layer of insulating material, very thin meaning less than 5 nm. Now in a superconductor there always exists a mixture of superconducting electrons and normal electrons. The

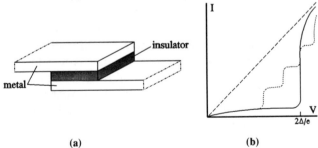

Figure 5.18 (a) SIS junction and (b) *I–V* characteristic.

superconducting electrons always move in pairs, the so-called Cooper pairs. The normal electron current needs an electric field to keep flowing, the Cooper pair current does not.

In an SIS junction both types of electrons can tunnel through the insulator if it is thin enough. The tunneling of normal electrons is well known and has led to the two kinds of tunnel diodes described earlier in this chapter. That Cooper pairs can tunnel was predicted theoretically by Josephson in 1963 and shortly afterwards verified. The *I–V* characteristic of an SIS junction therefore has two branches (Fig. 5.18). The superconducting so-called Josephson current needs no voltage, whereas the current carried by the normal electrons does, like it does in a p–n junction.

If one increases the current from zero, at first there will be only Josephson current. There is, however, a maximum for the superconducting current an SIS junction can sustain. If one tries to exceed this, the diode switches to normal current and a voltage appears across it. The maximum Josephson current decreases when a magnetic field is applied.

In microwave technology the normal current characteristic of an SIS junction is used for detection and mixing because the knee in the characteristic is very sharp. To this end the Josephson current has to be suppressed and this can be done with a magnetic field. Josephson junctions can also be used as fast switching elements. This is employed in digital systems and will be discussed in Chapter 11.

5.8 NEGATIVE-RESISTANCE DIODES

5.8.1 Tunnel diodes

The tunnel diodes of the p–n junction and double-barrier types have been discussed in sections 5.2.4 and 5.6.2. These devices have a negative-slope region in their *I–V* characteristics. When they are biased in this region the differential resistance, as it is experienced by an a.c. modulation around the bias point, is negative. This can be used for amplification or generation of signals (Chapters 9 and 10). The frequency range over which the negative resistance is experienced is limited for two reasons. The fundamental reason is that the tunneling process needs time and when the signal period is of the order of this time constant transit-time effects occur as in a vacuum triode, and the real part of the diode impedance becomes positive. This cutoff frequency can be very high. It is estimated to be in the order of 1 THz (1000 GHz). A more mundane limitation is caused by the fact that these diodes always have parasitic series resistance and parallel capacitance. This gives the equivalent circuit of Fig. 5.19. From this circuit it can easily be seen that the real part of the impedance is only negative if $R_d > R_s$ and becomes zero when

$$(\omega C R_d)^2 = \frac{R_d}{R_s} - 1.$$

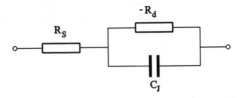

Figure 5.19 Equivalent circuit of a tunnel diode.

Since R_d in these devices cannot be made very large it is important to keep the parasitics small.

5.8.2 Gunn diodes

The name Gunn diode is given to all two-terminal devices whose operation depends on the transferred-electron effect, described in Section 5.4, although the Gunn effect occurs only in diodes of sufficient length.

In very long diodes (order of magnitude 100 µm) so-called Gunn domains will result, i.e. a combination of an accumulation layer and a depletion layer (Fig. 5.20(a)) that moves with constant velocity. In short diodes this does not happen because a disturbance will reach the anode and disappear before it has had time to grow into a fully developed domain. The domain absorbs a large part of the voltage over the device, so that the field strength outside the domain is below the threshold value. The external current is determined

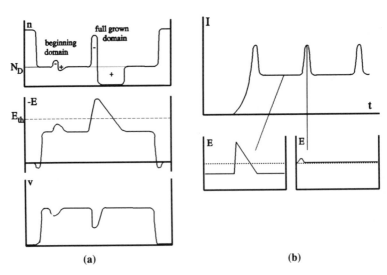

(a) (b)

Figure 5.20 Gunn domains: (a) density, field and velocity in the presence of a domain; (b) current spikes due to domain formation and annihilation.

by this field strength. When the domain arrives at the anode (the positive contact) the space charge suddenly disappears and with it the voltage drop over the domain. The field strength in the whole diode then rises and with it the external current. When the field strength rises above the threshold value an unstable situation occurs again and a new domain starts to form, usually in the neighborhood of the cathode (the negative contact). This domain travels towards the anode and the whole story repeats itself. So even with a constant voltage over the diode a periodic phenomenon occurs, consisting of short current peaks (Fig. 5.20(b)). This is what I.B. Gunn discovered in 1963, when he was investigating the noise of semiconductors at high field strengths, and which is called the *Gunn effect* after him. It had been predicted, however, a few years before by Ridley and Watkins and Hilsum [7]. Extensive discussions of the Gunn effect are given in refs [5] and [8].

When we put such a diode in a resonant circuit tuned to the frequency of the current pulses, the circuit will be excited by the current and it will oscillate. The oscillation frequency is roughly equal to the transit frequency $1/T_d$, T_d being the transit time of a domain, equal to L/v_d where the domain velocity v_d lies between 1×10^5 and 2×10^5 m/s. For instance, a diode length of 10 µm gives a transit frequency of about 1 GHz. The whole phenomenon now becomes more complicated, because the diode voltage is no longer constant but also has a periodic variation. This is further discussed in Chapter 10.

In short diodes (<10 µm for frequencies of 10 GHz and up) as used in modern Gunn oscillators, there is no time for full-grown domains to be formed. What happens in these devices is that periodically an accumulation layer forms at the cathode that travels to the anode. Depending on the doping and length ($N_D L < 10^{12}$ cm^{-2}) it is also possible that the electron density rises continuously towards the anode. The d.c. characteristic of these devices shows no negative slope but the a.c. impedance has a negative real part in a wide band around the transit frequency. They are called stable Gunn diodes and can be used for small-signal amplification.

The bandwidth of the Gunn effect is limited by the speed with which the electrons can transfer from one valley to another and back. This is a scattering process which is characterized by a time constant. It turns out to be shorter in InP than in GaAs. In GaAs the Gunn effect can be used up to about 80 GHz; in InP frequencies about twice as high can be reached.

5.8.3 Transit-time diodes

Charge carriers need time to traverse a diode and this *transit time* can have an influence on the high-frequency properties. We have already encountered examples of this in the discussions on tunnel diodes and the Gunn effect above. However, even in materials without a negative slope in the v–E

characteristic transit time can be used to create an impedance with a negative real part. The reason is that the finite transit time of the carriers causes the current to be delayed with respect to the voltage. For the alternating component of the current this delay means a phase shift that is greater the higher the frequency is. When the phase shift is between 90° and 270° the real part of the diode impedance is negative. To understand this let us have a look at Fig. 5.21.

For the analysis we split the device into an injection region, which is very short, and a drift region, where the transit-time delay occurs. From the injection region charge is injected into the drift region. The amount of injected charge will in general be determined by the electric field in the injection region. We shall come back to this later. For the moment let us assume that somewhere in the positive period of the device voltage a spike of charge is injected into the drift region (Fig. 5.21(b)). By virtue of the Ramo–Shockley theorem (Appendix 2.A) external current will flow as long as this charge travels through the drift region. Assuming a constant drift velocity the external current will then have the form of a rectangular pulse, the length of which is determined by the transit time. It will be clear from looking at the figure that the fundamental harmonic component of the current is delayed with respect to the voltage and with a suitable combination of the injection point of time and the transit delay the phase shift may be more than 180°.

All this is very qualitative and approximate (for instance, the time of injection is not determined by the total device voltage but by the part that

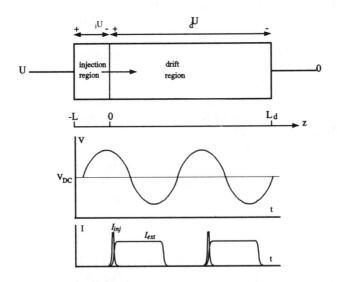

Figure 5.21 Transit-time diode: (a) division into injection and drift regions; (b) simplified picture of voltage and current.

drops over the injection region) so let us apply some small-signal theory to obtain a more accurate picture. As usual we assume that the injected charge can be split up into a constant and a time-dependent part as follows:

$$\rho(0, t) = \rho_0 + \text{Re}(\rho_1 e^{j\omega t}). \tag{5.29}$$

The factors that determine the amplitude and phase of ρ_1 will be dealt with later. We now assume that this charge travels with a constant velocity v through the drift region. The charge density at an arbitrary position is then easy to calculate:

$$\rho(x, t) = \rho(0, t - x/v) = \rho_0 + \text{Re}[\rho_1 e^{j\omega(t - x/v)}]. \tag{5.30}$$

In the following we will skip the direct current components and limit ourselves to a description of the alternating components. Furthermore we use complex calculus, so the operator $\text{Re}(\ldots e^{j\omega t})$ will be assumed to be implied and not be written explicitly.

Integration of the Poisson equation using the charge density given by the a.c. part of equation (5.30) yields the a.c. electric field:

$$E(x) = E_1 + j\frac{v}{\omega\varepsilon}\rho_1(e^{-j\omega x/v} - 1) \tag{5.31}$$

where E_1 is the field at $x = 0$. Integrating once more yields the voltage over the drift region:

$$U_d = E_1 L_d - j\frac{v\rho_1}{\omega\varepsilon}L_d - \frac{v^2}{\omega^2\varepsilon}\rho_1(e^{-j\omega L_d/v} - 1). \tag{5.32}$$

The current density in the drift region is the sum of the conduction current and the dielectric current densities:

$$J_d = J_c(x) + j\omega\varepsilon E(x)$$

$$= v\rho_1 e^{-j\omega x/v} + j\omega\varepsilon\left[E_1 + j\frac{v}{\omega\varepsilon}\rho_1(e^{-j\omega x/v} - 1)\right]$$

$$= j\omega\varepsilon E_1 + v\rho_1. \tag{5.33}$$

In agreement with a theorem from electromagnetic field theory, discussed in Appendix 2.A, J_d is divergence free, that is independent of x. The impedance of the drift region now is found by dividing U_d by J_d and the diode area A:

$$Z_d = \frac{U_d}{J_d A} = \frac{L_d}{j\omega\varepsilon A} - \frac{v^2}{\omega^2\varepsilon A}\frac{\rho_1}{j\omega\varepsilon E_1 + \rho_1 v}(e^{-j\omega L_d/v} - 1). \tag{5.34}$$

This still contains the boundary values E_1 and ρ_1. We now have to specify the relation between these two. It is reasonable to assume that ρ_1 is determined by the electric field in the injection region and, when the latter is thin, the a.c. component of the field may be assumed to be constant and equal to E_1, so that ρ_1 is completely determined by E_1. Since we are using small-signal theory there should be a linear relationship. We therefore define a complex constant η as follows:

$$\eta = \frac{\omega \varepsilon E_1}{J_c(x = 0)} = \frac{\omega \varepsilon E_1}{v \rho_1}. \tag{5.35}$$

Now if we define also the transit angle $\Phi = \omega L_d / v$ and the capacitance of the drift region $C_d = \varepsilon A / L_d$ we can write equation (5.34) in the form

$$Z_d = \frac{1}{j\omega C_d}\left(1 - \frac{1}{1 + j\eta}\frac{1 - e^{-j\Phi}}{j\Phi}\right). \tag{5.36}$$

The function of Φ in this formula is the transit-time function which we already have encountered in section 2.3. Whether $\mathrm{Re}(Z_d)$ is negative is now determined by η and Φ. Suppose that η is real; then we have

$$\mathrm{Re}(Z_d) = \frac{1}{\omega C_d}\frac{1 - \cos\Phi + \eta\sin\Phi}{(1 + \eta^2)\Phi}. \tag{5.37}$$

Since $1 - \cos\Phi$ is always positive, $\sin\Phi$ should be negative, so Φ must lie between 180° and 360°. The optimal angle is 270°; in other words, the transit time L_d / v must be 3/4 of the signal period. In this case the optimal value of η is about 2.5 and we find for the so-called *negative quality factor* $Q_n = |\mathrm{Im}\, Z_d / \mathrm{Re}\, Z_d|$ a maximum value of *ca.* 20. This is rather large for a negative-resistance device. Preferably it should be below 10. Evidently large oscillator power should not be expected from a device with in-phase injection. Better performance is obtained when η is positive imaginary. In that case we find

$$\mathrm{Re}(Z_d) = \frac{1}{\omega C_d}\frac{1 - \cos\Phi}{(1 - |\eta|)\Phi} \tag{5.38}$$

Now $\mathrm{Re}(Z_d)$ and Q_n can reach very large negative values, and the negative Q factor can be quite small.

These two examples for η are not chosen arbitrarily. They represent cases that are realized in practice. A real value for η is found in the punch-through diode treated in section 5.5. When it is used as a transit-time device it is called a BARITT diode (from barrier injection transit time). The forward-biased junction now serves as the injection region and the rest of the depletion layer as the drift region. For a superimposed small a.c. voltage we can apply a Taylor series expansion to equation (5.24):

$$J_1 = -\frac{q}{kT}J_s \exp\left[\frac{q(V_{bi} - U_{B0})}{kT}\right]U_{B1} = -\frac{q(J_0 + J_s)}{kT}U_{B1}. \tag{5.39}$$

When we now assume that the a.c. field strength over this region is uniform we can replace U_{B1} by $- E_1 L_i$ (the minus sign is because when E increases U_B decreases) and we find for η

$$\eta = \frac{\omega \epsilon k T}{q(J_0 + J_s)}. \tag{5.40}$$

This is a real quantity. BARITT diodes are therefore not efficient. The achieved oscillator power lies in the range 5–100 mW at 10 GHz and the highest frequency is about 30 GHz. However, they are low-noise devices. Noise measures of about 15 dB have been obtained. Also, the so-called *self-rectification* (section 5.8.4) is large. This makes them very suitable as self-oscillating mixers in small Doppler radars.

An example of inductive injection (η positive imaginary) is found in IMPATT diodes (from impact avalanche transit time). Here an avalanche diode (section 5.2.5) is used as the injection region. The structure of the IMPATT diode as conceived by its inventor Read [10] is sketched in Fig. 5.22. The electric field profile is so designed that in the p–n junction avalanche multiplication occurs, while in the drift region the electron velocity reaches its saturation value. As a result of the typical properties of the multiplication process the injection current lags 90° with respect to the voltage over the avalanche region.

To show this let us start again from equations (5.10) of section 5.2.5. We split all quantities in a d.c. and (small-signal) a.c. part, e.g.

$$J = J_0 + \text{Re}[J_1 \exp(j\omega t)].$$

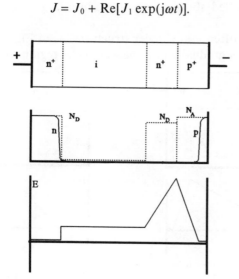

Figure 5.22 Read-type IMPATT diode.

Making the same assumptions as before we arrive at an equation for the a.c. part of the difference current:

$$\frac{j\omega}{qv_s}J_1 + \frac{\partial}{\partial z}K_1 = 2\alpha_0 J_1 + 2\alpha' J_0 E_1 \tag{5.41}$$

where $\alpha_0 = \alpha(E_0)$ and $\alpha' = d\alpha/dE$. We now approximate $\partial K_1/\partial z$ by its average over the avalanche zone:

$$\frac{\partial}{\partial z}K_1 = \frac{K_1(a) - K_1(-a)}{2a} = \frac{J_1}{a} \tag{5.42}$$

where a and $-a$ are the limits of the avalanche zone and the same boundary conditions as for the d.c. values have been used. Substituting this in equation (5.41) we obtain

$$J_1 = -j\frac{v_s}{\omega}2\alpha' J_0 E_1 = -j\frac{\omega_a^2}{\omega^2}\omega\varepsilon E_1 \tag{5.43}$$

where the *avalanche frequency* $\omega_a = (2J_0\alpha' v_s/\varepsilon)^{1/2}$ has been introduced. Note that the $-j$ factor in this equation implies that J_1 lags E_1 by 90°, in other words the a.c. avalanche current shows an inductive behavior. For the injection parameter η defined by equation (5.35) we find

$$\eta = j\frac{\omega^2}{\omega_a^2} \tag{5.44}$$

and, as already stated above, this can produce large values of the negative resistance.

In practice one often uses simpler structures, as in Fig. 5.23(a), where the drift region merges gradually into the avalanche region. At millimeter wave

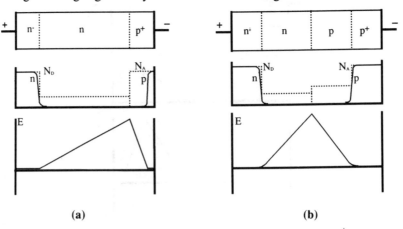

(a)   (b)

Figure 5.23 Alternative types of IMPATT diodes: (a) simple p⁺–n structure; (b) double drift region structure.

frequencies sometimes a *double-drift region* (DDR) structure is used where both electrons and holes are allowed to drift. Clearly the widths of the two regions must be matched to the different drift velocities of electrons and holes, in order that they both have the same transit delay. This structure is shown in Fig. 5.23(b).

IMPATT diodes are efficient converters of d.c. energy into a.c. energy. The highest efficiencies (up to 35%) are attained with GaAs diodes because in GaAs a lower field strength in the drift region is necessary to reach the saturated drift velocity. The highest frequencies are reached with Si diodes. They can be used to *ca.* 200 GHz. On the other hand, they are rather noisy, owing to the multiplication process (section 4.4.5). Si IMPATTs typically have noise measures of 40 dB; GaAs IMPATTs, because here the ionization coefficients are more equal, reach about 30 dB.

5.8.4 Large-signal properties

When negative-resistance diodes are used in oscillators or power amplifiers they operate under conditions of large signal amplitude and so it is important to know their impedance properties under large-signal condi- tions. In general this is not an easy task since even in the simplified models presented previously one runs into non-linear differential equations which are not analytically solvable. So we have to satisfy ourselves here with an approximate analysis which gives an idea how Re(Z) varies for the different devices when the a.c. amplitude increases.

For a tunnel diode it is expected by looking at the *I–V* characteristic that as the voltage amplitude increases the current amplitude will increase less than linearly (Fig. 5.24). This means that in an equivalent circuit consisting

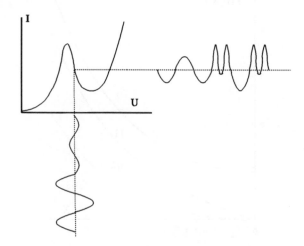

Figure 5.24 Tunnel diode under small-signal and large-signal conditions.

of – R_d parallel with C_J as in Fig. 5.19 the magnitude of R_d will *decrease* as the signal amplitude increases. For Gunn diodes the situation is more complicated. However, the behavior of these devices is governed by the N-shaped v–E characteristic and to this we can apply the same reasoning as to the I–V characteristic of the tunnel diode. So it seems plausible that the large-signal behavior of Gunn diodes is similar to that of tunnel diodes.

For a large-signal analysis of BARITT diodes we can again start from equation (5.24), in which we now substitute $U_B = U_0 + U_1 \cos \omega t$, which gives as the injected current:

$$J_i = J_s \exp\left[\frac{q(V_{bi} - U_0)}{kT}\right] \exp\left(-\frac{qU_1}{kT} \cos \omega t\right). \tag{5.45}$$

In this equation we can use an expansion from the theory of Bessel functions, similar to the one we used in Chapter 2 for the analysis of the klystron:

$$\exp\left(-\frac{qU_1}{kT} \cos \omega t\right) = I_0\left(\frac{qU_1}{kT}\right) + 2 \sum_1^\infty I_n\left(-\frac{qU_1}{kT}\right)\cos(n\omega t) \tag{5.46}$$

where the I_n are modified Bessel functions of the first kind (Fig. 5.25). With this expansion we find for η after some manipulation

$$\eta = \frac{\omega \varepsilon kT}{q(J_0 + J_s)} \frac{\gamma I_0(\gamma)}{I_1(\gamma)} \tag{5.47}$$

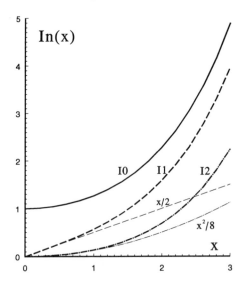

Figure 5.25 Modified Bessel functions of the first kind.

with

$$\gamma = \frac{qU_1}{kT}.$$

This formula means that η *decreases* with increasing signal amplitude and from equation (5.37) we can conclude that R_d also decreases.

Another conclusion we can draw from equation (5.46) is that the direct current also changes with a.c. amplitude thanks to the function I_0. This is the *self-rectification* effect, already mentioned above in the description of BARITT diodes.

For IMPATT diodes we can make a similar analysis. To make it tractable we approximate the field dependence of the ionization coefficient by

$$\alpha = \alpha_0 \exp\left(\frac{E - E_0}{E_A}\right) \tag{5.48}$$

where E_0 is the value of the d.c. component of the electric field and E_A is chosen to match the curve of Fig. 5.8 in a limited range. For a sinusoidal a.c. field ($E = E_0 + E_1 \cos \omega t$) we can now apply the expansion

$$\alpha = \alpha_0 \exp\left(\frac{E_1}{E_A} \cos \omega t\right) = \alpha_0 \left[I_0\left(\frac{E_1}{E_A}\right) + 2 \sum_1^\infty I_n\left(\frac{E_1}{E_A}\right) \cos(n\omega t) \right].$$

With this expansion we can now write the avalanche frequency as

$$\omega_a^2 = \frac{2 J_0 \alpha' v_s}{\varepsilon} \frac{2 I_1(E_1/E_A)}{E_1/E_A}. \tag{5.49}$$

This increases, so η *decreases*, as the signal amplitude increases and R_d decreases. The IMPATT diode thus shows similar large-signal behavior to that of the BARITT diode. It is noteworthy that tunnel and Gunn diodes, which might be called *voltage-controlled devices*, since for stable operation they should be biased from a voltage source, show an increase of R_d whereas BARITT and IMPATT diodes, which are *current-controlled devices*, since they should be biased from a current source, show a decreasing R_d.

REFERENCES

1. Sze, S. M. (1981) *Physics of Semiconductor Devices*, 2nd edn, Wiley, New York.
2. Shur, M. S. (1990) *Physics of Semiconductor Devices*, Prentice-Hall, Englewood Cliffs, NJ.
3. Neamen, D. A. (1992) *Semiconductor Physics and Devices: Basic Principles*, Irwin, Homewood, IL.
4. Kane, E. O. (1961) Theory of tunneling. *J. Appl. Phys.*, **32**, 83.
5. Shockley, W. (1954) Negative resistance arising from transit time in semiconductor diodes, *Bell Syst. Tech. J.*, **33**, 799–826.

6. Rose-Innes, A. C. and Rhoderick, E. H. (1987) *Introduction to Superconductivity*, 2nd edn, Pergamon, Oxford.
7. Carroll, J. A. (1970) *Hot Electron Microwave Generators*, Arnold, London.
8. Bulman, P. J. Hobson, G. S. and Taylor, B. C. (1972) *Transferred Electron Devices*, Academic Press, London.
9. Howes, M. J. and Morgan, D. V. (1976) *Microwave Devices, Device–Circuit Interactions*. Wiley and Sons, New York.
10. Read, W. T. (1958) A proposed high-frequency negative resistance diode, *Bell Syst. Tech. J.*, **37**, 401–46.

FURTHER READING

1. Watson, H. A. (ed.) (1969) *Microwave Semiconductor Devices and Their Circuit Applications*, McGraw-Hill, New York.
2. White, J. F. (1982) *Microwave Semiconductor Engineering*, Van Nostrand Reinhold, New York.
3. Garver, R. V. (1976) *Microwave Diode Control Devices*, Artech, Dedham, MA.
4. Kollberg, E. L. (ed.) (1984) *Microwave and Millimeter Wave Mixers*, IEEE, New York.

PROBLEMS

5.1 Calculate the built-in voltage and the capacitance per unit area for an n^+–p junction diode with $N_D = 10^{20}/cm^3$ and $N_A = 10^{18}/cm^3$ at zero bias. At what bias will this capacitance (at least according to the simple theory) become infinite? Explain why this should happen in the model used and give reasons why it does not happen in reality.

5.2 Draw the equivalent of Fig. 4.1 for a p–i–n diode. When the p and n dopings are both $10^{18}/cm^3$ and the i region is 30 µm wide, calculate

- the built-in voltage,
- the electric field in the i region (approximately) at zero bias, and
- the width of the depletion regions in the n and p layers.

At what bias voltage will the field in the i region become zero?

5.3 Calculate the differential resistance and the capacitance of a GaAs Schottky diode with the following data at a forward bias of 0.3 V: bandgap, 1.4 eV; $\Phi_B = 0.7$ eV; $N_D = 10^{16} cm^3$; area, 400 µm². Calculate the admittance at 5 GHz with a series resistance of 5 Ω included.

5.4 An often-used analytic approximation for the v–E characteristic of GaAs is

$$v = \frac{\mu E + v_s(E/E_A)^4}{1 + (E/E_A)^4}$$

where μ, the low-field mobility, is a decreasing function of temperature and doping concentration. Give an expression for E_A when μ, v_s and the threshold field E_{th} are known. Calculate E_A for $\mu = 0.5 \, m^2/(V \, s)$, $v_s = 9 \times 10^4$ m/s, $E_{th} = 4$ kV/cm. For the same parameters calculate the negative differential mobility at a field of 6 kV/cm. If the doping is $10^{16}/cm^3$, calculate the dielectric relaxation time at this field strength. For a diode of 10 µm length, estimate whether a charge disturbance has sufficient time to grow. Take a d.c. operating point at 6 kV/cm and apply a field modulation with an amplitude of 2 kV/cm. Estimate the amplitude of the fundamental component of the current and use this to make an estimate of the maximum efficiency of a Gunn diode.

5.5 Design a Read-type IMPATT diode with the structure and field profile of Fig. 5.22 with the following conditions:

- the p^+ and n^+ contact regions are partly depleted and the doping concentrations in these regions are $10^{19} cm^3$;
- the n region is fully depleted;
- the electric field in the drift region is 50 kV/cm;
- the operating frequency range is centered around 10 kV/cm;
- the ionization coefficients of electrons and holes are equal and can be approximated by $\alpha = a \exp(E/E_0)$ with $a = 200/cm$, $E_0 = 80$ kV/cm;
- $\varepsilon_r \varepsilon_0 = 10^{-10}$ A s/V m.

Calculate the width of the n and i regions, the peak electric field and the diode voltage. Calculate the impedance at 10 GHz for a diode diametre of 50 μm and a direct current of 100 mA.

6

Transistors

6.1 SILICON BIPOLAR TRANSISTORS

6.1.1 Structure and technology

The bipolar transistor has since its invention in 1948 undergone a continuous development, and its frequency limits have been pushed upward steadily. In particular the invention of planar silicon technology has increased the possibilities for reducing device dimensions. As we will see this is essential for reaching higher frequencies. In this technology the impurities are diffused via a mask into the surface of the silicon. The diffusion depth can be controlled by time and temperature and is of the order of 1 μm. By two subsequent diffusions of acceptors and donors a cross-section such as that in Fig. 6.1(a) is obtained. The doping profiles have the form of Fig. 6.1(b). Since bipolar transistors are treated extensively in many textbooks we will here only discuss those properties that are essential for microware applications.

The frequency limit of a bipolar transistor is for a large part determined by the time the carriers injected from the emitter need to cross the base region. To keep this transit time short we should, to begin with, use the carriers with the highest mobility, which in all practical semiconductors are the electrons. This is why microwave transistors are always of n–p–n type. Next we must keep the base as thin as possible, a precaution that also keeps the electron–hole recombination low (it is roughly proportional to the ratio of transit time and electron lifetime). It turns out that as an emitter dopant arsenic is the best because it has a steeper diffusion profile than the other n dopants (mainly phosphorus) so that the transition from emitter to base is sharper. It also helps to introduce the dopants by ion implantation instead of diffusion, which offers shallower emitter depths and better control of the junction positions. In this way base thicknesses of about 0.1 μm can be obtained.

6.1.2 Basic principles

The band diagram of an n–p–n transistor is shown in Fig. 6.2. Transistor operation is based on the fact that when the emitter–base junction is biased

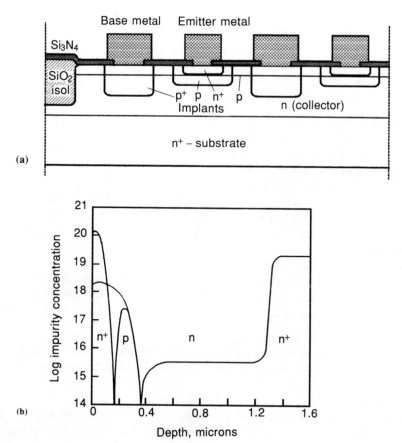

Figure 6.1 Bipolar transistor: (a) structure; (b) doping profile.

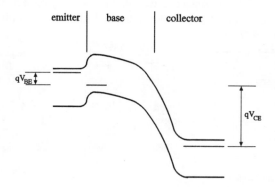

Figure 6.2 Band diagram of a bipolar transistor.

in the forward direction the barrier for electrons from the emitter to the base is lowered so more electrons are injected into the base. Because of the short distance to the collector most of these electrons diffuse to the collector where they are picked up by the electric field that exists in the reverse-biased base–collector junction. In principle no base current is needed, only base–emitter voltage. In practice base current is unavoidable, mainly for three reasons.

- Injection of holes (in the case of an n–p–n transistor) takes place from the base into the emitter. The holes in the base see a potential barrier of the same magnitude as the electrons in the emitter. By doping the emitter much higher than the base one can nevertheless achieve a hole current that is much lower than the electron current. Unfortunately the high doping in the emitter reduces the bandgap and this shows up as a reduction of the barrier in the valence band, thus increasing the hole current. A way to counteract this is to make use of a heterojunction (section 6.2).
- Recombination occurs in the base.
- Base layer resistivity exists: the center of the emitter–base junction, being farther away from the base contact, is biased less far in the forward direction than the edges owing to the voltage drop the base current causes.

The last-mentioned effect causes most of the current to flow through the emitter edges. At microwave frequencies this will be enhanced further by the skin effect. The center of the emitter therefore contributes more to the emitter–base capacitance than to the base current. For this reason it is necessary to make the emitters of microwave transistors as narrow as possible. To increase the collector current and the output power many emitter fingers are connected in parallel. This leads to layouts as in Fig. 6.3. A problem is connected with this: a bipolar transistor is inherently unstable. When its temperature is raised electron–hole pairs are thermally generated in the base leading to increased base current. This gives an increase of the collector current which raises the temperature further. At high temperatures

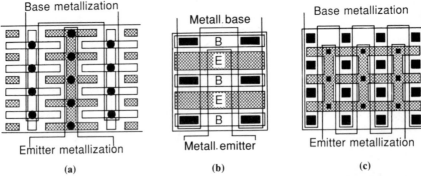

Figure 6.3 Layouts of bipolar transistors: (a) fishbone; (b) overlay; (c) mesh.

this can lead to thermal runaway and destruction of the transistor. In the case of parallel connected emitters it can lead to an unbalance in the current distribution. If one emitter draws more current than the others it becomes hotter by which its current further increases. To counteract this a technique called *emitter ballasting* is applied: the emitters are not connected directly but via resistors. An increase of emitter current will now produce an increased voltage drop over this resistor so that the emitter–base voltage is reduced. This counteracts the apparent increase in base current caused by thermal generation.

The frequency response of a bipolar transistor is limited by

- the transit time in the base and the collector depletion layer (this gives a cutoff frequency $f_\tau = 1/2\pi\tau$),
- the emitter–base capacitance C_{eb} (this produces a capacitive emitter current which does not reach the collector, and as a result α becomes frequency-dependent: $\alpha = \alpha_0/(1 + jf/f_\alpha)$, $f_\alpha = 1/2\pi R_{be} C_{be}$), and
- the base resistance R_b (together with the emitter capacitance this forms a low-pass filter in grounded-emitter configuration).

The transit-time delay can be split up in two main components.

- The first is the transit time in the non-depleted part of the base. Here the principal driving force for the electron transport is diffusion. From the diffusion equation one can derive that the time then is equal to $t_b = L_b^2/2D_{el}$, where L_b is the base thickness and D_{el} the electron diffusivity. For a base thickness of $0.15\,\mu m$ and a diffusivity of $0.001\,m^2/(V\,s)$ we find a time of about 12 ps. This would severely limit the cutoff frequency. In practice therefore one arranges the doping profile to have a gradient in the base, as in Fig. 6.1. By outdiffusion of the holes an electric field is created which counteracts the outdiffusion by pushing the holes towards the emitter. The same field will push the electrons towards the collector giving them an extra velocity. Under favorable circumstances this velocity can come close to the saturation velocity which is about 10^5 m/s. At the above-mentioned base thickness this yields a transit time of 1.5 ps.
- The second is the transit time in the base–collector depletion region. We may assume that here the field strength is high enough so that the electrons attain their saturation velocity, *ca.* 10^5 m/s. A layer thickness of $1\,\mu m$ would result in a transit time of 10 ps.

Including both of these two times, we obtain an f_τ value of about 10 GHz

In microwave transistors f_α is of the same order of magnitude as the transit-time frequency. As already stated, to keep C_{eb} small, we should use narrow emitter fingers. We cannot do much about the base resistance: in an n–p–n transistor the base current is carried by holes which have a lower mobility than the electrons; furthermore the base doping cannot be too high because the emitter doping has to be greater. The only thing is to keep the emitter fingers narrow and to put the base metallization as close as possible to the emitter.

The technology described above evidently has its limitations. A disadvantage not yet mentioned is that the high emitter doping causes a narrowing of the bandgap, leading to increased injection of holes from the base into the emitter and reduced current gain. The usable frequency range of Si bipolar transistors is therefore restricted to about 6 GHz for amplifiers and 12 GHz for oscillators. On the bright side we have the fact that reasonable power output can be obtained by putting a large number of emitters in parallel: 20 W at 2 GHz.

For further improvements we have to apply new techniques such as the following.

- *(Hetero)epitaxy*. If we grow the emitter and base layers epitaxially the base doping can without any problem be made higher than the emitter doping; also one can with modern growth techniques such as MBE produce very thin and highly doped base layers ($0.01\ \mu m$, $N = 10^{20} cm^3$); also with epitaxial techniques we can make sure that the emitter has a greater bandgap than the base by using different materials; this reduces the injection of holes into the emitter.
- *Materials with higher electron mobility*. Here GaAs is a promising material; for this material the heteroepitaxy technique is well developed.
- *Electron beam lithography*. With this technique even narrower emitter fingers and smaller emitter–base distances can be made.

Heterojunction bipolar transistors are treated in section 6.2.

6.1.3 Circuit properties and noise

The simplest small-signal equivalent circuit for a bipolar transistor is given in Fig. 6.4.

As already mentioned in the previous section the frequency dependence of the current amplification in the grounded base configuration can approximately be described by

$$\alpha = \frac{\alpha_0}{1 + jf/f_\alpha} \qquad \text{with } f_\alpha = \frac{1}{2\pi R_{be} C_{be}} \tag{6.1}$$

where α_0 is the direct current amplification. For the grounded emitter current amplification we find

$$\beta = \frac{\beta_0}{1 + jf/f_\beta}, \qquad \beta_0 = \frac{\alpha_0}{1 - \alpha_0} \tag{6.2}$$

where f_β reflects the combined effects of f_α and f_T. At low collector currents R_{be} is high and f_α is low. At high currents the effective base width increases

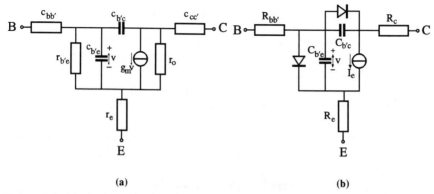

(a)   **(b)**

Figure 6.4 Equivalent circuit of a bipolar transistor: (a) small signal; (b) large signal.

by the Kirk effect and f_τ decreases. The cutoff frequency f_T is defined as the frequency where $|\beta| = 1$. It has the highest value at intermediate collector currents.

The maximum frequency of oscillation is that for which the power amplification is equal to 1. This is found to be $f_{max} = (f_T/8\pi r_b C_c)^{1/2}$.

For the noise figure F, defined as the degradation of the signal-to-noise ratio (Chapter 4), one finds, assuming that the main noise source is thermal noise in the resistances,

$$F = 1 + \frac{r_b}{R_s} + \frac{r_e}{2R_s} + \frac{(r_b + r_e + R_s)^2}{2\alpha_0 r_e R_s}\left[\left(\frac{f}{f_\alpha}\right)^2 + \frac{1}{\beta_0} + \frac{I_{co}}{I_e}\right] \qquad (6.3)$$

where R_s is the source resistance connected to the input and I_{co} the collector leakage current. From this formula the great interest of a small r_b and r_e and a high f_α is evident. At high frequencies we find that F increases as f^2 i.e. with 6 dB/octave.

6.1.4 Applications

Silicon bipolars are mainly used in the lower microwave ranges. Since their power capability is quite good but the noise figures are inferior to those of GaAs MESFETs (section 6.3) at frequencies above 500 MHz, their main uses are in power amplifiers and oscillators. They may also be used in small-signal amplifiers when the noise performance is not critical. Since the emitter–base junction has a diode characteristic they can be used as mixers by feeding the signal and local oscillator into the base with the emitter grounded. The intermediate frequency can be taken off the collector. Because of the amplification of the transistor this mixer can show conversion gain.

6.2 HETEROJUNCTION BIPOLAR TRANSISTORS

Recent years have shown a strong development in the heterojunction bipolar transistor (HBT). Above we have already mentioned the possibility of limiting the injection of holes into the emitter by the use of an emitter with a larger bandgap than the base. The bandgap difference manifests itself as a discontinuity in the conduction band or the valence band, or both.

For n–p–n HBTs we need a discontinuity in the valence band such that the holes have a higher energy in the emitter than in the base (Fig. 6.5). As

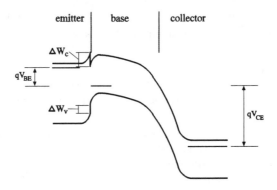

Figure 6.5 Band diagram of a heterojunction bipolar transistor.

pointed out in Chapter 4, to make good quality heterojunctions the two materials should be lattice matched. Some possible combinations that fulfill the lattice-matching condition are GaP/Si and AlGaAs/GaAs. It has been found that when one of the layers is very thin lattice matching is not absolutely necessary: the thinnest layer is stretched or compressed to adapt itself to the crystal lattice of the other materials. Fortunately one wants the base of a bipolar transistor to be thin so it becomes possible to use a so-called *strained layer* as a base. This opens a much wider field: we now have combinations such as AlGaAs/InGaAs and Si/SiGe. In the combinations mentioned here the first-mentioned material has the highest bandgap.

6.2.1 AlGaAs–GaAs HBTs

One of the most researched combinations is AlGaAs–GaAs, with an AlAs fraction of 20–30%. A disadvantage of GaAs is its short electron lifetime, 10^{-9} s as compared with 10^{-6} s for Si. However, if the base is thin enough, the transit time through the base is small compared with the lifetime and recombination is negligible.

GaAs-based HBTs have the following advantages compared with silicon:

- a greater electron mobility, resulting in higher cutoff frequencies;
- a greater bandgap, and hence less thermal generation of charge carriers;
- the possibility of using a semi-insulating substrate which eases the isolation of devices;
- the possibility of integration with optoelectronic components.

Compared with GaAs MESFETs (section 6.3) GaAs HBTs also have some advantages:

- higher transconductance and output current;
- more uniform threshold voltages (important for digital ICs);
- less low-frequency noise, absence of backgating (section 6.3.1).

Recently with GaAs HBTs cutoff frequencies of 100 GHz, have been reached comparable with those of the best MESFETs. Figure 6.6 shows the structure of a recent AlGaAs HBT, from which it appears that these are not easy structures to fabricate.

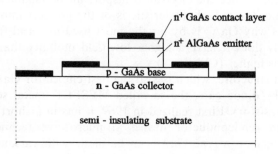

Figure 6.6 AlGaAs–GaAs HBT structure.

6.2.2 Si–SiGe HBTs

In this transistor, which has been receiving much attention recently, the base is made of an alloy of silicon with about 10% germanium. Since Ge has a bandgap of 0.7 eV, as compared with 1.1 eV for Si, the alloy will have a lower gap than the Si emitter, even if one takes bandgap lowering by the emitter doping into account. A problem is that the lattice constant of Ge differs strongly from that of Si. This could result in crystal defects at the interfaces. However, if the Ge content is not too high and the base layer not too thick, say 0.1 µm or less, the SiGe layer is strained to match the Si crystal lattice and defects are absent. The bandgap difference can therefore not be more than about 0.04 eV, just a little more than kT at room temperature. Fortunately the bandgap difference is entirely spent on the discontinuity in the valence band, where it is needed.

As an added bonus it turns out that the SiGe alloy can be p doped to a higher degree than pure Si, about 10^{20} atoms/cm³. This helps greatly in achieving a low base resistance. The SiGe HBT technology has the great advantage of being compatible with standard Si technology. A possible disadvantage could be that the lower bandgap in the base means higher thermal generation of electron–hole pairs, making the transistor more sensitive to temperature variations.

6.3 FIELD EFFECT TRANSISTORS: GaAs METAL SEMICONDUCTOR FIELD EFFECT TRANSISTOR

6.3.1 Structure and technology

The structure of field effect transistors is favorable for microwaves. All contacts are on the surface so that the parasitic capacitances can be kept small. The cutoff frequency is thus mainly determined by the transit time of the electrons under the gate and we therefore should aim for short gate lengths ($\leq 1 \mu$m). Since the electron transport in the channel is mainly by the electric field, the v–E characteristic is of the greatest importance. This partly explains why GaAs is by far the most used material for microwave FETs: not only does it have a higher low-field mobility than Si, also the peak velocity is higher (Chapter 4).

The simplest structure, which is also the best choice for GaAs, is the one with a Schottky diode gate, the so-called MESFET (metal semiconductor field effect transistor). First realized in 1969, it has in a short time become the most used semiconductor device in microwave technology. Good Schottky barriers on n-type GaAs can be made with various metals, e.g. Au, Al, Ti. The v–E characteristic is not the whole explanation for the much better performance of GaAs compared with Si MESFETs. For one thing, in MESFETs the channel is doped with a density of 10^{17}/cm³ or higher, and this reduces the peak velocity considerably, as can be seen from Fig. 4.1. Furthermore, in normal operation most of the channel is in the saturated-velocity regime and in only a small part of it is the low-field mobility important. So on the basis of the v–E characteristic alone one would expect a performance advantage for GaAs of at most a factor of 2.

However, early experiments [1] had already showed a 4 times higher cutoff frequency in GaAs MESFETs, compared with Si devices of the same dimensions. The main reasons for this are as follows. First, the source and drain parasitic resistances are made up by regions where the electric field is low and consequently the mobility dominates. Second there is the semi-insulating substrate of GaAs, something which is not possible in Si. This keeps the parasitic capacitances to ground of the contact pads low. So in the end the GaAs MESFET does perform much better than its

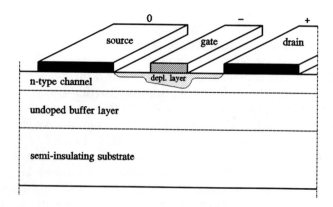

Figure 6.7 Cross-section of a MESFET.

Si counterpart, but this is as much thanks to the parasitics as to the channel.

The structure of a MESFET is shown in cross-section in Fig. 6.7 and layouts in Fig. 6.8. The layout in Fig. 6.8(b) is for power devices and essentially consists of a number of transistors in parallel. In this case air bridges have to be made to connect all the sources.

The MESFET is made in an n-type layer ($N_D = 10^{17}$–10^{18} atoms/cm^3 thickness, 0.1–0.2 µm) on a semi-insulating substrate ($\rho = 10^8$ Ω cm). For the ohmic source and drain contact metallizations mostly a combination of $Au_{0.88}Ge_{0.12}$ with Ni is used, which after evaporation is alloyed at a temperature of about 400 °C. Nowadays this is usually done by so-called *rapid thermal annealing*, in which process the temperature is increased very

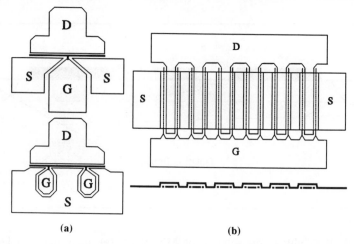

Figure 6.8 Possible layouts of MESFETs: (a) low power; (b) high power.

rapidly to the desired value, held there for a short time (usually less than a minute) and then brought down again rapidly. This produces a shallow alloyed layer with a smooth interface. The Ge in this mixture serves as an n-type dopant and the Ni as a wetting agent.

The Schottky gate contact is nowadays mostly a three-layer metallization of Ti–Pt–Au, where the Ti forms the Schottky barrier with good adherence to the GaAs surface, the Au is the low-resistance connection to the outside world and the Pt acts as a diffusion barrier to prevent the Au from diffusing into the GaAs. Also alloys such as TiW and WN are used.

The great advantage of the semi-insulating substrate is that the connecting leads can be laid over this substrate without introducing large parasitic capacitances. In fact, by metallizing the bottom side of the substrate, microstrip lines (section 7.2) can be made, which offers the possibility of integrating passive microwave circuits with transistors into MMICs (monolithic microwave integrated circuits) (Chapter 11).

Originally the n layer was made by epitaxial growth on the substrate, often preceded by an undoped buffer layer. The latter is necessary because the substrate contains lots of impurities which tend to diffuse into the n

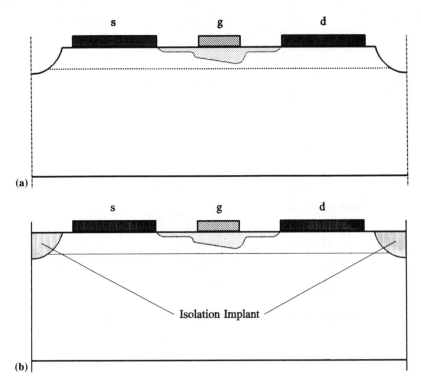

Figure 6.9 Possibilities for isolation of MESFETs: (a) mesa etch; (b) local ion implantation.

layer, causing recombination centers. In the mean time the quality of the substrates is improved to a point where it is feasible, at least for digital applications, to make the n layer by direct ion implantation in the substrate. For analog low-noise applications one still prefers an epitaxial buffer layer. Besides, it is now possible to grow the epilayer by MBE or MOVPE (Chapter 3). These will give very sharp transistions between the epilayer and the substrate. The influence of the sharpness of the channel–substrate transition will be discussed below.

Isolation of the devices is in the case of epitaxial n layers done by etching away the n layer (mesa isolation) between the devices and in the case of ion implanted layers by local implantation via a mask (Fig. 6.9). Local ion implantation also offers the possibility of enhancing the conductivity under the source and drain contacts (Fig. 6.9(b)), giving lower contact resistances.

The specific properties of GaAs (Chapter 3) can also cause problems. The surface pinning of the Fermi level at midgap energy creates a depletion layer also outside the gate region. Because of the small thickness of the n layer this leads to an increase of the series resistances at the source and drain sides. Moreover the noise also increases. As a remedy one has resorted to using a thicker n layer which in the gate region is etched thinner again, the so-called gate recess (Fig. 6.10(a)). This gives also the possibility of compensating for batch-to-batch variations in the as-grown n layer thickness by adjusting the etch depth. Furthermore, it is possible to improve the source and drain contact resistances by an extra epitaxial n^+ layer, as is shown in the figure. A very interesting application of the gate recess technique is first to deposit the source and drain metallizations and to use these as an etch mask for the recess. Since most etch processes give some underetching the recess will always be wider than the source–drain spacing. In a second step one can now use the source and drain as a mask for the gate metal evaporation (Fig. 6.10(b)). This *self-aligned* process produces the shortest possible gate–source and gate–drain spacings.

Another problem is formed by substrate impurities such as Cr. These often pile up at the interface between substrate and n layer and form there recombination centers that can trap electrons from the channel. The degree to which this happens depends on the local position of the Fermi level which in turn depends on V_{DS}. Because the charging and discharging of these traps occur with long time constants one often sees the occurrence of hysteresis in the I_D–V_{DS} characteristic. Another thing is that the Fermi level also depends on the substrate potential meaning that the latter can influence the drain current. This is called *backgating* and causes problems when the source is not at ground potential, e.g. in a source follower. Also, contacts or interconnections on the surface close to the transistor can influence the substrate potential. This is called *sidegating*. A third effect is the occurrence of low-frequency noise (quasi-1/f noise) (Section 6.3.5).

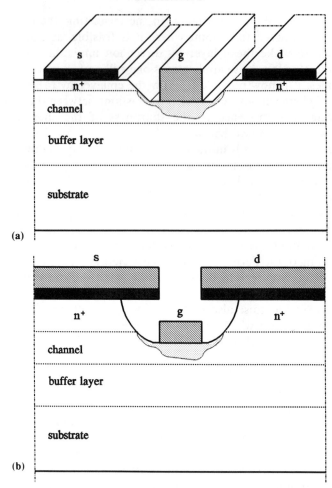

Figure 6.10 MESFETs with recessed gate: (a) standard; (b) self-aligned.

To eliminate or at least to reduce these detrimental effects it has become good practice to insert an undoped buffer layer between the substrate and the channel.

6.3.2 Dual-gate MESFETs

The planar structure of a MESFET makes it easy to add a second gate, (Fig. 6.11). This can be used to control the amplification of the transistor. Figure 6.12(b) gives the transconductance as a function of both gate voltages. Applications include heterodyne mixers, automatic volume control circuits and amplitude modulation of oscillators.

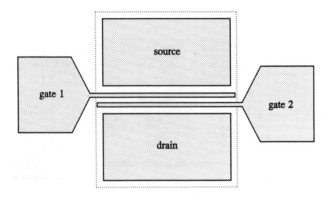

Figure 6.11 Layout of dual-gate MESFET.

In the application as a heterodyne mixer, where each of the two signals is fed to one of the gates, one can have gain, in contrast to passive diode mixers. The mutual coupling by the gate–gate capacitance is small so the isolation between local oscillator and signal ports is good.

6.3.3 Principles of operation of the MESFET

The characteristics of a MESFET are in principle the same as those of other field effect transistors. One can distinguish a linear part at low drain voltages, where the current is determined by the low-field mobility, and a saturated part where the current becomes more or less independent of drain voltage owing to a combination of velocity saturation and pinch-off of the channel near the drain.

Some insight into its operation can be obtained from Shockley's long-channel model for a field effect transistor which is illustrated in Fig. 6.12. It uses two assumptions:

- the variation of the channel width over its length is small compared with the width itself;
- the channel is in a low-field regime with a constant mobility.

We consider the intrinsic transistor, that is we place a virtual source and drain at the beginning and end of the channel. The voltage drops over the ohmic regions between the source or drain and the channel can be accounted for later. If the drain is at zero bias with respect to the source the channel will have the same width everywhere (Fig. 6.12(a)). In the following ε is the permittivity of the semiconductor and μ the mobility of the electrons, w the width, a the height and L_{ch} the length of the channel (Fig. 6.7).

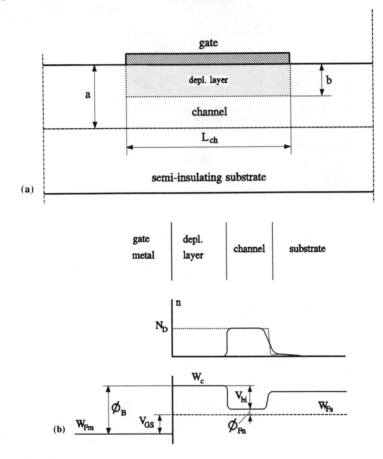

Figure 6.12 (a) Schematic of a MESFET with important parameters and (b) cross-section of the channel.

A cross-section through the gate gives the picture of Fig. 6.12(b) and from this picture we see that the voltage drop between source and gate is composed of the following components:

$$V_{GS} = V_{bi} - V_{depl} \tag{6.4}$$

where V_{depl} is the voltage drop over the depletion region and the value of V_{depl} at $V_{GS} = 0$ is called the *built-in voltage* V_{bi}:

$$V_{depl} = \frac{qN_D}{2\varepsilon} b^2 \tag{6.5a}$$

$$V_{bi} = \phi_B - \phi_F. \tag{6.5b}$$

Here b is the thickness of the depletion layer (Fig. 6.12(a)), ϕ_B is the Schottky barrier potential also encountered in Chapter 5 and ϕ_F is the position of the Fermi level at the source contact, all given in volts. We can express the latter in terms of known parameters of the transistor:

$$\phi_F = \frac{kT}{q} \ln(N_C/N_D). \tag{6.6}$$

The resistance of the channel is now

$$R_{ch} = \frac{L_{ch}}{q\mu N_D(a-b)w}. \tag{6.7}$$

This describes the linear part of the I_D–V_{GS} characteristic. When the gate voltage is increased in the negative direction a point comes where $b = a$, that is the channel disappears and R_{ch} becomes infinite. This is called *pinch-off* and the gate voltage then equals the *threshold voltage*

$$V_T = V_{bi} - V_{po} \tag{6.8a}$$

with

$$V_{po} = \frac{qN_D}{2\varepsilon} a^2 \tag{6.8b}$$

the *pinch-off voltage*. The picture changes when V_{DS} has an appreciable value. A potential gradient appears over the length of the channel, the thickness of the depletion layer is position dependent (Fig. 6.13) and equation (6.4) acquires an extra term:

$$V_{GS} = \phi_B - \phi_F - V_{depl} + V_{ch}. \tag{6.9}$$

Using equations (6.8) we can write this equation as

$$V_{GS} = V_T + \frac{qN_D}{2\varepsilon} (a^2 - b^2) + V_{ch}. \tag{6.10}$$

The drain current is, assuming constant mobility in the channel, given by

$$I_D = q\mu_n N_D w(a-b)E_z$$

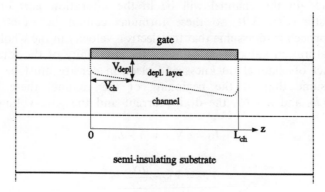

Figure 6.13 Depletion layer in a MESFET.

where E_z is the longitudinal electric field in the channel. E_z and b are now both functions of z. Using equation (6.10) we can write the drain current as

$$I_D = \frac{2\varepsilon\mu w}{a+b}(V_{GS} - V_T - V_{ch})\frac{dV_{ch}}{dz}.$$

For gate voltages close to V_T the channel is narrow and we can approximately replace b by a. The above equation can then be integrated along the z coordinate. Since V_{ch} changes from O to V_{DS} as z goes from O to L_{ch} the result is

$$I_D = \frac{\varepsilon\mu w}{aL_{ch}}[(V_{GS} - V_T)V_{DS} - \tfrac{1}{2}V_{DS}^2]. \tag{6.11}$$

I_D saturates when the channel thickness becomes zero at the drain end. The value of V_{DS} at which this happens can be found from equation (6.10).

$$V_{GS} = V_T + V_{DSsat}.$$

Substituting this in equation (6.11) we obtain

$$I_{Dsat} = \frac{\varepsilon\mu w}{2aL_{ch}}(V_{GS} - V_T)^2. \tag{6.12}$$

In an ideal transistor I_D does not increase any more when V_{DS} is raised further so we can use this expression for the whole saturation range. The transconductance in this range then is given by

$$g_m = \frac{dI_{Dsat}}{dV_{GS}} = \frac{\varepsilon\mu w}{aL}(V_{GS} - V_T). \tag{6.13}$$

This all looks very nice but the typical microwave MESFET is characterized by a very short channel in which the electric field can become very high. At drain voltages of 2 V or more and gate lengths of 1 µm or less the average electric field in the channel will be in the saturation part of the v–E characteristic (Fig. 3.9), so these formulas cannot be exact. Another simple approach is to assume that the electron velocity in the whole channel is at its saturation value v_s. Because of the continuity of the total current the product of channel thickness and electron density must be constant. If we assume that at the beginning of the channel these have the values $b = b_0$ and $n = N_D$ the drain current and the gate voltage can be written as

$$I_D = qN_D v_s w(a - b_0) \tag{6.14}$$

and

$$V_{GS} = V_{bi} - \frac{qN_D}{2\varepsilon}b_0^2 \tag{6.15}$$

from which it follows for the transconductance that

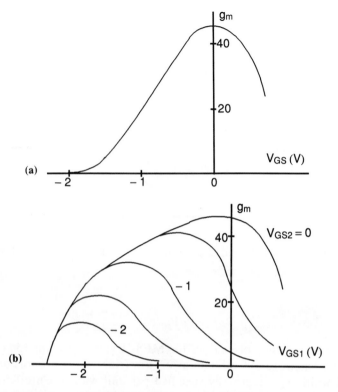

Figure 6.14 Typical g_m–V_{GS} characteristics of MESFETs: (a) single gate; (b) dual gate.

$$g_m = \frac{\varepsilon v_s w}{b} = v_s w \left[\frac{q \varepsilon N_D}{2(V_{bi} - V_{GS})} \right]^{1/2}. \tag{6.16}$$

This also predicts an increase of g_m with increasing V_{GS} but at a quite different rate. The actual behavior of g_m is illustrated in Fig. 6.14. Somewhat surprisingly, in view of all the assumptions involved in deriving equation (6.13), it turns out that a linear dependence of g_m on V_{GS} is not a bad approximation over the lower part of the range. Also, it is found that g_m increases with decreasing gate length although not as much as predicted by formula (6.13). When V_{GS} becomes positive the gate starts to conduct and the model is not applicable any more. Instead of going to infinity g_m decreases at positive gate voltages because the gate current goes at the cost of the drain current.

Shur [2] has given an expression, based on the assumption that the drift velocity is saturated in part of the channel, that unites the 'linear' and 'saturation' models:

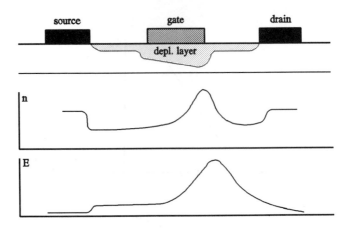

Figure 6.15 Electric field and electron density profiles in the channel of a typical MESFET.

$$I_{\mathrm{D}} = \frac{2\varepsilon\mu w}{3aL + \mu a V_{\mathrm{po}}/v_{\mathrm{sat}}}\,(V_{\mathrm{GS}} - V_{\mathrm{T}})^2. \qquad (6.17)$$

Typical transconductance values for MESFETs range from 150 mS/mm at 1 μm gate length to 300 mS/mm at 0.3 μm (since g_{m} is proportional to the channel width w, it is usually specified per unit width). Cutoff frequencies for these gate lengths are about 30 GHz and 70 GHz, respectively.

The actual behavior of the inside of a MESFET can only be found by two-dimensional numerical modeling. A sample is shown in Fig. 6.15 for a negative gate bias and a drain bias in the saturation region. Because of the narrowing of the channel and the decrease of velocity with electric field an electron accumulation occurs near the drain end of the channel and this produces a strong increase of the electric field. This field peak is the cause of electrical breakdown by avalanche multiplication at high drain voltages. If one wants high breakdown voltages, e.g. in power transistors, measures have to be taken to reduce this field peak (section 6.3.4).

In view of the v–E characteristic of GaAs (Chapter 4) one might expect a negative slope of the I_{D}–V_{DS} characteristic in the saturation region. In practice this is seldom observed because of the very inhomogeneous field distribution in the channel. In Chapter 5, when discussing Gunn diodes, it has already been explained that in a structure with a negative differential mobility an I–V characteristic with a positive slope may result from an inhomogeneous charge and field distribution. Also, injection into the substrate contributes to a positive slope. This is explained in the following section.

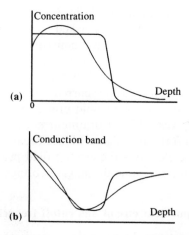

Figure 6.16 (a) Possible doping profiles in MESFETs and (b) corresponding potential profiles, ———, epitaxial, ·····, implanted.

6.3.4 Influence of the doping profile

The doping profile can have a great influence on the characteristics of a MESFET. Consider two profiles, one with a sharp transition from channel to substrate and the other more gradual (Fig. 6.16(a)). The transverse potential profile near the drain then looks approximately like that in Fig. 6.16(b).

One must remember now that the electrons near the drain become hot because of the high electric field there, and the more so the higher the drain voltage is. These hot electrons are at a higher energy level and can diffuse somewhat further into the substrate than electrons at the bottom of the conduction band. Obviously this effect is greater at a more gradual doping profile so that in this case we will have more substrate current. The results are a more gradual pinch-off at V_T and a stronger increase of current with drain voltage in the saturation region. So we will find the I_D–V_{DS} curves of

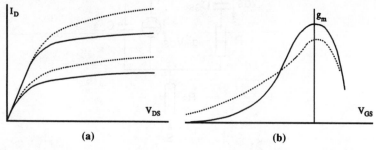

Figure 6.17 Influence of doping profile on (a) I_D and (b) g_m: ———, sharp transition; ·····, gradual transition.

Fig. 6.17(a) and the g_m–V_{DS} curves of Fig. 6.17(b). Also possible effects of negative differential mobility will be masked by the extra current.

Ladbrooke [3] has argued that this has consequences from an applications point of view. For a small-signal low-noise transistor one would prefer the g_m–V_{DS} characteristic given by the full line in Fig. 6.17(b) because it maintains a high g_m at low drain current where the noise is lowest. Conversely, for power devices one would prefer the other case, the gradual transition, because it gives a higher drain current at positive gate bias (the dotted line in Fig. 6.17(a)) and also a higher breakdown voltage: because of the spreading of the electrons the electric field peak at the drain end of the gate is less sharp for the same drain–gate voltage.

6.3.5 Small-signal equivalent circuit and cutoff frequency

For small-signal applications the equivalent circuit of Fig. 6.18 is used. The figure also indicates with which physical regions of the device the various circuit elements can be identified. This is the so-called intrinsic equivalent circuit. In practice we have to add parasitic circuit elements arising from the package in which the transistor is mounted and bond wires by which the gate, drain and source are connected.

In principle all element values can be determined by measuring the S-parameters (Chapter 8) over a wide frequency band and matching these to values calculated from the equivalent circuit. In practice this can be a dangerous procedure: since the equivalent circuit is itself an approximation it can never produce the exact frequency dependence of the S-parameters. In combination with the large number of parameters that has to be extracted there is a great risk that the best match yields physically impossible parameter values. Therefore the best procedure is to measure as many parameters as possible with other methods and then to do the wide band matching.

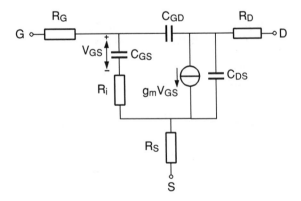

Figure 6.18 Small-signal equivalent circuit of a MESFET.

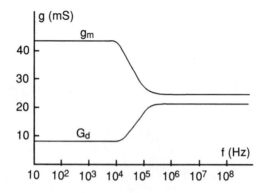

Figure 6.19 Low-frequency dispersion of g_m and G_d.

The transconductance g_m and the drain conductance G_d can in principle be determined from the d.c. characteristics. Some care has to be taken here: the traps discussed in Chapter 3 that cause hysteresis in the $I-V$ characteristic can also lead to low frequency dispersion in g_m and G_d (Fig. 6.19). Therefore it is wise to measure these quantities at a frequency of 1 MHz or higher. C_{GS} can be determined from $C-V$ measurements at 1 MHz. R_s can be measured by taking recourse to the Shockley model discussed earlier. Substitute b from equation (6.5a) and V_{depl} from equation (6.4) into equation (6.7). This gives

$$R_{ch} = \frac{L_{ch}}{q\mu N_D a w} X, \text{ where } X = \left[1 - \left(\frac{V_{bi} - V_{GS}}{V_{bi} - V_T} \right)^{1/2} \right]^{-1}. \qquad (6.18)$$

Now the resistance between the source and drain contacts is equal to $R_s + R_{ch} + R_d$, so, if one measures it as a function of gate voltage, a plot of the measured R against X will give a straight line whose intersection with the vertical axis yields $R_s + R_d$. Another method consists of biasing the source–gate diode in the forward direction and using the drain as a potential probe to measure the voltage drop over R_s. Interchanging the source and the drain yields R_d.

To apply this method V_{bi} and V_T have to be known. V_{bi} is found from ϕ_B and N_D according to equation (6.5b); V_T can be found by plotting $(I_D)^{1/2}$ against V_{GS}. This should give a straight line in at least part of the range and it is in this part that equation (6.18) can be applied. Extrapolation of the straight line to $I_D = 0$ gives V_T.

The cutoff frequency, as in the case of the bipolar transistor, is defined as the frequency at which the current amplification becomes equal to unity. In a simplified equivalent circuit where we keep only the most important elements, i.e. the gate–source capacitance and the current source, the current amplification is given by

$$\frac{i_D}{i_G} = \frac{g_m V_{GS}}{j\omega C_{GS} V_{GS}} = \frac{g_m}{j\omega C_{GS}}. \tag{6.19}$$

Now, if we use the saturated-velocity approximation, C_{GS} can be expressed in terms of g_m using equation (6.16):

$$C_{GS} = \frac{\varepsilon w L_{ch}}{b} = \frac{g_m L_{ch}}{v_s} \tag{6.20}$$

so that we find for the cutoff frequency

$$f_T = \frac{g_m}{2\pi C_{GS}} = \frac{v_s}{2\pi L_{ch}} = \frac{1}{2\pi \tau} \tag{6.21}$$

where τ is the transit time from source to drain.

When this simplified equivalent circuit is extended slightly by including the gate resistance the frequency dependence of the power gain can be calculated. The power dissipated in the gate port is

$$P_G = \tfrac{1}{2} |I_G|^2 R_G. \tag{6.22}$$

The drain current is controlled by the voltage over the gate–source capacitance C_{GS} so that we have

$$I_D = g_m V_C = \frac{g_m}{j\omega C_{GS}} I_G \tag{6.23}$$

and the power delivered into a drain load resistance R_L is

$$P_L = \tfrac{1}{2} |I_D|^2 R_L = \tfrac{1}{2} \left(\frac{g_m}{\omega C_{GS}} \right)^2 |I_G|^2 R_L \tag{6.24}$$

from which follows the power amplification

$$G = \frac{P_L}{P_G} = \left(\frac{g_m}{\omega C_{GS}} \right)^2 \frac{R_L}{R_G} = \left(\frac{f_T}{f} \right)^2 \frac{R_L}{R_G}. \tag{6.25}$$

This expression should not be taken too seriously as far as the dependence on R_L is concerned. It suggests that the power can be raised indefinitely by increasing the load resistance. However, this is all assuming that the current source keeps working no matter how high the drain voltage gets. In reality there is only a limited range of load resistances over which this expression can be applied.

More important is the amplification as a function of frequency that it predicts. This decreases at 6 dB/octave, as in the case of the bipolar

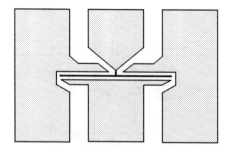

Figure 6.20 Layout for on-wafer microwave characterization.

transistor. This behavior can be modified by the other elements of the equivalent circuit and especially by the parasitic reactances of the transistor itself and of its mounting. The result can be that the amplification first decreases at 6 dB/octave, then at a slower rate and then faster again. If one tries to extrapolate the cutoff frequency from amplification measurements at lower frequencies as is often done one should take care to extrapolate the 6 dB/octave part of the characteristic. In the past often unrealistically high cutoff frequencies and channel velocities have been obtained by extrapolating from the part with a less steep slope [4].

To avoid the problems with mounting parasitics measurement of the microwave characteristics directly on the wafer is nowadays common practice. This can be done by using a microwave probe consisting of a piece of coplanar transmission line ending in three small spring contacts which are pressed on the contact pads of the transistor. The layout of the transistor has to be adapted to fit these probes (Fig. 6.20).

In spite of the questionable validity of equations (6.16) and (6.20) they are often used to calculate the (average) channel velocity from measured data. Perhaps the most reliable way to do this is to make use of the fact that, in fabrication, all technological parameters, in particular the gate length, show some spread. If one plots $1/f_T$ against gate length the slope of this curve should give the velocity.

6.3.6 Large-signal equivalent circuits

At large signal amplitudes the strong non-linearities of the MESFET become manifest and a linear equivalent circuit such as that of Fig. 6.18 is no longer applicable. For applications where the signals are more or less sinusoidal, such as power amplifiers and oscillators, it still can be used, with the provision that the element values are now dependent on the signal amplitude and therefore should also be measured as a function of the latter. Using this technique oscillators can be designed obtaining accurate predictions of power output and load-pull characteristics. Since one assumes a

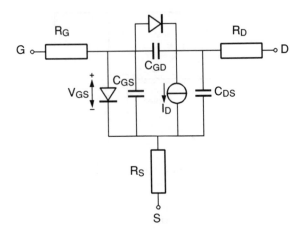

Figure 6.21   Large-signal equivalent circuit of a MESFET.

sinusoidal waveform it is not possible to obtain information about the
higher harmonic content of the signal. For pulse-shaped signals (digital
applications) this will not work and one has to use a non-linear equivalent
circuit such as that of Fig. 6.21 where a number of voltage-dependent
elements appear, including a diode between the gate and the source. This
diode is necessary for the case that the gate is biased so far in the forward
direction that it starts to conduct. If we are sure that this situation will
never occur the diode can be left out. If on the contrary the possibility exists
that the drain becomes negative with respect to the gate a diode should be
included between the gate and drain also.

In this circuit the non-linear dependences of $I_D(V_{GS}, V_{DS})$ and $C_{GS}(V_{GS})$
are especially important. For these a fairly large number of models exists,
most of which can be traced back to that of Curtice [5]:

$$I_D = \beta(V_{GS} - V_T)^2(1 + \lambda V_{DS}) \tanh(\alpha V_{DS}) \tag{6.26}$$

$$C_{GS} = C_1 + C_2(1 - V_{GS}/V_{bi})^{-1/2}. \tag{6.27}$$

In the first formula the V_{GS} dependence is described by a function similar
to equation (6.12) and clearly is only of limited applicability. The tanh
function describes the transition from the linear region to the saturation
region and although it has no physical basis it often describes the measured
curves very well. The factor between these takes care of a finite drain
conductance in the saturation region. The formula for C_{GS} is that of a plane
depletion layer whose thickness is given by equations (6.4) and (6.5) plus
edge effects. This is only correct for uniformly doped channels as long as
the channel is not fully depleted. When the depletion edge reaches the

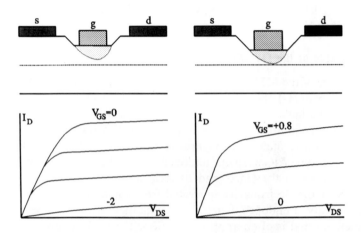

Figure 6.22 Depletion and enhancement FETs.

substrate junction it cannot move downward any more and only the sidewalls can be displaced by an increasing gate voltage. The result is that when V_{GS} drops below V_T the gate–source capacitance suddenly drops to a low value. If the channel doping is not uniform, as in ion-implanted channels, the capacitance also can deviate strongly from the formula above.

Many refinements and extensions of the Curtice model have been developed in the course of the years. Recent reviews and comparisons of the various models have been given in refs 6 and 7.

6.3.7 Applications of MESFETs

We distinguish two types of MESFETs, the depletion type and the enhancement type (Fig. 6.22). In the first the channel is open at $V_{GS} = 0$ V and is cut off at *ca.* -2 V. This is the standard transistor for analog microwave applications. Nowadays MESFETs are also used in fast digital ICs (Chapter 10) and in these also enhancement types are used. These have a cutoff channel at $V_{GS} = 0$ and start to conduct at a positive gate voltage. V_{GS} can in these devices not be much higher than $+0.8$ V because otherwise the gate starts to conduct. For this reason enhancement FETs are sometimes made as junction FETs. In this case the gate voltage can be increased to 1 V.

The main microwave applications of MESFETs are in small-signal amplifiers with low noise figures (Chapter 8) and in oscillators (Chapter 9). Furthermore, they are used as power amplifiers mainly at frequencies above 4 GHz. Dual-gate MESFETs can be used as mixers (with conversion gain) and for control purposes.

6.4 FIELD EFFECT TRANSISTORS: HIGH ELECTRON MOBILITY TRANSISTOR

6.4.1 Principle of the HEMT

A disadvantage of the MESFET is that at the fairly high doping in the channel (10^{17}–10^{18} atoms/cm^3) the mobility is degraded (Fig. 3.9). We cannot fully benefit from the high mobility of GaAs. The electrons are hindered in their movements by the ionized impurities in the channel. A way to avoid this is the application of a heterojunction. As already discussed in Chapter 3, if we make a junction of n-type AlGaAs with undoped GaAs, the electrons tend to move from the AlGaAs to the GaAs and to form a conducting channel near the interface. Now the electrons are separated from the donors and the electrons have the mobility of undoped material although their density can be quite high.

A refinement that is often applied is to dope the AlGaAs not over the full thickness but to leave a thin undoped layer of *ca.* 6 nm adjacent to the GaAs. This is because the electrons in the GaAs can feel the Coulomb field of the ionized donors in the AlGaAs close to the interface. With this *spacer layer* ionized impurity scattering is further reduced. We thus obtain a transistor structure such as that in Fig. 6.23. The width and doping of the

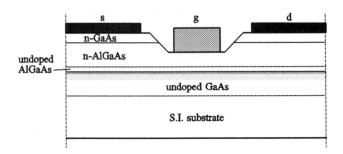

Figure 6.23 Structure of an HEMT or a MODFET.

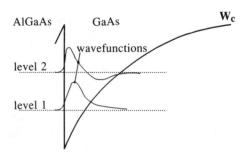

Figure 6.24 Discrete energy levels in an HEMT.

AlGaAs have to be designed carefully so that at all operating bias points the AlGaAs layer under the gate is fully depleted. If this is not the case, a parasitic channel in the AlGaAs results with low electron velocity and consequently bad high-frequency properties.

The situation is further improved by a quantum mechanical effect: at the bottom the potential well is so narrow that the electron wavefunction is confined between the walls (Fig. 6.24), and the electrons can only move in two dimensions. This is called a *two-dimensional electron gas*. These electrons are scattered less strongly by lattice vibrations (phonons) and can reach higher velocities. The effects of reduced scattering are not very strong at room temperature but they show up quite clearly at lower temperatures.

This transistor has different names: HEMT (high electron mobility transistor), TEGFET (two-dimensional electron gas FET), MODFET (modulation-doped FET) and SDHT (selectively doped heterojunction transistor). Recently the name HFET (heterojunction FET) has come into use. As Fig. 6.21 shows, their noise performance is quite impressive, especially at the higher microwave frequencies.

6.4.2 Theory

The analysis of an HEMT is simple and complicated at the same time. The simple part is evaluated with the help of Fig. 6.25 where the electric field profile and the conduction band are shown at a positive gate voltage. We have assumed, as is usually done but not always explicitly mentioned, that

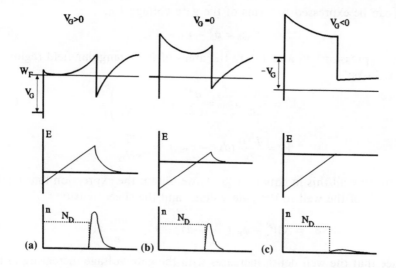

Figure 6.25 Cross-section of an HEMT, showing (a) charge density, (b) electric field and (c) conduction band.

the Fermi level jumps an amount qV_G at the metal–AlGaAs junction. Another condition we presuppose is that the AlGaAs layer is fully depleted. If it were not we would have a parasitic channel in the AlGaAs, where the electron velocity is very low and the performance of the device would be strongly degraded.

Since the conducting channel is very thin (in the figure it is drawn wider for clarity) the electric field jumps from zero in the GaAs substrate to a finite value at the AlGaAs side of the heterojunction. With the help of Poisson's equation the junction field E_j can be related to the electron density integrated over the width of the channel, the *sheet density* n_s:

$$E_j = \frac{qn_s}{\varepsilon}. \tag{6.28}$$

In the AlGaAs layer, where we have left out the undoped AlGaAs spacer to keep the analysis simple, the electric field has a linear variation:

$$E = E_j + \frac{qN_D}{\varepsilon}(x - d_{Al}) \tag{6.29}$$

so the position of the potential minimum is given by

$$d_{Al} - x_{min} = \frac{\varepsilon E_j}{qN_D} = \frac{n_s}{N_D}. \tag{6.30}$$

Now the depth of the potential well below the Fermi level is

$$\phi_w = \Delta W_c - V_{min} - V_2. \tag{6.31}$$

V_{min} can be expressed in terms of the gate voltage V_G:

$$V_{min} = \phi_B - V_G - V_1. \tag{6.32}$$

In the expressions V_1 and V_2 are the areas of the triangular field regions in the AlGaAs:

$$V_1 = \frac{qN_D}{2\varepsilon} x_{min}^2 = \frac{qN_D}{2\varepsilon}\left(d_{Al} - \frac{n_s}{N_D}\right)^2 \tag{6.33}$$

$$V_2 = \frac{qN_D}{2\varepsilon}(d_{Al} - x_{min})^2 = \frac{qn_s^2}{2\varepsilon N_D}. \tag{6.34}$$

Substituting all this in equation (6.31) we obtain the expression that relates the depth of the well to the gate voltage and the sheet density:

$$\phi_w = \Delta W_c - \phi_B + \frac{qN_D}{2\varepsilon} d_{Al}^2 + V_G - \frac{qn_s}{\varepsilon} d_{Al}. \tag{6.35}$$

We see that the well depth increases with the gate voltage increasing in the positive direction, but this increase is counteracted by the sheet density as the well fills up with electrons.

On the other hand, and this is the complicated part, the sheet density in the channel is a function of the depth of the potential well. To calculate this dependence is a complicated matter because one has to solve Poisson's and Schrödinger's equations simultaneously. When the electron charge is small enough to be ignored a *triangular well* results and for this case an analytic solution in terms of Airy functions is possible. For the positions of the subbands then the following approximate formula can be derived:

$$W_i \approx \left(\frac{\hbar^2}{2m^*}\right)^{1/3} [\tfrac{3}{2}\pi qE_j(i + \tfrac{3}{4})]^{2/3}. \tag{6.36}$$

As one would expect, the subband levels move up when the field increases, i.e. when the well becomes narrower. If we assume that only the first subband level is below the Fermi level the sheet density is given by equation (3.35):

$$n_s = \frac{1}{\pi}\frac{m_c^* kT}{\hbar^2} \ln\left[1 + \exp\left(\frac{\phi_w - W_1}{kT}\right)\right]. \tag{6.37}$$

Here too we see a negative feedback of the sheet density via E_j and W_1. Nevertheless, the result is what one would intuitively expect, a monotonic increase of n_s with ϕ_w. Since equation (6.35) gives a decrease of ϕ_w with n_s, the crossing of the two curves (Fig. 6.26) will yield the values of n_s and ϕ_w.

If we rearrange equation (6.31) a bit, using equation (6.34), we obtain

$$\phi_w + \frac{qn_s^2}{2\varepsilon N_D} = \Delta W_c - V_{min} \tag{6.38}$$

and we see that, if we want to maximize the sheet density n_s, we have to make $\Delta W_c - V_{min}$ as large a possible. Now V_{min} has to have a minimum positive value, say $4kT/q$, otherwise electrons will assemble in this minimum and form a parasitic channel. Thus, the maximum value of n_s depends mainly on ΔW_c.

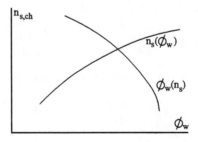

Figure 6.26   Mutual dependence of $n_{s, ch}$ and ϕ_w.

Another way to look at it is to realize that the HEMT as a whole stays neutral. The electrons diffusing out of the n-AlGaAs layer distribute themselves over the channel and the gate metal in a ratio that depends on the gate voltage. So we can write

$$n_s + n_{gate} = N_D d_{Al} \tag{6.39}$$

where n_{gate} is the density per unit area on the gate. Clearly n_s reaches a maximum when n_{gate} is zero, that is when the potential minimum in the AlGaAs is at the gate. For the gate voltage we have in this case (equation (6.32))

$$V_{GS} = \phi_B - 4kT/q. \tag{6.40}$$

This gives an upper limit for the positive gate voltage that can be applied without the risk of drawing gate current. $\phi_B \approx 0.8$ V for most gate metals on AlGaAs. The channel is pinched off when $n_s \approx 0$, i.e. $E_j \approx 0$. In this case W_1 is small and ϕ_w is close to zero, so from equation (6.35) the threshold value of the gate voltage is

$$V_T = \phi_B - \Delta W_c - \frac{qN_D}{2\varepsilon} d_{Al}^2. \tag{6.41}$$

This can be positive or negative. When it is negative one speaks of a *depletion HEMT (DHEMT)* and when zero or positive of an *enhancement HEMT (EHEMT)*. These names are borrowed from the MESFET but for the HEMT they are not very appropriate since in this device in both cases we simultaneously have depletion (in the AlGaAs) and enhancement (in the GaAs).

The drain current is also simply found if we assume that the saturated-velocity approximation may be used:

$$I_D = q w_{ch} n_s v_{sat}. \tag{6.42}$$

If we now change gate voltage E_j will change according to

$$\Delta V_G = d_{Al} \Delta E_j \tag{6.43}$$

so that for the transconductance we find

$$g_m = \frac{\Delta I_D}{\Delta V_G} = \frac{w_{ch} \varepsilon v_{sat}}{d_{Al}} = \frac{C_G v_{sat}}{L_{ch}} \tag{6.44}$$

from which also follows, in agreement with equations (6.35) and (6.41),

$$I_D = \frac{w_{ch} \varepsilon v_{sat}}{d_{Al}} (V_G - V_T). \tag{6.45}$$

From these equations it would appear that g_m changes very little with gate voltage, which is a nice property from an applications point of view.

Unfortunately in practice things are not so nice and g_m does change, mainly because of two effects [4].

● The channel is not infinitely thin, as we have supposed, but has a non-negligible and, worse, a non-constant width. In the equations above we should replace d_{Al} by $d_{Al} + d_{ch}$, where d_{ch} is an effective width characterizing the spatial spread of the electrons in the channel. When V_G approaches V_T, E_j decreases and the channel widens. Consequently we should replace equation (6.43) by

$$\Delta V_G = (d_{Al} + d_{ch}) \, \Delta E_j + E_j \, \Delta d_{ch}. \tag{6.46}$$

The increase of the last term makes g_m decrease at low V_G.

● The potential minimum in the AlGaAs will always contain some electrons and, even though these are not noticed in the d.c. drain current, they have an influence on the transconductance. The variations of the gate voltage will not only modulate the main channel but also the parasitic channel, so the modulation of the former will be reduced. It is as if the parasitic channel screens the main channel from the gate voltage. In particular at positive gate voltages, when V_{min} becomes small, this has a strong effect.

The overall result is that the g_m–V_G curve of a HEMT does not look so much different from that of a MESFET, although it is wider and its peak is higher, mainly because of the higher channel velocity and because the AlGaAs layer in a HEMT can be thinner than the channel layer of a MESFET.

6.4.3 HEMT technology

The question now is how to maximize n_s. It turns out that the Al-GaAs/GaAs HEMT performs somewhat disappointingly in this respect, for the following reasons. One of the less pleasant effects in AlGaAs is that the Si atoms used as donors can form so-called *DX centers* which have an energy level near the middle of the forbidden gap. This gives rise to low-frequency relaxation phenomena and noise, and also to *persistent photoconductivity*, meaning that under illumination the conductivity increases and stays high when the light is switched off again. The latter effect occurs mainly at low temperatures.

To avoid these problems the Al content should be kept below 25% which means that ΔW_c cannot be higher than about 0.23 V. The result is that the highest sheet density one can obtain is 10^{12}/cm^3. Thus, the AlGaAs HEMT can be a high-transconductance device but not a high-current device. It therefore performs excellently as a low-noise, small-signal amplifier but not in power applications. It looks as if the HEMT will replace the MESFET in amplifier applications for frequencies above *ca.* 30 GHz.

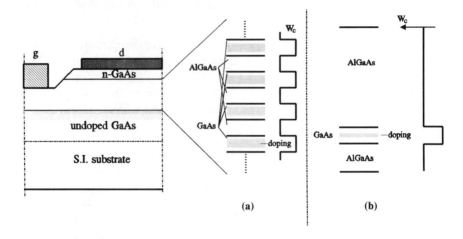

Figure 6.27 Avoiding DX centers in HEMTs: (a) modulation-doped superlattice; (b) delta-doped quantum well.

A number of solutions have been thought out to avoid the problems with DX centers in HEMTs. They all try to keep the Si donors away from the Al atoms. One consists of replacing the AlGaAs by a superlattice of layers AlAs and GaAs (thickness *ca.* 5 nm each) of which only the GaAs is doped (Fig. 6.27(a)). Since the AlAs barriers are so thin the electrons can tunnel easily through them so that from the outside the superlattice behaves as a doped AlGaAs layer. Another method puts a thin layer (10 nm) of GaAs in the AlGaAs layer close to the channel. The GaAs layer has a very thin very highly doped layer in the middle. This is called *delta doping* (Fig. 6.27(b)).

Even with these tricks the conduction band discontinuity cannot be raised very much because of the multivalley structure of GaAs (Chapter 3). If ΔW_c is made too high hot electrons in the channel (which will appear at the drain side) can transfer to the upper valley where their mobility is low. So a ΔW_c value of 0.3 eV is the maximum anyway.

To increase the sheet density further another possibility is to put under the GaAs layer a second n-type AlGaAs layer, so that one obtains a double channel (Fig. 6.28(a)). This increases the channel conductivity by a factor of 2. An extension of this idea is a channel consisting of a multi-quantum-well structure (Fig. 6.28(b)). In this case the barriers have to be so thin ($\leqslant 3$ nm) that the electrons can easily tunnel through, otherwise we obtain separate channels, which can give rise to strange phenomena, such as a g_m–V_G characteristic with two peaks.

Since in the AlGaAs/GaAs combination the band discontinuity cannot be very high one has in recent years looked for other material combinations,

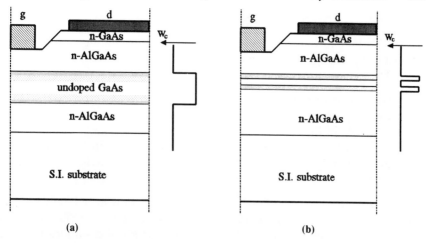

Figure 6.28 Increasing the sheet density of HEMTs: (a) double-sided channel; (b) multiple quantum well.

e.g. $In_xGa_{1-x}As$ for the channel. This material has several nice properties: it has a lower electron mass and a greater separation between the central and satellite valleys than GaAs. As a result the electron mobility and peak velocity are greater than those of GaAs (because it suffers less from intervalley scattering). The greater separation between the central and satellite valleys also means that when it is used as the channel material in an HEMT a greater conduction band discontinuity can be allowed.

The lattice of $In_xGa_{1-x}As$ does not fit that of GaAs, but when $x = 0.53$ it fits InP, so this material can be used as a substrate. As the high bandgap material one can use InP or better $In_xAl_{1-x}As$ which fits the lattice of InP at $x = 0.48$. If, however, the InGaAs layer is not too thick (<10 nm) and the In content not too high it can be sandwiched without dislocations between two GaAs layers. The InGaAs layer is then stretched out somewhat, a so-called *strained layer*. The combination strained $In_xGa_{1-x}As/AlGaAs$ with $x \leqslant 0.25$ can be grown on GaAs substrates. These devices are often called pseudomorphic HEMTs.

Both InGaAs/InAlAs and InGaAs/AlGaAs can provide larger conduction band discontinuities. Consequently higher sheet densities are obtained: $2 \times 10^{12}/cm^2$ for InGaAs/AlGaAs and 3×10^{12} for InGaAs/InAlAs. The transistor data reached so far are quite impressive: g_m values of 500 mS/mm and 1000 mS/mm and cutoff frequencies of 110 GHz and 170 GHz for InGas/AlGaAs and InGaAs/InAlAs, respectively. Perhaps more important are the noise figures: 1.6 dB at 60 GHz and 2 dB at 94 GHz. The fabrication of MESFETs in InGaAs has not been found to yield acceptable results because on this material no good Schottky barriers can be made.

6.5 NOISE OF MESFETS AND HEMTS

MESFETs had low noise figures from the very beginning and with the advancement of the technology they have been improved steadily (Fig. 6.29). The main sources of noise are

- thermal noise of the parasitic resistances, mainly R_S and R_G,
- thermal velocity fluctuations in the channel,
- fluctuations in the thickness of the channel, and
- scattering of electrons between the different valleys in the conduction band (section 4.3); this will only occur in regions of high field strength, i.e. at the drain side of the channel.

At low frequencies there is extra noise caused by traps in bulk and surface states. Traps occur especially at the free surfaces and in the interface between the substrate and the channel, so to have low noise it is a good idea to put a buffer layer in between the latter two. These interface or surface traps capture and release electrons with time constants of micro- to milliseconds. The result is noise with a quasi-$1/f$ character in the frequency region below 1 MHz. This noise is especially troublesome for oscillators, where it gives rise to high sideband noise levels, and for mixer circuits with a low intermediate frequency.

The simplest noise model, due to Fukui [8, 9], assumes that the main noise source is the thermal noise of the gate and source resistances. This leads to the following formula for the minimum noise figure (the definition of noise figure is given in Chapter 4 and that for minimum noise figure in Chapter 9):

$$F_{\min} = 1 + k_1 C_{GS} f \left(\frac{R_S + R_G}{g_m} \right)^{1/2} = 1 + K_F \frac{f}{f_T} \left[g_m (R_S + R_G) \right]^{1/2} \quad (6.47)$$

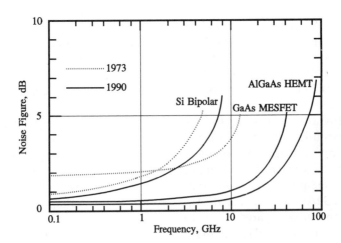

Figure 6.29 Evolution of microwave transistor noise figures.

where k_1 is a fitting factor which was given by Fukui as 0.016. In practice different values are found, depending on the particular transistor studied. To obtain the second expression equation (6.21) has been used. In this form it is used mostly. K_F is known in the literature as the Fukui factor and is evidently equal to $k_1/2\pi$. It has been found that for HEMTs the same noise model is applicable, but with a lower k_1 factor than in MESFETs.

A different approach, based on the assumption that the physical origin of noise is diffusion noise in the source–gate and source–drain regions, has been proposed by Pospieszalski [10]. It is assumed that the gate–source region can be characterized by a small-signal resistance r_{gs} (this is the internal resistance between gate and source in Fig. 6.18) with an associated noise voltage source given by

$$S_{V, gs} = 4kT_g r_{gs}. \tag{6.48a}$$

Similarly, the drain–source region is characterized by the drain conductance g_{ds} (the conductance parallel to the current source in Fig. 6.18) with an associated noise current source

$$S_{I, ds} = 4kT_d g_{ds} \tag{6.48b}$$

where T_g and T_d are the noise temperatures of these resistances.

In practice it is found that T_g and T_d can have values in a wide range, depending on the transistor studied. The intuitive assumption that these temperatures should be equal, or at least related, to the electron temperatures calculated with a hydrodynamic or Monte Carlo method (Chapter 4) seems not always to be justified.

6.6 FIELD EFFECT TRANSISTORS: Si MOSFET

Silicon MOSFETs are not really microwave transistors. They are mainly included here because of their importance for large-scale integrated circuits. GaAs ICs, to be treated in Chapter 11, will always be compared with NMOS and CMOS. Another reason is that their operation resembles very much that of high electron mobility transistors, described in the previous section.

The great advantage of MOS structures is that the gate can be biased in the forward direction without drawing current. In addition, their linearity is good. Compared with the GaAs MESFET, however, there are distinct disadvantages: the electron mobility is a factor of 10 smaller, resulting in much longer transmit times and greater parasitic resistances, giving higher noise figures. Also, the Si substrate cannot be made semi-insulating because of the smaller bandgap of Si. Connection lines therefore have to be laid over an oxide layer and consequently have larger parasitic capacitances. The result is that MOSFETs do not perform very well above 1 GHz. Below this frequency there are r.f. applications, mainly as power amplifiers.

Much effort has been spent to make the channel length as short as possible and at the same time to keep the peak electric field at the drain side low to maintain a reasonable drain–source breakdown voltage. Thanks to these efforts the upper frequency limits of MOSFETs have in the course of time moved up steadily and are now above 1 GHz for NMOSTs, mainly because of the reduced gate length. NMOSTs with gate lengths shorter than 0.1 μm can be made nowadays and there are indications that here too velocity overshoot can be observed.

6.7 OTHER HETEROJUNCTION FETS: MISFET AND SISFET

After the success of the MESFET people have for some time tried to also make MOSFETs on GaAs, mainly with digital applications in mind. This has never been a success because, no matter which oxide one uses, the Fermi level at the GaAs surface is always pinned at *ca.* 0.7 eV below the conduction band, resulting in a depletion layer at the surface. The channel is therefore difficult to modulate. To create an inversion one therefore always needs a positive threshold voltage, which is not very useful in digital technology.

On the other hand, the structure and band diagram of an HEMT resemble very much those of a MOSFET. It is, however, not possible in a HEMT to bias the gate in the forward direction because, owing to the doping in the AlGaAs, the gate can draw current. The obvious thing to do is to use undoped (or compensated) AlGaAs. Because of its large bandgap it forms a good insulator. The problem is then to get electrons in the channel. One method is to rely on diffused or implanted N^+ contacts, so that the electrons are injected from the source. At small gate lengths this works well for digital circuits where the demands on current capability are not strong. With P^+ contacts one can in the same way make p channel transistors. By selecting a suitable metal, or even a semiconductor, for the gate the threshold voltage can be adjusted to the desired value. In the case of a semiconductor gate one speaks of a SISFET (semiconductor–insulator–semiconductor). Another method consists of doping the GaAs, thereby losing the high-mobility property of the HEMT. The loss of mobility can be offset by an increase of the carrier density.

On InP the Fermi level pinning problem does not occur so on this material MISFETs can be made. There may be an application as power amplifiers.

6.8 TRANSISTOR PACKAGING

To be used in circuits transistors have to be mounted in some way. The simplest way of mounting is to take a chip and to mount it directly into a microstrip circuit. The connections are made by bonding gold wires from

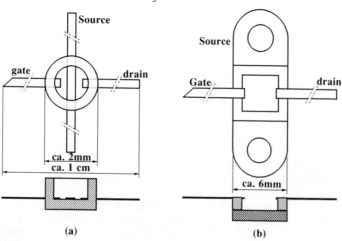

Figure 6.30 Microwave transistor packages: (a) low power; (b) high power.

the bonding pads of the transistor to the microstrip lines. To protect the device the whole chip except the bonding pads is covered with an insulating material, such as Si_3N_4 or polyimide. Better protection against environmental influences and handling damage is obtained by mounting each transistor individually in a package. These packages consist of a ceramic ring on a base of the same material (Fig. 6.30(a)). The chip is soldered onto this base and the electrodes are connected to thin strips of gold-plated metal sticking through the ceramic. The layout and the strip dimensions are designed to match alumina microstrip circuits. When the transistor is a MESFET the gate and drain pads are connected to opposing strips and the source pads are connected to the other two. In the case of bipolar transistors read 'base' for 'gate', 'collector' for 'drain' and 'emitter' for 'source'. The base of the package may be mounted on a screw with which the package can be secured to the ground plane of the microstrip for better thermal contact. In the case of power transistors, which have a larger chip size, the package is often made in the form of a solid metal bar on which a rectangular ceramic ring is mounted (Fig. 6.30(b)). This provides better heatsinking than the package of Fig. 6.30(a). The source is now connected directly to the ground plane.

REFERENCES

1. Baechtold, W., Daetwyler, K., Forster, T., Mohr, T. O., Walter, W. and Wolf, P. (1973) Si and GaAs 0.5 μm-gate Schottky-barrier field-effect transistors. *Electron. Lett.*, **9**, 232–4.
2. Shur, M. S. (1990) *Physics of Semiconductor Devices*, Prentice-Hall, Englewood Cliffs, NJ.
3. Ladbrooke, P. (1986) GaAs MESFETs and high electron mobility transistors (HEMT), (1991) in *Gallium Arsenide for Devices and Integrated Circuits* (eds

H. Thomas, D. V. Morgan, B. Thomas, J. E. Aubrey and G. B. Morgan), IEE Electronic Materials and Devices Series 3, Peregrinus, London, Chap. 10.

4. Tasker, P. (1991) High electron mobility transistors, in *Physics and Technology of Heterojunction Devices* (eds D. V. Morgan and R. H. Williams), IEE Electronic Materials and Devices Series 8, Peregrinus, London, Chap. 5.
5. Curtice, W. (1980) A MESFET model for use in the design of GaAs integrated circuits. *IEEE Trans. Microwave Theory Tech.*, **28**, 448–56.
6. Curtice, W. R. (1988) GaAs MESFET modeling and nonlinear CAD. *IEEE Trans. Microwave Theory Tech.*, **36**, 220–30.
7. Corsi, F., Perri, A. G., Armenise, M. N. and Andresciani, V. (1988) A comparison of non-linear circuit simulation models of GaAs MESFETs by DC/AC characterization. *Alta Freq.*, **57**, 321–9.
8. Fukui H. (ed.) (1981) *Low-Noise Microwave Transistors and Amplifiers*, IEEE, New York.
9. Fukui, H. (1979) Design of microwave GaAs MESFETs for broad-band low-noise amplifiers. *IEEE Trans. Microwave Theory Tech.*, **27**, 643–50.
10. Pospieszalski, M. W. (1989) Modeling of noise parameters of MESFETs and MODFETs and their frequency and temperature dependence. *IEEE Trans. Microwave Theory Tech.*, **37**, 1340–50.

FURTHER READING

1. de Cogan, D. (1987) *Solid State Devices: A Quantum Physics Approach*, Springer, New York.
2. Neamen, D. A. (1992) *Semiconductor Physics and Devices: Basic Principles*, Irwin, Homewood, IL.
3. Sze, S. M. (1981) *Physics of Semiconductor Devices*, 2nd edn, Wiley, New York.
4. Hess, K. (1988) *Advanced Theory of Semiconductor Devices*, Prentice-Hall, Englewood Cliffs, NJ.
5. Pengelly, R. S. (1986) *Microwave Field Effect Transistors – Theory, Design and Applications*, Wiley, Chichester, New York.
6. Soares, R., Graffeuil, J. and Obregon J. (eds) (1983) *Applications of GaAs Mesfets*, Artech, Dedham, MA.
7. Thomas, H., Morgan, D. V., Thomas, B., Aubrey, J. E. and Morgan G. B. (eds) (1986) *Gallium Arsenide for Devices and Integrated Circuits*, IEE Electronic Materials and Devices Series 3, Peregrinus, London.
8. Morgan, D. V. and Williams R. H. (eds) (1991) *Physics and Technology of Heterojunction Devices*, IEE Electronic Materials and Devices Series 8, Peregrinus, London.

PROBLEMS

6.1 A bipolar transistor has an interdigitated layout with ten emitter fingers. What is the number of base fingers? The emitter diffusions each have an area of $200 \times 4\,\mu m^2$. Assume that the emitter and base dopings are uniform and equal to $10^{20}/cm^3$ and $10^{18}/cm^3$, respectively. Estimate the capacitance of the emitter-base diode, assuming $V_{BE} = 0\,V$ and neglecting edge effects. Use the equations for diode built-in voltage and capacitance of Chapter 4. Estimate the base resistance, assuming a hole mobility of $300\,cm^2/(V\,s)$ and a base thickness of $0.1\,\mu m$. What is the cutoff frequency given by this RC filter? Repeat the calculations but now for $V_{BE} = 0.1\,V$ in the forward direction.

6.2 In the previous example the collector doping is $10^{16}/cm^3$. Calculate the width of the collector depletion layer at a bias $V_{CB} = 5$ V. Calculate the transit time through this layer assuming the electrons travel at their saturated velocity. An estimate for the base transit time is given by $t_b = W^2/2D$, where W is the base thickness and D the diffusion coefficient of electrons in the base. Calculate this transit time for $D = 2 \times 10^{-3}$ m^2/s. Compare the cutoff frequency given by these transit times with that of Problem 6.1.

6.3 Give three reasons why GaAs MESFETs perform better then their Si counterparts.

6.4 A GaAs MESFET has a channel length of 1 μm, a channel thickness of 0.2 μm and a doping of 2×10^{17} cm^3. Calculate the saturated drain current and the transconductance at $V_{GS} = 0$ V. Now increase the doping to 10^{18} cm^3. Adapt the channel thickness so that I_D remains the same and see whether g_m can be improved this way. Remember that the mobility decreases with increasing doping. Use Hilsum's formula (Chapter 3). What possible other advantage and disadvantages does this procedure have?

6.5 A MESFET shows low-frequency dispersion so that the d.c. transconductance is 40 mS but at frequencies above 1 MHz it has dropped to 20 mS. If the transistor is biased so that the direct current is 40 mA, and a voltage step of $+0.1$ V is applied at the gate with a rise time of 1 ns, what will be the initial current step and to which value will the current eventually rise?

6.6 A MESFET has 300 μm gate width and a 4 μm source–drain spacing. The chip size is 600×600 μm^2 and the substrate thickness is 100 μm. Calculate the thermal resistance under the following assumptions:

- the heat is carried away uniformly over the whole chip;
- only the area under the source–drain channel is removing heat.

For the second case calculate the temperature rise at $V_{DS} = 3$ V, $I_D = 300$ mA.

6.7 Find an expression for the *saturation voltage* V_{DSsat} (the 'knee' voltage in the I_D–V_{DS} characteristic) for a MESFET, for each of the three models given in section 5.3.3 for the saturation current. How does it depend on V_{GS}? Calculate an estimate for V_{DSsat} at $V_{GS} = 0$ V, for the original MESFET of Problem 6.3, using $\phi_B = 0.8$ V.

6.8 Design an HEMT for a maximum sheet density of 10^{12} cm^3, given that the doping concentration in the AlGaAs cannot be made higher than 10^{18} cm^3. Take an undoped AlGaAs spacer of 6 nm (this has proved to be a good value for HEMTs). Take the channel thickness d_{ch} equal to 6 nm. Calculate the necessary minimum thickness of the doped AlGaAs. Calculate the theoretical g_m and the minimum and maximum gate voltage. A spacer thickness of more than 6 nm has been found to increase the mobility in the channel considerably, especially at low temperatures. What could be the reason(s) that one nevertheless does not use thicker spacers in HEMTs?

7

Transmission lines and microwave circuits

7.1 TRANSMISSION LINE THEORY

7.1.1 Propagation coefficient and characteristic impedance

A transmission line is called uniform when its parameters do not vary in the direction of propagation. If we assign the z coordinate to the propagation direction then the propagation of a sinusoidal wave can be described by a function $\exp(j\omega t - \gamma z)$ which is common to all field components. The constant γ is called the *propagation exponent* and is in general a complex quantity. Its real part (α) is called the *damping coefficient* and its imaginary part (k) the *propagation coefficient* or *wave number*. For lossless lines the t–z dependence simplifies to $\exp[j(\omega t - kz)]$. Another important quantity that characterizes a transmission line is the *characteristic impedance Z_0*, in principle also a complex quantity.

The full solution of the wave pattern is found from Maxwell's equations:

$$\nabla \times E = -\frac{\partial B}{\partial t}$$

$$\nabla \times H = \frac{\partial D}{\partial t} + J_c.$$

Usually the waveguides we are talking about are filled with a uniform isotropic medium such as air or a simple dielectric. Then the conduction current J_c is zero and B and D are related to E and H as follows:

$$D = \varepsilon_r \varepsilon_0 E, \ B = \mu_r \mu_0 H.$$

Sometimes a medium can be anisotropic. In particular for a ferrite in a constant magnetic field the r.f. magnetic permeability exhibits a tensor character and B has a direction different from H.

The field pattern in the transverse plane obtained from Maxwell's equations is determined by the boundary conditions and depends strongly on the geometry of the waveguide. Each of the possible transverse field profiles defines a *waveguide mode*. In every waveguide an infinite number of modes is possible, each with its own frequency-dependent propagation coefficient and characteristic impedance. In the limited frequency band for which the waveguide is intended a finite number of modes exists. Each mode has a cutoff frequency below which the propagation coefficient becomes imaginary meaning that no propagation is possible. In some special cases, as discussed below, the cutoff frequency is zero.

7.1.2 TEM Lines

Lines that consist of two metallic conductors in a uniform medium are called TEM (transverse electric and magnetic) lines. In this class belong two-wire or Lecher lines, coaxial lines and striplines (Fig. 7.1). In TEM lines both the electric field and the magnetic field of the *fundamental* propagating mode are transverse to the direction of propagation which means that a local line voltage can be defined as the integral of the electric field in the transverse plane from one conductor to the other. This integral is independent of the integration path since no magnetic field lines are crossing the transverse plane. Also, the cutoff frequency of TEM modes is zero.

The local voltages and currents on TEM transmission lines can be written in terms of left- and right-travelling waves:

$$V(z) = V^+ \exp(-\gamma z) + V^- \exp(\gamma z) \tag{7.1a}$$

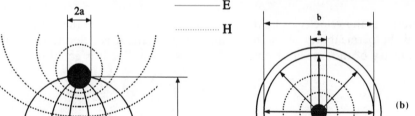

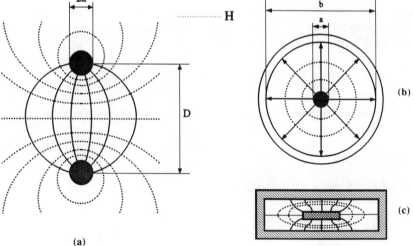

Figure 7.1 TEM transmission lines: (a) two wire (Lecher); (b) coaxial; (c) stripline; ——, electric field lines; ·····, magnetic field lines.

$$I(z) = \frac{V^+}{Z_0} \exp(-\gamma z) - \frac{V^-}{Z_0} \exp(\gamma z) \qquad (7.1b)$$

and conversely

$$V^+ \exp(-\gamma z) = \tfrac{1}{2}[V(z) + Z_0 I(z)] \qquad (7.2a)$$

$$V^- \exp(+\gamma z) = \tfrac{1}{2}[V(z) - Z_0 I(z)]. \qquad (7.2b)$$

A short line section of length Δx can be represented by a T-network with series impedance $(R + j\omega L)\,\Delta x$ and parallel admittance $(G + j\omega C)\,\Delta x$ where R, L, G and C are given per unit length. The propagation exponent and characteristic impedance are then given by the following equations:

$$\gamma = [(R + j\omega L)(G + j\omega C)]^{1/2} \qquad (7.3)$$

$$Z_0 = \left(\frac{R + j\omega L}{G + j\omega C}\right)^{1/2}. \qquad (7.4)$$

For lossless lines, as well as in the limit of infinitely high frequencies, these formulas simplify to

$$\gamma = jk = j\omega(LC)^{1/2} \text{ and } Z_0 = R_0 = \left(\frac{L}{C}\right)^{1/2}. \qquad (7.5)$$

In terms of the medium filling the line they can also be written as

$$k = \frac{\omega}{c} (\varepsilon_r \varepsilon_0 \mu_r \mu_0)^{1/2} \text{ and } R_0 = K(\mu_0/\varepsilon_0)^{1/2}, \qquad (7.6)$$

where c is the velocity of light in vacuum, and ε_r and μ_r are the permittivity and the permeability of the medium. The constant K is determined by the geometry of the transmission line. For a two-wire line it is equal to

$$K_{2w} = \frac{1}{\pi} \cosh^{-1}\left(\frac{D}{2a}\right) \qquad (7.7)$$

where D and a are given in Fig. 7.1(a). For a coaxial line we have

$$K_{coax} = \frac{1}{2\pi} \ln\left(\frac{b}{a}\right) \qquad (7.8)$$

where a and b are shown in Fig. 7.1(b).

7.1.3 Non-TEM lines

Non-TEM transmission occurs in hollow waveguides and in two-conductor transmission lines where the dielectric medium is non-uniform, e.g. micro-strip lines and finlines (Fig. 7.2). Here at least one of the fields has a

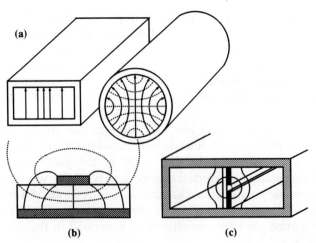

Figure 7.2 Non-TEM transmission lines: (a) waveguide; (b) microstrip; (c) finline;
——, electric field lines; ·····, magnetic field lines.

component in the propagation direction and a unique definition of line
voltage or current is not possible.

Hollow waveguide modes can also occur in coaxial lines and striplines.
They are characterized by a cutoff frequency below which no propagation
can occur. Figure 7.3 shows the ω–k diagram for a few modes of propagation. It is seen that the TEM modes are non-dispersive, that is their phase
velocity ω/k and group velocity $\mathrm{d}\omega/\mathrm{d}k$ are independent of frequency and
both equal to the velocity of light.

In spite of the non-TEM character of waveguide and microstrip it would
be very handy for circuit analysis if also in these cases one could use
transmission line theory. This can be done by defining as the line voltage
the integral of the electric field along a well-chosen path. In microstrip the
integration path is from the middle of the strip down to the ground plane

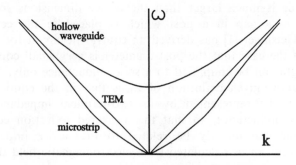

Figure 7.3 ω–k diagrams of some waveguides.

and the line current is the current through the strip. From Fig. 7.3 it can be seen that the ω–k diagram of microstrip does not differ greatly from that of TEM line. For this reason microstrip is often classified as a quasi-TEM line and its properties are summarized in an *effective permittivity* which is defined in such a way that equations (7.4) and (7.6) remain valid. Microstrip lines are treated more extensively in Section 5.2.

In a rectangular waveguide one uses the fact that for the fundamental TE$_{01}$ mode the electric field lines in the transverse plane are vertical and this field has its maximum in the center of the guide (Fig. 7.2a)), so the integration path is the vertical in the center of the waveguide. In this case it is possible to define a characteristic impedance in different ways but having defined the line voltage as above we can use the fact that the power transported on a transmission line, expressed in the line voltage, is equal to

$$P = \frac{1}{2} \frac{V^2}{Z_0}.$$

This can be inverted to obtain a definition of Z_0. For the TE$_{01}$ mode the result is

$$Z_0 = \frac{2b}{a} \frac{\lambda_g}{\lambda_0} \left(\frac{\mu_0}{\varepsilon_0} \right)^{1/2}. \tag{7.9}$$

where a and b are given in Fig. 7.2(a), λ_0 is the free space wavelength and λ_g the wavelength in the guide given by

$$\lambda_g = \frac{\lambda_0}{(1 - f_c^2/f^2)^{1/2}}; \quad f_c = \frac{1}{a^2} \left(\frac{1}{4\mu_0\varepsilon_0} \right)^{1/2}. \tag{7.10}$$

The parameter f_c is called the *cutoff frequency*. This definition of Z_0 is especially useful in cases where a two-terminal device, e.g. a Gunn or an IMPATT diode, is mounted in a waveguide. Normally the height of the waveguide is much larger than the device dimensions so the device is mounted in a gap in a post which is placed in the centre of the waveguide. Getsinger [1] has derived an equivalent circuit for this mounting which in the case that the post diameter is very small compared with the wavelength can be simplified to a series inductance only. The advantage of the above-given definition of Z_0 is that in the equivalent circuit the diode can be represented by its two-terminal impedance in series with the post inductance, so that the measured reflection coefficient in the waveguide can directly be related to the device impedance. The post inductance can be calculated from electromagnetic field theory but it can also be measured by inserting a known impedance in the device position.

7.2 MICROSTRIP LINES

7.2.1 Structure and properties

A microstrip line is sketched in Fig. 7.4. The dielectric medium is now non-uniform and therefore this is not a TEM line. However, the greatest part of the electric field is in the dielectric, so that the axial component of the field is not large and we may speak of a quasi-TEM line. The formulas given above for γ and Z_0 are now not valid any more, but since they are so handy we keep using them by the introduction of an effective permittivity ε_{eff}, that is defined from the capacitance per meter, related to that of the same line with air as a dielectric:

$$\varepsilon_{\text{eff}} = \frac{C(\varepsilon_r)}{C(\varepsilon_r = 1)}. \tag{7.11}$$

At large strip width ($w/h \gg 1$) most of the electric field is in the dielectric and ε_{eff} approaches the permittivity of the material, ε_r. At very small strip width ($w/h \ll 1$) ε_{eff} approaches $(\varepsilon_r + 1)/2$. In many handbooks (e.g. ref. 2)

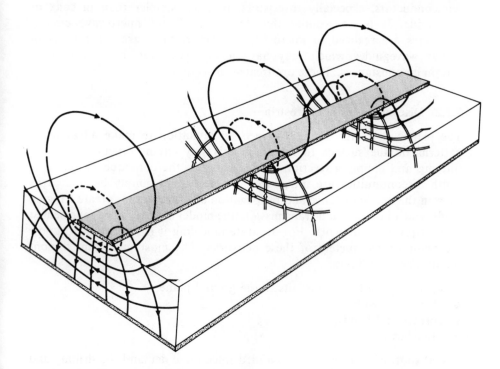

Figure 7.4 Microstrip line. Redrawn from Van Heuven, J. H. C. and Van Nie, A. G., Geintegreerde Microgolfschakelingen in *Philips Technisch Tijdschrift*, **32**, 161–73; published by Philips Technisch Tijdschrift, 1971.

approximate formulas are given with which ε_{eff} can be calculated as a function of the dimensions and substrate permittivity. At high frequencies a microstrip line becomes dispersive [2], that is its propagation coefficient and characteristic impedance become frequency-dependent quantities. This can be represented by allowing ε_{eff} to be a function of frequency. The general tendency is that ε_{eff} goes from its low-frequency limit given above to the high-frequency limit ε_r. It follows from what is said above that strips with a large w/h ratio are less dispersive than narrow ones.

As materials for microstrip aluminum oxide, Teflon and polyester–glass fiber are used, and sometimes also ferrite material which gives the possibility to integrate non-reciprocal components such as circulators. At frequencies above 30 GHz the losses of these materials begin to increase strongly and quartz is the preferred material. Besides these dielectrics, high-resistivity semiconductors can also be used, in particular GaAs, which can have a resistivity of $10^8\ \Omega$ cm. This offers the possibility to integrate diodes and transistors monolithically with microstrip circuits (Chapter 11).

Microstrip lines are small and light and can be made with printed-circuit techniques. This makes them relatively cheap. Moreover, the mounting of semiconductors, especially transistors, is much simpler than in coax or waveguide. It is no wonder that by far the most microwave circuits nowadays are realized in microstrip. However, there are other types of planar waveguides which may have advantages over microstrip under certain circumstances. These are treated in section 7.3.

7.2.2 Discontinuities in microstrip

The planar character of microstrip is an invitation to make all kinds of patterns on the surface to realize in this way circuit functions such as filtering and coupling (section 7.2.3). This has the consequence that line width discontinuities are introduced. These cannot simply be treated as steps in the line impedance and propagation coefficient; they create fringing fields which in terms of a transmission line model have to be be represented by an equivalent network. For accurate modeling it is important to know the two-port parameters of these networks. The most important discontinuities are as follows:

- open ends and short circuits to the ground plane;
- changes of width;
- corners and bends;
- branches.

Real short circuits are to be avoided since these demand the drilling and metallization of holes in the substrate material which is a costly procedure. Also, such a feed-through creates fringing fields which cause it to have a finite impedance. It is better to simulate a short circuit by an open piece of

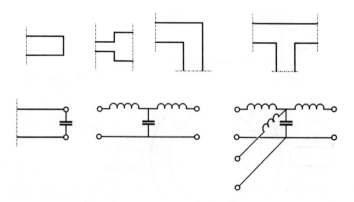

Figure 7.5 Microstrip discontinuities and their equivalent circuits.

line a quarter-wavelength long. The input impedance of this is given by equation (7.18), assuming the impedance of the open end is infinite. Unfortunately, however, an open end has fringing fields too which can be represented by a capacitance (Fig. 7.5). We can improve this by making the line as wide as possible so that its characteristic impedance is low. Also, the capacitance can be compensated at one frequency by making the line a little longer than a quarter-wavelength, which adds an inductance. The other elements mentioned produce electric and magnetic stray fields and have equivalent circuits with inductances and capacitances. In refs 2 and 3 formulas can be found with which these elements can be calculated.

7.2.3 Elements of microstrip circuits

For the realization of circuits in microstrip a great variety of basic elements is available, to be distinguished in four groups:

- line pieces of different width, straight or curved;
- interruptions;
- coupled lines;
- discrete elements.

With pieces of microstrip line we can for instance make open or closed resonators (Fig. 7.6). With alternating wide and narrow line pieces of less than a quarter-wavelength we can simulate a ladder LC circuit. The wide pieces behave as capacitances to ground and the narrow ones as series inductances (Fig. 7.7), so the whole acts as a low-pass filter.

Another way to make filters is with coupled lines. When two line pieces run partially parallel (Fig. 7.8), they behave as a combination of a coupler and a resonator. At the resonance frequency the field amplitudes in the

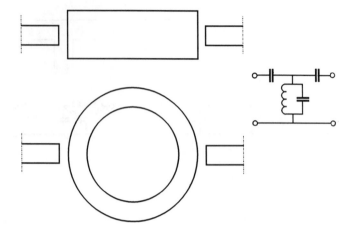

Figure 7.6 Microstrip resonators.

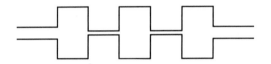

Figure 7.7 Low-pass filter.

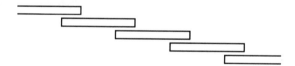

Figure 7.8 Filter with coupled lines.

resonator become high so we have a strong through-coupling. A cascade of identical sections acts as a band filter, similar to coupled LC circuits at low frequencies. Directional couplers will be treated in section 7.4.4.

Discrete elements used in microstrip circuits are resistors, capacitors and inductors. Resistors can be made by deposition of a resistive material by means of thin film or thick film techniques. The standard technique to make microstrip circuits is to metallize the substrate on both sides and then to etch the desired pattern in the metal on one side. Resistors can be included in this technique by making the metallization from a high-resistivity nichrome layer, on top of which a gold layer is deposited. By etching away both layers one can make microstrip patterns and by etching only the gold one can make resistors (Fig. 7.9). With three resistors in a T-junction a

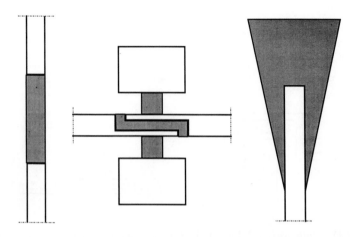

Figure 7.9 Resistor, attenuator and matched load.

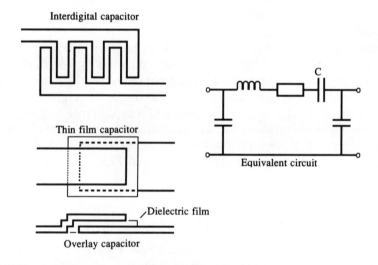

Figure 7.10 Capacitors: interdigital capacitor, thin film capacitor, overlay capacitor and equivalent circuit.

matched attenuator can be realized. A reflection-free load can be made by an NiCr layer of increasing width at both sides of the strip.

Series capacitors of small value can be made simply by interrupting the strip. The capacitance can be increased by making the cut in the parallel direction (Fig. 7.10). If greater capacitances are desired one has to use overlapping metallizations with a thin dielectric between them. Another solution is the mounting of so-called chip capacitors.

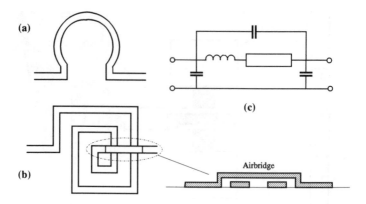

Figure 7.11 Inductors: (a) loop inductor; (b) spiral inductor; (c) equivalent circuit.

Inductances can be made as narrow line pieces, curved line pieces and so-called spiral inductors (Fig. 7.11).

Discrete elements behave only as such as long as they are very small compared with a wavelength. When this condition is not fulfilled they become electromagnetic structures. The resistance of Fig. 7.9 for instance is really a piece of lossy transmission line. The simplest equivalent circuit for this contains at least a series inductance and a capacitance to ground. This holds for all discrete elements: as the frequency goes up their parasitic elements become more important and have to be accounted for when designing a circuit. In the equivalent circuits of Figs. 7.10 and 7.11 this is already accounted for.

7.2.4 Transitions to other waveguides

In practice microstrip circuits never work alone. They are always connected to other circuits, measuring equipment, antennas etc., which mostly are made in coax or waveguide. This makes it necessary to have transitions to these media. Microstrip-to-waveguide transitions are difficult to make so they are seldom used. Moreover, in the frequency range where microstrip is normally used (< 30 GHz) coax works well too.

A coaxial-to-microstrip transition can be in plane or at a right angle (Fig. 7.12). The right-angle transition is difficult to make and has rather marked parasitic reactances. Therefore usually the in-plane transition is used which can be made reasonably reflection free. The taper at the end of the coaxial line inner conductor seems like a minor detail but it really helps in keeping the reflection low.

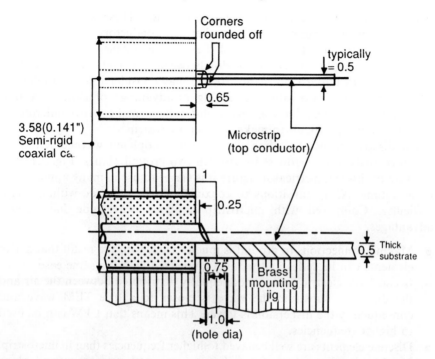

Figure 7.12 Coax-to-microstrip transition. Reproduced with permission from England, E. H. (1976) *A Coaxial to Microstrip Transition*, IEEE Trans. Microwave Theory and Techniques, **MTT-24**, 47–8. © 1976 IEEE Press.

7.3 OTHER PLANAR WAVEGUIDES

7.3.1 Slotline and coplanar waveguide

One disadvantage of microstrip is that it has the ground plane at the bottom side of the substrate. This becomes especially clear when one has to make a short circuit to ground as has been explained before. Another, more economical, disadvantage is that one has to start from a substrate metallized on both

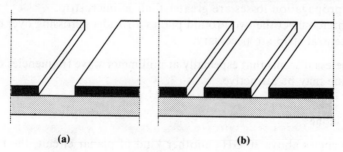

Figure 7.13 (a) Slotline and (b) coplanar waveguide.

sides and etch away most of the metal on one side. These disadvantages are avoided by planar transmission lines where all conductors are on the same side of the substrate. The simplest of these is the slotline (Fig. 7.13(a)).

A slotline simply consists of a slot between two metal planes, one of which serves as the ground plane. So one may consider it as a topological deformation of a two-wire line. A big disadvantage of slotline is that transitions to coaxial lines are very difficult to make. Also, microwave circuitry is more difficult to make than in microstrip.

The disadvantages of slotline are avoided in coplanar waveguide (CPW) which consists of a narrow strip between two ground planes (Fig. 7.13(b)). Because of this symmetrical structure it is easier to use in many applications than slotline. Also, transitions to coaxial line can be made without great difficulty. Compared with microstrip coplanar waveguide has several advantages.

- Making connections to ground is much easier, with the result that active elements can be series and parallel connected with the same ease.
- Because the electric field is more equally distributed between the air and the dialectric substrate the wave is closer to a true TEM wave and consequently the dispersion is lower. This means that CPW can be used to higher frequencies.
- Discrete elements are well behaved to higher frequencies than in microstrip.
- Neighboring lines do not couple (as long as there is a ground plane between them) so that circuit designs can be more compact.
- Points within a circuit are accessible for measurement with coplanar probes (section 7.5.7).

Of course there are also disadvantages.

- Besides the desired mode where the two outer conductors are at the same potential (even mode) there is also a mode of propagation possible where the outer conductors are at different potentials and where the field pattern is similar to that of the slotline (odd mode), and yet another one where all three conductors are at the same potential and form together with the outer casing a kind of stripline configuration.
- The propagation losses are greater than in microstrip.
- To interconnect the two ground planes airbridges crossing over or under the central strip are necessary.

Nevertheless it seems that especially at millimeter wave frequencies coplanar waveguide may be attractive.

7.3.2 Finline

At frequencies above 30 GHz another kind of planar circuit, the finline, is used whose cross-section is sketched in Fig. 7.14(a). In fact it is a combina-

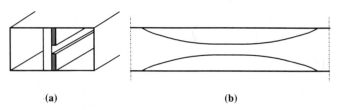

Figure 7.14 Finline: (a) cross-section; (b) side view.

tion of slotline with rectangular waveguide. The name finline derives from the taper that is made in the slot (Fig. 7.14(b)) to give a smooth transition to the rectangular waveguide. The basic idea behind finline is to keep as much as possible of the circuit in a waveguide which has the lowest loss, but at the same time to provide a place where semiconductor devices can be mounted easily.

7.4 MICROWAVE NETWORKS

7.4.1 Standing waves, reflection coefficients, Smith chart

When a wave propagates down a transmission line a reflected wave is excited at every discontinuity of the line to satisfy the local boundary conditions. The most prominent and always-present discontinuity is the end of the line. When an impedance Z_L is mounted here the boundary condition is $V(L) = Z_L I(L)$. The *reflection coefficient* is defined as the ratio of the voltage of the reflected wave to that of the incident wave. For an arbitrary position z it follows from equations (7.2) that the reflection coefficient is given by

$$\Gamma_z = \frac{V(z) - Z_0 I(z)}{V(z) + Z_0 I(z)}. \tag{7.12}$$

In particular, at the end of the line we have

$$\Gamma_L = \frac{Z_L - Z_0}{Z_L + Z_0}. \tag{7.13}$$

A terminating impedance equal to Z_0 gives zero reflection and is called a *matched load*.

From equations (7.2) it also follows that we can express the reflection coefficient at an arbitrary place in terms of the one at the end:

$$\Gamma_z = \Gamma_L \exp[2\gamma(z - L)]. \tag{7.14a}$$

For a lossless line this becomes

$$\Gamma_z = \Gamma_L \exp[2jk(z - L)]. \tag{7.14b}$$

The incident and reflected waves give rise to a standing wave pattern on the line whose local amplitude is given by

$$|V^+ \exp(-jkz) + V^- \exp(jkz)| = |V^+| \, |1 + \Gamma_L \exp[2jk(z - L)]| . \quad (7.15)$$

In the complex plane the last factor describes a circle with centre at 1 and radius $|\Gamma_L|$ so its maximum is $1 + |\Gamma_L|$ and its minimum $1 - |\Gamma_L|$. The ratio of these two is called the *voltage standing wave ratio*:

$$\text{VSWR} = \frac{1 + |\Gamma_L|}{1 - |\Gamma_L|} .$$

So, if the VSWR is known, the magnitude of Γ_L is known. The phase of Γ_L can be found from the position of the minima of the line voltage. From equation (7.15) it can be seen that a minimum occurs wherever $\phi_r + 2k(z - L) = (2n + 1)\pi$, so

$$\phi_r = (2n + 1)\pi - 2k(z_{\min} - L)$$

where n is zero for the minimum closest to the end of the line. It follows that a measurement of the VSWR and the position of a minimum is sufficient to determine the reflection coefficient. This will be discussed further in section 7.5.

The input impedance of a line with characteristic impedance Z_0 and length L, that is terminated with an impedance Z_L, can from equations (7.12) and (7.14) be calculated as

$$Z_i = Z_0 \frac{Z_L \cosh(\gamma L) + Z_0 \sinh(\gamma L)}{Z_0 \cosh(\gamma L) + Z_L \sinh(\gamma L)} \quad (7.16)$$

or, for the lossless case,

$$Z_i = Z_0 \frac{Z_L + jZ_0 \tan(kL)}{Z_0 + jZ_L \tan(kL)} . \quad (7.17)$$

When L is made a quarter of a wavelength, $kL = \pi/2$ and this equation simplifies to

$$Z_i = \frac{Z_0^2}{Z_L} . \quad (7.18)$$

So an impedance that is small in magnitude is transformed to a large one and vice versa. This is the famous *quarter-wavelength transformer*, which is much used for impedance matching. It transforms an open circuit into a short and vice versa.

A much-used diagram in microwave technology is the Smith chart. The original purpose of this was to convert impedances into reflection coefficients and vice versa. The reflection coefficient at the end of a transmission line with (real) characteristic impedance Z_0 that is terminated with a (complex) impedance Z_L is given by equation (7.13). If we now make a

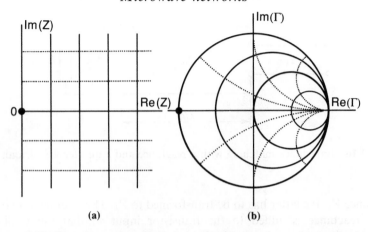

Figure 7.15 Graphic representation: (a) impedance plane; (b) Smith chart.

conformal representation of the ζ plane (where $\zeta = Z_L/Z_0$) onto the Γ plane using this formula then the lines of constant $\mathrm{Re}(Z_L)$, and of constant $\mathrm{Im}(Z_L)$ are transformed into circles. When we plot Z_L/Z_0 on these circles then we can read Γ in rectangular or polar coordinates (Fig. 7.15). If we move to another point on the transmission line, then Γ is given by equations (7.14), so Γ describes a circle around the origin. The value of Γ at an arbitrary point of the line is then easy to determine if we know it at the end of the line. On the impedance coordinates we can then also read the impedance at this point. This diagram is called the Smith chart after its inventor.

For the admittance Y a transformation similar to equation (7.13) is valid:

$$\Gamma = - \frac{Y_L - Y_0}{Y_L + Y_0}$$

so the admittance plane can also be represented on the Smith chart.

A difficulty arises with impedances that have a negative real part, since normally only the right-half of the ζ plane is represented on the Smith chart. In this case we have to extend the working area to a line $\mathrm{Re}(Z) = -C$ in the Z plane, corresponding to a circle outside the unit circle in the Γ plane. This is indicated by dotted lines in Fig. 7.15.

7.4.2 Impedance matching

A quarter-wavelength transformer can be used for impedance matching, a task that is often necessary in microwave technology. For instance, to obtain optimal transfer of power from a signal source to a transistor the output impedance of the source should be matched to the input impedance of the transistor. Supposing the former is equal to the characteristic

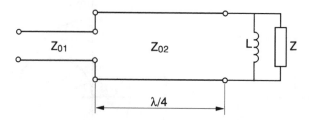

Figure 7.16 Impedance matching with a reactance and a quarter-wavelength transformer.

impedance Z_0, the latter has to be transformed to Z_0. This is done in two steps: first a reactance is added to the transistor input so that the total input impedance becomes real, say Z_i; then Z_i is transformed to Z_0 by a quarter-wavelength of transmission line with characteristic impedance $Z_{02} = (Z_0 Z_i)^{1/2}$. This is illustrated by Fig. 7.16 where it is assumed that the imaginary part of the transistor input impedance is capacitive so that an inductor is necessary to compensate it. Another'way to compensate the imaginary part of an impedance is to use a so-called *parallel stub*, that is a piece of open- or short-circuited transmission line connected in parallel to the complex impedance Z_L. By varying the length of the stub its reactance can be varied from $-\infty$ to $+\infty$ as equation (7.9) shows. The real part of Z_L can also be changed by inserting the stub in the transmission line at a distance l_1 from the end (Fig. 7.17(a)): by choosing l_1 we can make $\mathrm{Re}(Z_i)$ is equal to Z_0, and by choosing the length of the stub l_2 we can compensate the imaginary part of Z_i.

In coax and waveguide it is preferable to use short-circuited stubs. This prevents radiation from the end and a short can also easily be made in

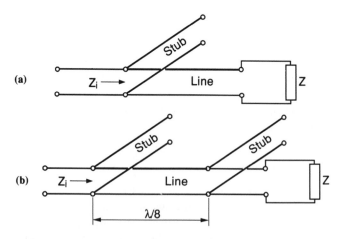

Figure 7.17 Impedance matching: (a) single stub; (b) double stub.

moveable form. In coax a problem could arise when there is a d.c. voltage on the line which would be short circuited. In this case a d.c. blocking capacitor has to be inserted at a suitable place.

With a single stub not all impedances can be matched. Moreover, whereas the stub length can be adjustable, the length l_1 most of the time cannot. With two stubs as in Fig. 7.17(b) we have an extra degree of freedom and the range of impedances that can be matched is greater. The distance between the stubs usually is taken as $\lambda/4$ or $\lambda/8$. It can be shown that in this case the greatest range is obtained. In waveguide very often a hybrid tee (section 7.4.4) with moveable shorts in arms 3 and 4 is used as a double-stub tuner.

7.4.3 S-parameters

Semiconductor devices are small compared with a wavelength in air, even at microwave frequencies, and so can be described in terms of voltages and currents. Indeed, if they were not small, they would not work well as microwave devices owing to transit-time effects. However, this does not hold for most passive components and certainly not for the measuring equipment. Measuring voltages and currents at microwave frequencies therefore leads to insurmountable difficulties. On the other hand, the amplitude and phase of waves on transmission lines can be measured, (section 7.3), so a description of twoports that is adapted to the wave picture is more appropriate. This leads to the concept of scattering parameters (S-parameters) (Fig. 7.18). These are relations between incoming and outgoing sinusoidal waves on transmission lines connected to a linear twoport. In the case of a non-linear twoport we can use S-parameters when we consider small a.c. quantities that are superimposed on the d.c. quantities (small-signal approximation).

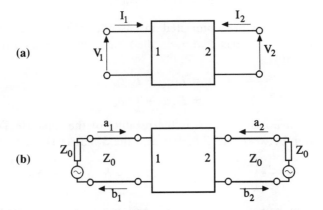

Figure 7.18 Two representations of a twoport.

The current–voltage relationship can be expressed in the impedance matrix or the admittance matrix, i.e.

$$V_1 = Z_{11} + I_1 + Z_{12}I_2$$
$$V_2 = Z_{21}I_1 + Z_{22}I_2 \tag{7.19}$$

$$I_1 = Y_{11}V_1 + Y_{12}V_2$$
$$I_2 = Y_{21}V_1 + Y_{22}V_2 \tag{7.20}$$

or by the mixed-mode h parameter matrix:

$$V_1 = h_{11}I_1 + h_{12}V_2$$
$$I_2 = h_{21}I_1 + h_{22}V_2. \tag{7.21}$$

The relations between waves are given by the S-parameter matrix. We define first the normalized complex wave amplitudes a_i and b_i (these contain the amplitude and phase information), where a_i indicates waves towards the twoport and b_i waves traveling away from it. The index i ($i = 1, 2$) indicates the port. The definitions are (with real $Z_0 = R_0$)

$$a_i = \frac{1}{2}\left(\frac{V_i}{(R_0)^{1/2}} + (R_0)^{1/2}I_i\right), \qquad b_i = \frac{1}{2}\left(\frac{V_i}{(R_0)^{1/2}} - (R_0)^{1/2}I_i\right). \tag{7.22}$$

This is based on equation (7.2). On comparison we see that a_i and b_i are proportional to the voltages of the left- and right-travelling waves, respectively. The normalization is chosen in such a way that $\frac{1}{2}|a_i|^2$ and $\frac{1}{2}|b_i|^2$ are equal to the power that the waves indicated by a_i and b_i, respectively, transport. With these definitions we can now describe a twoport by its S-matrix:

$$b_1 = S_{11}a_1 + S_{12}a_2$$
$$b_2 = S_{21}a_1 + S_{22}a_2. \tag{7.23}$$

When more twoports are connected in cascade it is handier to use the transfer matrix or T-matrix that gives the input quantities as a function of the output quantities:

$$b_1 = T_{11}a_2 + T_{12}b_2$$
$$a_1 = T_{21}a_2 + T_{22}b_2. \tag{7.24}$$

The T-matrix of a cascade now is the product of the separate T-matrices. The connection between the T-matrix and the S-matrix is

$$T = \frac{1}{S_{21}}\begin{bmatrix} -D_s & S_{11} \\ -S_{22} & 1 \end{bmatrix} \quad \text{with } D_s = S_{11}S_{22} - S_{12}S_{21}. \tag{7.25}$$

The connection between the S-matrix and other four-pole parameters is given by the following formulas, where $z_{ij} = Z_{ij}/Z_0$, $y_{ij} = Y_{ij}Z_0$:

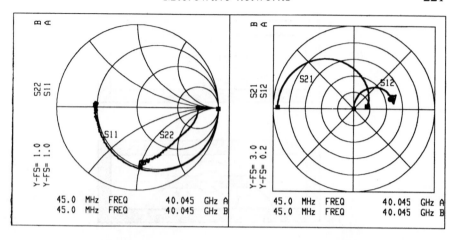

Figure 7.19 *S* parameters of a microwave transistor.

$$z_{11} = \frac{1 + S_{11} - S_{22} - D_s}{1 - S_{11} - S_{22} + D_s}, \qquad z_{12} = \frac{2S_{12}}{1 - S_{11} - S_{22} + D_s}$$

$$z_{21} = \frac{2S_{21}}{1 - S_{11} - S_{22} + D_s}, \qquad z_{22} = \frac{1 - S_{11} + S_{22} - D_s}{1 - S_{11} - S_{22} + D_s} \qquad (7.26)$$

$$h_{11} = Z_0 \frac{1 + S_{11} + S_{22} + D_s}{1 - S_{11} + S_{22} - D_s}, \qquad h_{12} = \frac{2S_{12}}{1 - S_{11} + S_{22} - D_s}$$

$$h_{21} = \frac{-2S_{21}}{1 - S_{11} + S_{22} - D_s}, \qquad h_{22} = Y_0 \frac{1 - S_{11} - S_{22} + D_s}{1 - S_{11} + S_{22} - D_s}. \qquad (7.27)$$

Since S_{11} and S_{22} have the character of reflection coefficients we can represent them in a Smith chart. On the impedance scale we can then directly read the input impedance on either side of the twoport if the other side is terminated reflection free. Figure 7.19 shows how the frequency-dependent parameters of a transistor can be represented in a Smith chart.

The Smith chart can in this case also be used to find a circuit that matches the input impedances to a transmission line. One has to remember, however, that this can be done only for one frequency at a time. Since the Smith chart is calibrated in propagation angle (kz) the corresponding length is different for each frequency. The other two S-parameters, S_{12} and S_{21}, are often plotted in the same diagram, although this is an improper use of the Smith chart, since the impedance coordinates in this case have no meaning. Moreover, for S_{21}, which in a transistor can be greater than 1, the scale has to be adapted.

7.4.4 Coupled transmission lines

When the EM fields of two transmission lines overlap energy transfer can occur. We speak then of coupled lines. In Fig. 7.20 an example is sketched

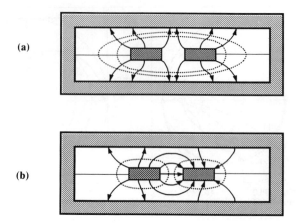

Figure 7.20 Coupled transmission lines (stripline): ——, electric field lines; ·····, magnetic field lines.

of coupled TEM lines. In the figure two field patterns are sketched, one in which the two conductors have the same potential, which is called the even mode, and one in which they have opposite potentials, the odd mode. An arbitrary distribution of waves on the two lines can always be written as the sum of an even and an odd mode. This simplifies the analysis, since for each mode we can consider the system as one single transmission line with a definite characteristic impedance and propagation coefficient. For a TEM structure like the one depicted here both modes have the same propagation coefficient (i.e. propagation velocity), but different characteristic impedances.

When we couple two transmission lines over a certain length we have a directional coupler. Two stripline realizations are shown in Fig. 7.21. The theory of TEM directional couplers is not difficult, but cumbersome. It is treated in ref. [2]. The coupling is strongest when the length of the coupler is a quarter-wavelength. Then, for the couplers in Fig. 7.21, with port 1 as the signal input, in the ideal case the output of port 4 is zero, and that of

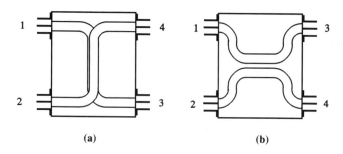

Figure 7.21 Directional couplers in stripline.

port 2 is 90° out of phase with that from port 3. The strength of the signal from port 4 depends on the distance between the lines. One strip above the other gives a stronger coupling than strips side by side. Couplers are named after their attenuation from port 1 to 4, e.g. when $S_{41} = S_{23} = 0$ and $S_{21} = S_{43} = -10$ dB it is called a 10 db coupler.

Strictly speaking a coupler like this works only at one frequency, namely that for which the length is one-quarter wavelength. In practice it is also usable in a frequency band around it. With the help of special tricks, such as a non-uniform distance between the conductors, the bandwidth can be increased to about one octave. In practical directional couplers there is also a signal emerging from port 4, and when port 3 is excited from port 2. The ratio of the desired signal coupling (i.e. from port 1 to port 2) to the unwanted one (from 1 to 4), in other words S_{21}/S_{41}, expressed in decibels, is called the *directivity*. It measures how well a directional coupler is able to distinguish the incoming and reflected waves and is an important figure of merit. The directivity of stripline couplers is seldom greater than 30 dB, of microstrip couplers 20 dB.

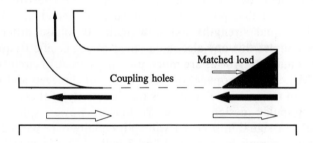

Figure 7.22 Waveguide directional coupler.

Waveguide directional couplers are made by soldering together the broad sides of two waveguides over a distance of several wavelengths and connecting the two guides by a large number of small holes in the interface (Fig. 7.22). Since the coupling holes are small compared with the guide wavelength the coupling is independent of frequency so these couplers work over the full waveguide band. Directivities of 40 dB can be achieved and even 50 dB in a limited frequency range.

A special type of (non-directional) coupler is the hybrid. A general principle, that can be used with different transmission lines, is the so-called *rat-race coupler* (Fig. 7.23(a)). By virtue of the phase delays introduced between the arms the coupling from arm 1 to 3 is zero while arms 2 and 4 receive equal signal strengths but with 180° phase difference. Of course this is exact at one frequency only but the coupler will work with acceptable

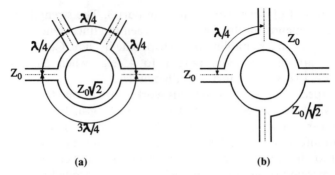

(a)   (b)

Figure 7.23 (a) Rat-race coupler and (b) 90° hybrid.

performance in a band of ± 10% around the design frequency. Even at the latter it will not work perfectly since the corners introduce reflections which couple into the nominally decoupled ports. This type of coupler can be made in rectangular waveguide, stripline and microstrip. Another type of hybrid is the 90° branch line (Fig. 7.23(b)). It uses a ring of four quarter-wave transformers, two of which have a lower characteristic impedance. The result is again that port 2 is decoupled from 1 and that 3 and 4 are coupled with equal strengths but now with 90° phase difference. Like the rat-race coupler this one also is designed for a specific frequency. Rat-race couplers and 90° hybrids are much used in microstrip circuits. Contrary to true TEM lines the propagation coefficients of the even and odd modes are different in microstrip. This makes the directivity of these couplers low.

Another popular design is the Wilkinson power splitter–combiner (Fig. 7.24(a)). A signal into port 1 will split equally over parts 2 and 3. On the other hand, signals from parts 2 and 3 will add up in port 1 when they have equal phases and amplitudes, and will cancel in port 1 and be dissipated in the resistor when the have equal amplitudes and opposite phases. In rectangular waveguide a hybrid coupler can take the form of a so-called *hybrid tee* (Fig. 7.24(b)). It consists of a piece of waveguide to

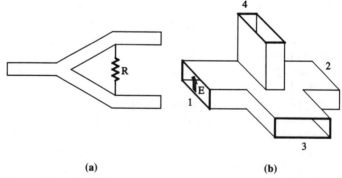

(a)   (b)

Figure 7.24 (a) Wilkinson splitter and (b) waveguide hybrid tee.

which two side-arms are made, one in the plane of the electric field of the TE_{10} mode and one in the plane of the magnetic field. Consequently these arms are called the E-arm and the H-arm, respectively. Following the field patterns it is not difficult to see that signals coming from the main ports (ports 1 and 2 in the figure) which are in phase, add up in the H-arm (port 3) and cancel in the E-arm (port 4), whereas signals 180° out of phase cancel in the H-arm and add up in the E-arm. Conversely this implies that a signal coming into port 3 will couple into ports 1 and 2 but not into port 4, and vice versa. Because no phase delays are involved this coupler operates over the full waveguide band. Also, since the decoupling properties depend entirely on the symmetry of the structure they can be very good for a carefully constructed device. Sometimes tuning screws are placed in the symmetry plane of the junction to obtain low reflection coefficients in all four arms. Such a device is called a *magic tee*.

7.5 MICROWAVE MEASUREMENTS

7.5.1 Introduction

Measuring the impedance of a one-port device such as a diode means at microwave frequencies that we place it at the end of a transmission line and measure its reflection coefficient. This can be done in different ways: by measuring the standing wave pattern on the line, by using directional couplers or by comparison with another reflection coefficient in a bridge. To measure the S-parameters of a transistor we must mount it in a transmission line with on one side a signal source and on the other side a matched load. A measurement of the reflection at the source side and the transmission at the other side yields S_{11} and S_{21}. By interchanging source and load we can determine S_{22} and S_{12}.

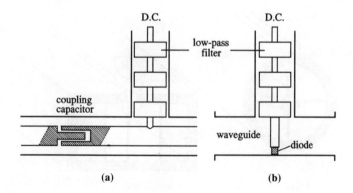

Figure 7.25 (a) Coaxial bias tee and (b) waveguide diode mount.

7.5.2 d.c. biasing

Of course the diode or transistor under test has to be supplied with the necessary d.c. voltages. This is done via so-called *bias tees*, in which the microwave signal and the d.c. voltage are separated. A coaxial bias tee is shown in Fig. 7.25(a). The coupling capacitor holds off the d.c. bias from the rest of the circuit while passing through the microwave signals and the low-pass filter prevents the r.f. from leaking into the bias circuit. A similar construction can be applied to a post-mounted diode in a waveguide (Fig. 7.25(b)).

7.5.3 Slotted-line methods

A simple but reasonably accurate method is to use a so-called *slotted line* (Fig. 7.26). It consists of a coaxial line or a rectangular waveguide in which a narrow longitudinal slot is made in a wall that is orthogonal to the transverse electric field component. A small probe inserted in this slot picks up a signal proportional to the field and transports it to a detector. One should keep in mind that most detectors, in particular diodes, have a quadratic detection characteristic (section 8.1.2) so their output is proportional to the square of the line voltage and consequently the square of the VSWR is measured.

The proportionality between the electric field in the waveguide and the detector output is unknown but usually that is no problem because we are only interested in the VSWR, which is the ratio of two signals picked up at different places. Only when absolute power must be measured is it necessary to know the relationship between line voltage or electric field and the detected signal. The other quantity that we need is the position of the minimum of the standing wave pattern (the maximum would do too of

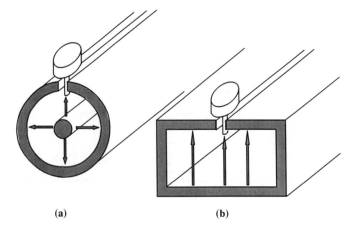

(a)   (b)

Figure 7.26   (a) Coaxial and (b) waveguide slotted lines.

course but usually the minima are much sharper and so their positions are better defined). This can also be disturbed by the probe since the latter, being a discontinuity in the waveguide, produces a reflected wave. This should be minimized by inserting the probe not deeper than is necessary to obtain good sensitivity.

The slotted-line method has been used very much in the past. It will be clear that it meets with difficulties when the magnitude of the reflection coefficient is close to unity, e.g. when an impedance is measured that is mostly reactive with a small resistive part. In this case the minimum of the standing wave pattern is very deep and difficult to measure accurately. As a result the real part of the load impedance is not measured accurately. Similarly great errors can result when the magnitude of the reflection coefficent is small. Then the VSWR is close to unity and small errors in the VSWR can cause large errors in the reflection coefficient. Also, because the slotted-line method needs a lot of manual work other methods have been thought out that give greater accuracy as well as ease of operation.

7.5.4 Directional coupler methods

Instead of directly measuring the standing wave pattern one can couple the incident and reflected wave out with the help of directional couplers (Fig. 7.27). Since at low frequencies it is much easier to measure phase and amplitude than at microwave frequencies the signals from the directional couplers are fed into mixers that downconvert them to intermediate frequency (i.f.) signals. When the same local oscillator signal is used for both mixers and the mixer diodes have identical characteristics the i.f. signals have the same phase and amplitude relationships as the output signals from the directional couplers. The latter in turn have fixed amplitude and phase relationships to the incident and reflected waves on the transmission line. With considerable algebra it is therefore possible to reduce the measured complex ratio of the two i.f. signals to the reflection coefficient of the device under test. Instead of trying to determine all the necessary relationships it

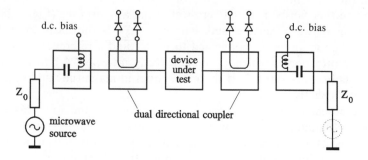

Figure 7.27 Reflection coefficient measurement using directional couplers.

is much simpler to do a calibration measurement first with a known impedance, for instance a short circuit. In the device measurement one then obtains the phase and amplitude relative to those of a short and from these the unknown reflection coefficient can easily be calculated. Measurement errors are introduced in this method by the finite directivity and unequal coupling factors of the directional couplers, and by unequal characteristics of the detection diodes. As far as these errors are linear, i.e. not dependent on the signal amplitude, they can be corrected by extra calibration measurements with two other load impedances. For these one usually takes an open circuit (infinite impedance) and a matched load (impedance equal to Z_0).

In coax or TE_{01} waveguide it is possible to make good short circuits, that is with a reflection coefficient of magnitude unity and located at a precisely defined plane in the transmission line. Since the electric field in these transmission lines is transverse to the propagation direction all one needs is a highly conducting plane transverse to the waveguide axis. Open circuits are much more difficult to make. A coaxial line or a waveguide that is open at the end radiates which means that the end presents an impedance with a resistive part. Therefore it is better to simulate an open circuit by a short at a quarter-wavelength distance. Actually the distance does not have to be exactly equal to a quarter-wavelength as long as it is known. For greatest accuracy, however, it should be close to a quarter-wavelength.

Matched loads, giving no reflection, are also difficult to make. These, however, one can simulate by using a moving load. If one moves this over a wavelength its reflection coefficient moves over a circle in the Smith chart whose radius is determined by the residual reflection and so is small. The results of the calibration maeasurements done with the load at different positions are also on a circle and the centre of this circle is what a perfect load would give.

7.5.5 Bridge methods

A third method to measure a reflection coefficient is to compare it with a known one in a microwave version of a Wheatstone bridge. Especially in waveguide this can be realized very well using a hybrid tee (Fig. 7.28). The decoupling between ports 3 and 4 in Fig. 7.21 is the basis for a bridge method. The unknown is connected to port 1 and a reference to port 2. A signal fed into port 3 will not couple directly into port 4 but only via the reflections in arms 1 and 2. If these are equal in magnitude and phase they will cancel exactly in port 3.

The reference in arm 2 can consist of an attenuator and a moveable short. The best attenuator for this purpose is the so-called *rotary-vane attenuator* which consists of a piece of round waveguide with transitions to rectangular waveguide at both ends. Inside it a dielectric sheet covered with lossy material on one side (graphite or nichrome) can be rotated. When the vane

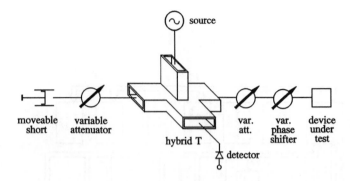

Figure 7.28 Waveguide bridge for measuring reflection coefficient.

is parallel to the transverse electric field the attenuation is at a maximum; when it is orthogonal to the field the attenuation is very low. A carefully constructed moveable short, driven by a micrometer screw, allows accurate phase determination.

7.5.6 Network analyzers

The measurement of S-parameters is in principle not different from that of a single reflection coefficient. In this case usually the directional coupler method is used since it can readily be automated. To extract from the signals produced by the directional couplers the phase and amplitude information they must be transposed to a lower frequency. This can for example be done by mixing them with a local oscillator signal as has been described above. To do this accurately over a wide frequency band is a big problem and therefore it is preferred to employ the sampling principle. Automated setups in which this is done are called network analyzers. The complete setup is shown in Fig. 7.29. For the signal source a sweep oscillator is used, that is an oscillator that can be tuned electronically over a wide band. In the past this always was a backward wave oscillator (Chapter 2) with a bandwidth of 1:2. Nowadays MESFET oscillators are used that cover the band of 2–18 GHz in one sweep. For accurate measurements the source must be frequency stabilized.

To measure transistors they have to be mounted in a special mount, that should have parasitic elements as small as possible. Figure 7.30 shows an example of a transistor mount that can be used for packaged transistors.

A network analyzer is a complicated system and, although every care has been taken to achieve a high accuracy, measuring errors are unavoidable. The main causes are the finite directivity of the directional couplers (40–45 dB), reflections of coaxial connectors and imperfections of the sampling

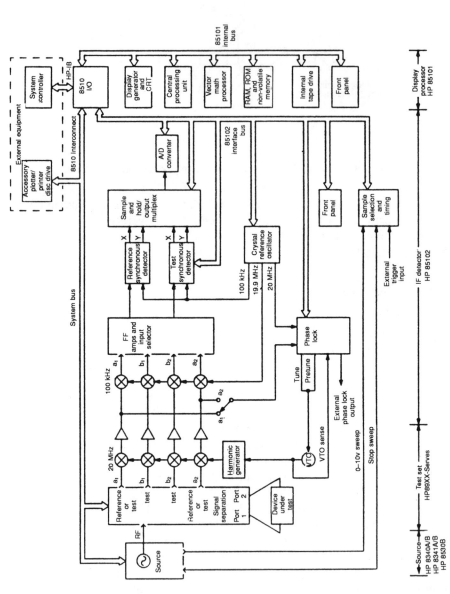

Figure 7.29 Network analyzer setup. Relabeled from *HP 8510B Network Analyzer, Operating and Programming Manual 1990;* published by Hewlett-Packard, 1990.

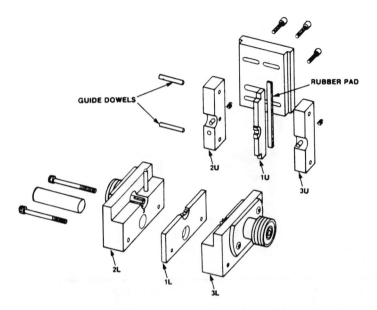

Figure 7.30 Transistor mount for *S* parameter measurements. Reproduced with permission from Lane, R. Q. and McCollum, G. L. (1979) *A New Self-Calibrating Test Fixture*, IEEE 1979 Microwave Symposium Digest, 99–101. © 1979 IEEE.

diodes that convert the signals to the lower frequency. We can represent these errors by error circuits that are inserted between the (ideal) network analyzer and the device under test. The S-parameters of these error circuits can be determined by a number of calibration measurements in which instead of the unknown device a known impedance or twoport is mounted. The device measurements can with the help of these error parameters be corrected using the computer that also takes care of the control of the apparatus.

One-port elements that can be used for calibration are a shorted and an open-ended transmission line and a matched (i.e. reflectionless) load. Since truly reflectionless loads are difficult to make and open-ended lines always have some stray fields, making their terminating impedance less than infinite, shorts are the preferred calibration elements. As calibration two-ports one uses two pieces of transmission line of different lengths. A two-port calibration method that uses a short and two through connections of different length is known as the through-reflect line (TRL) method.

7.5.7 Microwave wafer probers

At low frequencies it is quite common to characterize devices on the wafer directly after processing so that the good ones can be selected for mounting

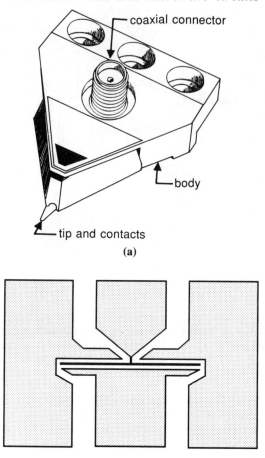

(a)

(b)

Figure 7.31 Microwave wafer probing: (a) probe. Redrawn from *Summit 9000 Analytical Probe Station Instruction Manual 1990*; published by Cascade Microtech, 1990. (b) device layout.

in a package. The long needles used in these probers restrict the frequency region to below 1 MHz. Using coaxial probes the upper limit can be extended to a few gigahertz. For higher frequencies it is necessary to have a transmission line that extends all the way to the device contact pads without reflections. A transmission line that accomplishes this is a coplanar waveguide that ends in small springs which can be put down on the contact pads (Fig. 7.31(a)). Of course this can only work when the device layout is adapted to this measurement technique (Fig. 7.31(b)).

This method makes it possible to characterize the microwave performance of a device without dicing and mounting it and so can save considerable time in a design cycle. An added advantage is that the parasitics introduced by the probe are much smaller than those of a microwave package, so a better picture of the intrinsic device properties is obtained and comparison with theoretical models is made easier.

REFERENCES

1. Getsinger, W. J. (1966) The packaged and mounted diode as a microwave circuit. *IEEE Trans. Microwave Theory Tech.*, **14**, 58–69.
2. Edwards, T. C. (1981) *Foundations for Microstrip Circuit Design*, Wiley, Chichester, New York.
3. Gupta, K. C., Garg, R. and Chadha, R. (1981), *Computer-Aided Design of Microwave Circuits*, Artech House, Dedham, MA.

FURTHER READING

1. Benson, F. A. and Benson, T. M. (1991) *Fields, Waves and Transmission Lines* Chapman & Hall, London.
2. Pozar, D. M. (1990) *Microwave Engineering*, Addison-Wesley, Reading, MA.
3. Liao, S. (1987) *Microwave Circuit Analysis and Amplifier Design*, Prentice-Hall International, Englewood Cliffs, NJ.
4. Laverghetta, T. S. (1988) *Modern Microwave Measurements and Techniques*, Artech, Norwood, MA.
5. Kurokawa, K. (1969) *An Introduction to the Theory of Microwave Circuits*, Academic Press, New York.
6. Matthaei, G. Young, L. and Jones M. E. T. (1980) *Microwave Filters, Impedance Matching Networks and Coupling Structures*, Artech, Dedham, MA.

PROBLEMS

7.1 A 50 Ω transmission line is connected to a negative-resistance device with an impedance of $-3 - j40\,\Omega$. Calculate the VSWR on the line and the point closest to the end of the line where the standing wave has its minimum. Do the same for an impedance of $+3 - j40\,\Omega$.

7.2 If the line in the previous problem is 10 cm long and the dielectric permittivity of the medium filling it is 3.5, calculate the input impedance at frequencies of 3 and 5 GHz, assuming the load impedance is equal to $-3 - j40\,\Omega$ at both frequencies.

7.3 A negative-resistance diode that can be represented by a parallel circuit of a resistance of $-1.5\,\Omega$ and a capacitance of 1 pF is matched to a 50 Ω coaxial line by a parallel connected inductance of 1 nH and a piece of coaxial line with a lower impedance. At what frequency is the impedance of device plus inductance real? Calculate the amplification at this frequency when the connecting line is a quarter-wavelength long at this frequency and its characteristic impedance is 9 Ω. Calculate the frequencies at which the amplification is down 3 dB from the previous value.

7.4 A microstrip transmission line is to be made on a substrate with a thickness of 1.6 mm and a dielectric permittivity of 2.3. What should the width be for a characteristic impedance of 100 Ω?

7.5 An impedance $Z = 25 - j70\,\Omega$ is to be matched to a 50 Ω transmission line. Determine the necessary parameters for a short-circuited series stub and for an open-circuited parallel stub. Note that there can be more than one solution for each case.

7.6   Design a quarter-wave transformer that matches a load of 133 Ω to a 50 Ω line. If the bandwidth is defined as the frequency range for which the VSWR is less than 2, determine this bandwidth in terms of the central frequency.

7.7   A transmission line, short circuited at both ends, is to be used as a half-wavelength resonator. The line wavelength is 5 cm and the damping of the line is 0.01 dB/m. Calculate the unloaded Q. (Hint: consider the impedance seen from a connection at the middle of the line.)

7.8   A microstrip ring resonator has a diameter of 1 cm and the effective permittivity is 3.5. Calculate the lowest resonant frequency of the resonator.

7.9   A directional coupler has a coupling of 10 dB, a directivity of 30 dB and an insertion loss of 0.5 dB. When power is fed into one port and all ports are matched calculate what fraction of the power is output at the other ports.

7.10   If the S-parameters of a transistor in grounded-emitter (source) configuration are given, outline how those of the same transistor in grounded-base (gate) configuration can be calculated.

7.11   To the intrinsic equivalent circuit of a transistor are added series inductances in the three connecting leads. Write down the steps by which the S-parameters of the complete circuit can be calculated from those of the intrinsic transistor.

8

Detection, modulation and frequency conversion

8.1 DETECTION AND MIXING

8.1.1 Detection diodes

The rectifying diode, formerly known as crystal detector, is the oldest microwave semiconductor device and is still in use as a point contact diode, (Fig. 8.1(a)). Its operation is based on the rectifying effect of a metal–semiconductor junction. Modern forms are the planar Schottky diode (Fig. 8.1(b)) and the beam lead diode (Fig. 8.1(c)). p–n junctions are not usable in the microwave range because of the charge storage under forward bias which slows down the response.

The cross-section of Schottky diodes used for detection at the lower microwave frequencies is shown in Fig. 8.1(b)). The substrate usually is about 400 µm thick and the chip size is of the same order. The Schottky contact itself is smaller: it has a diameter of the order of 50 µm microns for

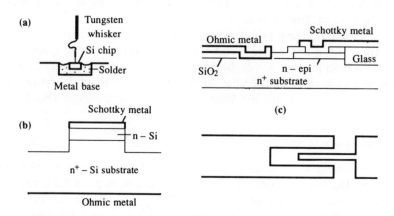

Figure 8.1 (a) Point contact diode, (b) mesa Schottky diode and (c) beam lead diode, cross-section and layout.

diodes used through the X-band. For higher frequencies the dimensions are scaled down to reduce the capacitance. The substrate contact functions as a low-resistance ohmic contact thanks to the high doping of the substrate. In the substrate the high-frequency current will suffer from the skin effect so that the current is pushed towards the sides of the chip. This increases the series resistance, which also includes the contact resistance of the substrate metallization and the resistance of the undepleted n layer. To reduce the latter one could increase the doping of this layer, but this will reduce the width of the depletion layer and so increase the capacitance of the diode. Therefore it is preferred to use a material with high electron mobility such as GaAs or to design the n layer width and doping such that it is fully depleted at zero bias. From equation (5.15) we can infer that this means for the n layer width

$$w_n = \left[\frac{2}{\varepsilon e N_D} (\phi_B - W_F) \right]^{1/2} \quad \text{with } W_F = \frac{kT}{q} \ln\left(\frac{N_C}{N_D} \right).$$

For instance, if $\phi_B = 0.7$ V, $N_D = 10^{17}/cm^3$ and $N_C = 4.2 \times 10^{17}/cm^3$ (typical for GaAs), then w_n should be about 4 μm.

At millimeter waves one prefers to use the beam lead diodes of Fig. 8.1(c), which have lower parasitic inductance and capacitance.

The old point-contact diodes were mounted in coaxial packages of about 1 cm length (Fig. 8.2(a)). Nowadays at frequencies up to about 15 GHz detection diodes are usually mounted in so-called varactor packages (Fig. 8.2(b), top). The chip, which has the mesa form of Fig. 8.1(b), is soldered onto the pedestal and contact with the top is made with a gold bond wire, or preferably a ribbon which has lower inductance. At higher frequencies the packages become smaller, as in Fig. 8.2(b), bottom. These packages are well suited for mounting in a coaxial or waveguide circuit but not for microstrip. In microstrip one can use unpackaged chips which are often made in the beam lead technique (Fig. 8.2(c), top). At extremely high frequencies, i.e. over 100 GHz, the diode diameters become so small (5 μm or less) that contacting with a bond wire is not possible. In this case one resorts to the whisker technique (Fig. 8.2(c), bottom). A great number of small-diameter diodes are spaced close together on a chip and contact is made with a tungsten wire whose point has been sharpened to a radius of about 10 μm.

8.1.2 Detection theory

When we subject a diode to an alternating voltage then, by its non-linear properties, the current contains, in addition to the applied frequency, also a d.c. component and higher harmonics. The direct current can be used for detection of signals, the harmonics for generation of high frequencies. For an analysis of the detection process we take a look at the simple detection circuit of Fig. 8.3.

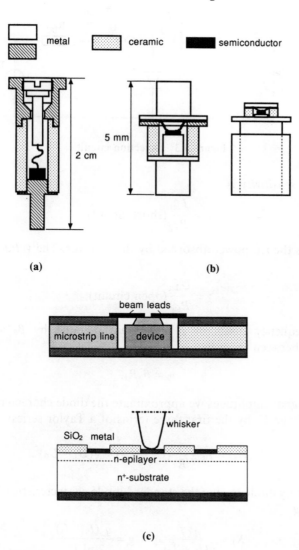

Figure 8.2 Diode packaging and mounting: (a) classical point-contact diode package; (b) modern diode packages (top, 1–15 GHz; bottom, 12–30 GHz); (c) beam lead diode in microstrip (top) and whisker contact (bottom).

Let us first make some definitions. The *junction resistance* is defined as

$$R_J = \frac{dU}{dI}.$$ (8.1a)

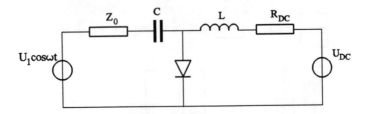

Figure 8.3 Detection circuit.

The *current sensitivity* is

$$\beta = \frac{I_{\text{det}}}{P_{\text{sig}}} \text{(short circuit)} \tag{8.1b}$$

where P_{sig} is the r.f. power absorbed by the detector. The *voltage sensitivity* is

$$\gamma = \frac{U_{\text{det}}}{P_{\text{sig}}} \text{(open circuit).} \tag{8.1c}$$

The low-frequency limits of β and γ are denoted by β_0 and γ_0. The connection between these two is

$$\gamma_0 = \beta_0 R_{\text{J}}. \tag{8.2}$$

For small signal amplitudes we approximate the diode characteristic around the bias voltage U_0 by the first three terms of a Taylor series:

$$I_{\text{D}}(U_{\text{J}}) = I_0 + \frac{1}{R_{\text{J}}}(U_{\text{J}} - U_0) + g(U_{\text{J}} - U_0)^2. \tag{8.3}$$

For a Schottky diode having the standard diode characteristic (5.17) we find for R_{J} and g

$$R_{\text{J}} = \frac{nkT}{q(I_0 + I_{\text{s}})}, \qquad g = \frac{q^2(I_0 + I_{\text{s}})}{2(nkT)^2}. \tag{8.4}$$

R_{J} is for small I_0 in the megohm range and is therefore large compared with R_{s} which is a few ohms.

For a tunnel diode we have, with the I–V characteristic (5.8),

$$R_{\text{J}} = \frac{V_{\text{P}}}{I_{\text{P}}} \frac{V_{\text{P}}}{V_{\text{P}} - U_0} \exp\left(\frac{U_0}{V_{\text{P}}} - 1\right) \tag{8.5}$$

$$g = -\frac{I_{\text{P}}}{2V_{\text{P}}^2}\left(2 - \frac{U_0}{V_{\text{P}}}\right)\exp\left(1 - \frac{U_0}{V_{\text{P}}}\right). \tag{8.6}$$

In this case g comes out as a negative number, at least for $U_0 < V_P$, which is the range for detection applications. However, if we turn the characteristic upside down so that it becomes similar to that of a Schottky diode, g is positive again. In the following we will use this convention.

With $U_J = U_0 + U_{J1} \cos(\omega t)$ we obtain for the current:

$$I_D = I_0 + \frac{1}{R_J} U_{J1} \cos(\omega t) + \tfrac{1}{2} g U_{J1}^2 [1 + \cos(2\omega t)]. \tag{8.7}$$

Because of the non-linearity an extra direct current $\tfrac{1}{2} g U_{J1}^2$ is induced by the modulation. Depending on the external load resistance this current changes the d.c. bias condition. For the d.c. circuit shown in Fig. 8.4, we have, when U_{DC} is the externally applied bias voltage,

$$U_{DC} = U_0 + (I_0 + \tfrac{1}{2} g U_{J1}^2)(R_{DC} + R_S)$$

or, after substituting equation (8.1a),

$$U_0 + (R_{DC} + R_S) I_0(U_0) = U_{DC} - \tfrac{1}{2} g U_{J1}^2 (R_{DC} + R_S). \tag{8.8}$$

This is an implicit equation the solution of which yields U_0. If we have a d.c. open circuit, there is no bias voltage and the total current is zero, which gives

$$I_0(U_0) + \tfrac{1}{2} g U_{J1}^2 R_{DC} = 0. \tag{8.9}$$

In this case the detected voltage is U_0 and can be calculated from this equation. If we have a d.c. short circuit then $R_{DC} = 0$ and equation (8.8) yields

$$U_0 + R_S I_0(U_0) = U_{DC} - \tfrac{1}{2} g U_{J1}^2 R_S. \tag{8.10}$$

Now the detected current is

$$I_{det} = I_0(U_0) - I(U_{DC}) + \tfrac{1}{2} g U_{J1}^2. \tag{8.11}$$

Both equations (8.9) and (8.10) indicate that the bias point is shifted towards negative voltages by the detection process.

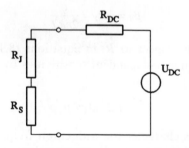

Figure 8.4 d.c. circuit.

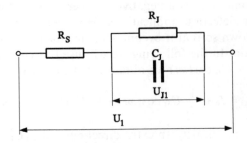

Figure 8.5   Equivalent circuit for detection diode.

For the fundamental harmonic the equivalent circuit, including the parasitic diode resistance and capacitance, is shown in Fig. 8.5. Using this circuit we find for the voltage over the junction

$$U_{J1} = \frac{R_J}{R_S + R_J + j\omega C_J R_J R_S} U_1 \tag{8.12}$$

and for the dissipated power, after some algebra,

$$P_1 = \frac{U_1^2}{2(R_J + R_S)} \frac{1 + \omega^2 C_J^2 R_J R_S R_J/(R_J + R_S)}{1 + [\omega C_J R_S R_J/(R_J + R_S)]^2}. \tag{8.13}$$

In the limit for zero frequency we have for the current sensitivity

$$\beta_0 = \frac{I_{det}}{P_1}$$

and for the voltage sensitivity

$$\gamma_0 = \frac{U_{det}}{P_1}.$$

For Schottky diodes where R_J is much larger then R_S the first two terms in the right-hand side of equation (8.11) cancel approximately and the current sensitivity comes out as

$$\beta_0 = gR_J = \frac{q}{2nkT}. \tag{8.14}$$

If we neglect R_S with respect to R_J in equations (8.12) and (8.13) we can approximate the frequency-dependent sensitivities by

$$\beta = \frac{\beta_0}{1 + \omega^2 C_J^2 R_J R_S}. \tag{8.15}$$

This yields a low-pass characteristic with cutoff frequency $1/2\pi C_J(R_J R_S)^{1/2}$. So for a good high-frequency performance a low junction capacitance as well as a low series resistance are important. Because of the high junction resistance (megohm range) a Schottky diode acts as a current source since in a microwave circuit the load resistance cannot be very high. This means that the current sensitivity is the only one that matters. Also, because of the high value of R_J the denominator of equation (8.15) is much greater than unity so that the microwave sensitivity of Schottky detectors is much lower than at direct current. From equation (8.14) it can be seen that the low-frequency sensitivity increases by reducing the temperature. At microwave frequencies, however, this is counteracted by the increase of R_J. On the other hand, R_J can be made smaller by increasing $I_s + I_0$. For this reason one sometimes gives the diode a forward d.c. bias. This makes an extra d.c. voltage source necessary and moreover the direct current increases the noise of the diode. A better way is to increase I_s. The theory of Schottky diodes given in Chapter 4 tells us that I_s can be increased by decreasing the barrier voltage. Therefore one has looked for metals that give a lower barrier, and hence the name *low-barrier diodes*.

For the backward diode the curvature in the origin is stronger than for a Schottky diode and therefore its sensitivity is greater. Looking at equation (8.4) and considering that V_P is in the millivolt range and I_P in the milliampere range, we see that R_J is small and the approximations used for the Schottky diode are not valid here. Also, the device looks more like a voltage source so that the voltage sensitivity is the quantity to look at. From equation (8.9), using equation (8.5) and (8.6), and assuming that $U_0 \ll V_P$, we obtain for the d.c. voltage sensitivity

$$\gamma_0 = g \frac{R_J^3}{R_S + R_J} = \frac{1}{eI_P} \frac{R_J}{R_S + R_J} \tag{8.16}$$

where e is exp(1). This can be large if I_P is made small by reducing the degeneracy of the p and n dopings.

8.1.3 Noise in detection

In both p–n junction diodes and Schottky diodes the current is given by the thermionic emission formula

$$I_0 = I_s \left[\exp\left(\frac{qU}{nkT}\right) - 1 \right]. \tag{8.17}$$

In fact this is made up of two components, the forward current

$$I_f = I_s \exp\left(\frac{qU}{nkT}\right) = I_0 + I_s$$

and the reverse current

$$I_r = -I_s.$$

Both components produce shot noise and since the noise contributions are uncorrelated the measured noise spectral density defined in Chapter 3 is equal to

$$S_I = 2q(I_0 + 2I_s). \tag{8.18}$$

At zero bias this goes to a limit $2qI_s \Delta f$ and not to zero as one would obtain by simply applying the shot noise formula to the bias current. The noise power spectral density associated with the current is according to equation (4.47)

$$S_P = \tfrac{1}{4} S_I R_J \tag{8.19}$$

where R_J is the differential resistance dU/dI_0 of the junction. It can be found from equation (8.4) and gives

$$S_P = \tfrac{1}{2} nkT \frac{I_0 + 2I_s}{I_0 + I_s}. \tag{8.20}$$

For an ideal diode $n = 1$ and at zero bias $S_P = kT$, i.e. the noise temperature is equal to the device temperature which is what one would expect from a device in thermal equilibrium. At higher bias the noise temperature becomes equal to half the device temperature.

There are three corrections to be made to this simple model. First, real diodes always have some series resistance due to the undepleted semiconductor regions and this series resistance produces thermal noise. Second, the saturation current I_s not only is produced by electrons injected over a barrier but also can contain contributions from other mechanisms, such as recombination of electrons and holes. This is particularly true for p–n junction diodes. In good quality Schottky diodes equation (8.18) may give a reasonably accurate description of the noise properties. Third, in Schottky barriers there is also a noise contribution from electrons tunneling through the barrier. The noise power from this source is not temperature dependent and it can be derived [1] that it contributes a noise temperature

$$T_{tunn} = \frac{\hbar q}{2k} \left(\frac{N_D}{\varepsilon m} \right)^{1/2}. \tag{8.21}$$

This will give a lower limit to the noise temperature when the device temperature is reduced. It is of the order of 50 K. Above this the detector noise can strongly be reduced by lowering the temperature.

8.1.4 Mixing

Detection diodes are also used for heterodyne mixing, that is the conversion of a signal to another frequency. A simple mixing circuit is drawn in Fig. 8.6. In addition to the signal with frequency f_s one also has a local oscillator with frequency f_{lo}. The amplitude of the local-oscillator voltage is much

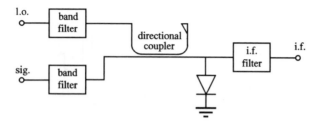

Figure 8.6 Mixer circuit.

greater than that of the signal. From equation (8.3) for the diode current we obtain then (bias point $U_0 = 0$)

$$I = \frac{1}{R_J} [U_{lo} \cos(\omega_{lo}t) + U_s \cos(\omega_s t)] + g[U_{lo} \cos(\omega_{lo}t) + U_s \cos(\omega_s t)]^2.$$

The last term yields a product $2gU_{lo}U_s \cos(\omega_{lo}t) \cos(\omega_s t)$ that we can write as

$$gU_{lo}U_s\{\cos[(\omega_{lo} - \omega_s)t] + \cos[(\omega_{lo} + \omega_s)t]\}.$$

In general we want to use the signal with the difference frequency, but it also happens that one wants to bring the signal to a higher frequency (upconversion). In all cases the converted signal amplitude is proportional to U_{lo}, which opens up the possibility of obtaining conversion gain.

If the difference frequency is zero (which means the signal modulation is converted to base band) one speaks of *homodyne mixing*, in the other case of *heterodyne mixing*.

In addition to the desired intermediate-frequency component other frequencies appear in the output signal and these must be filtered out. The spectrum is sketched in Fig. 8.7 for the case where the local-oscillator frequency is higher than the signal frequency. In communication systems,

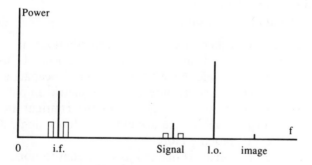

Figure 8.7 Frequency spectrum of a mixer.

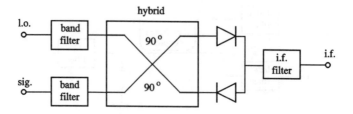

Figure 8.8 Balanced mixer circuit.

that which is called the 'signal' does not consist of a single frequency but of a central frequency with sidebands arising from modulation of the carrier wave. The sidebands contain the information that is to be transmitted. Since the frequency spectrum of the modulated signal is symmetric with respect to the central frequency often one sideband is filtered out before transmission to make more economic use of the electromagnetic spectrum. This is called *single-sideband transmission*, the other case *double-sideband transmission*.

In addition to the desired signal around frequency f_s, the noise at this frequency and furthermore signals and noise at a frequency $f_{im} = f_{lo} + f_{if}$, the *image frequency*, are converted to the intermediate frequency $f_{if} = f_{lo} - f_s$. On the other hand, part of the desired signal is *upconverted* to the image frequency by mixing with the second harmonic of the intermediate frequency.

Important figures of merit for mixers are the *conversion loss* and the *noise figure*. Conversion loss is defined as the signal power available at the input divided by i.f. power at the output. It consists of

- absorption in the non-linear element,
- power lost into harmonics,
- absorption in series resistance,
- reflected signal at the mixer input, and
- reflected signal at the i.f. output port.

In some types of mixers there are also gain contributions, in particular in superconductive mixers and mixers where a transistor is used as the non-linear element. In diode mixers the non-linear capacitance (if the non-linearity is strong) can contribute some parametric amplification. The conversion loss of diode mixers depends on the configuration used, as described below. Typical values are around 6 dB for a single-diode mixer and 4 dB for mixers employing more diodes.

The noise figure of a mixer is defined as the signal-to-noise ratio at the signal input divided by that at the i.f. output, where the input noise is supposed to come from a matched noise source with a noise temperature equal to a reference temperature T_0. One has to distinguish the *single-side-band noise figure* NF_{SSB} which is measured in a configuration where signal from only one sideband is allowed to reach the mixer from the *double-side-band noise figure* NF_{DSB} where both sidebands are included. Since a diode mixer does not remove any noise it will reduce the signal-to-noise ratio by at least its conversion loss and consequently the noise figure will at least be equal to the conversion loss.

The noise contribution of the mixer can come from the following sources:

- the non-linear element;
- the circuit (thermal noise from its lossy components);
- the local oscillator;
- the i.f. amplifier.

Also, in a single-sideband system the noise comes from both sidebands whereas the signal is present in only one sideband. This means that always $NF_{SSB} > NF_{DSB}$. When the signal and image channels are equivalent then the difference is exactly a factor of 2.

The unwanted signals at the image frequency that arrive via the signal line can be filtered out, in which case one speaks of an *image rejection mixer*. This reduces the noise figure (in the ideal case it would make NF_{SSB} equal to NF_{DSB}) as well as the interference from unwanted signals. The filter can be designed such that it reflects the upconverted signal back to the mixer diode. If the reflection has the correct phase it is converted to an i.f. voltage in phase with the main i.f. signal so the latter is enhanced. Consequently this is called an *image enhancement mixer*.

The noise sidebands of the local oscillator are not filtered out by the image rejection filter. To eliminate these one has to use a balanced mixer circuit, of which the principle is shown in Fig. 8.8. In one branch the local oscillator is given a phase shift of 90° and in the other the signal. Moreover, the diodes in the two branches are oppositely polarized.

For the i.f. component caused by the signal we have now after mixing

$$2U_{lo}U_s \cos(\omega_{lo}t - \pi/2) \cos(\omega_s t) - 2U_{lo}U_s \cos(\omega_{lo}t) \cos(\omega_s t - \pi/2)$$

$$= 2U_{lo}U_s[\sin(\omega_{lo}t) \cos(\omega_s t) - \cos(\omega_{lo}t) \sin(\omega_s t)]$$

$$= 2U_{lo}U_s \sin[(\omega_{lo} - \omega_s)t]$$

which is the desired signal at the intermediate frequency.

The sideband noise at the signal frequency yields as a mixing product

$$2U_{lo}U_s \cos(\omega_{lo}t - \pi/2) \cos(\omega_s t - \pi/2) - 2U_{lo}U_s \cos(\omega_{lo}t) \cos(\omega_s t)$$

$$= 2U_{lo}U_s[\sin(\omega_{lo}t) \sin(\omega_s t) - \cos(\omega_{lo}t) \cos(\omega_s t)]$$

$$= -2U_{lo}U_s \cos[(\omega_{lo} + \omega_s)t];$$

that is, there is no mixing product at the intermediate frequency.

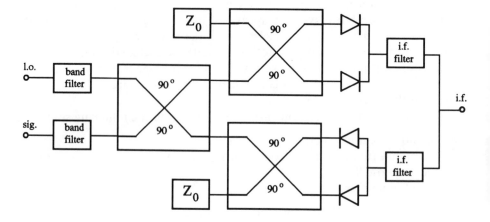

Figure 8.9   Double-balanced mixer.

The circuit with two diodes described above is called a *single-balanced mixer*. One can combine two of these with a phase difference of 180° into a *double-balanced mixer* (Fig. 8.9). Now the local oscillator and signal ports are completely decoupled so that no local signal power can leak out of the signal port and vice versa. Also, the even harmonics of signal and local oscillator are cancelled, leading to reduced intermodulation distortion.

Some practical realizations of microwave mixers are shown in Fig. 8.10. The balanced microstrip mixer in the left of the figure employs a rat-race hybrid (Chapter 6) to feed the signal and the local oscillator in the proper phase to both diodes. The circuit on the right is used at the high millimeter wave frequencies. It is a single-ended mixer in which the signal and the local oscillator are combined simply by radiation into a wide horn antenna which is connected to the diode by a stepped waveguide section acting as an impedence transformer. The diode is connected by a whisker as in Fig. 8.2(c).

8.2   MODULATION, SWITCHING AND CONTROL

8.2.1   Introduction

To transmit information, carrier waves have to be modulated. The properties that can be modulated are amplitude, frequency and phase. Switching a signal from one channel to another can be regarded as an extreme form of amplitude modulation. Control is also a form of modulation with the aim of keeping frequency or amplitude or phase constant. Amplitude can be modulated or controlled via the absorption of a network, i.e. via the real part of the impedance of a network element. Frequency and phase can be

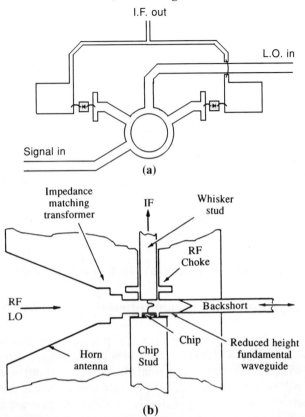

Figure 8.10 Microwave mixers: (a) balanced microstrip; (b) single-ended waveguide. Reproduced with permission from Clifton, B. J., Receiver technology for the millimeter and submillimeter wave regions in *Integrated Optics and Millimeter and Microwave Integrated Circuits*, **317**, 339–47; published by SPIE, 1981.

modulated through the reactive part of an impedance. So there is a need for impedances whose real or imaginary part can be varied without varying the other part. The most used are p–i–n diodes and varactor diodes.

8.2.2 Amplitude modulation and switching

For these applications p–i–n diodes are used. They are p–n junctions with an additional intrinsic layer and have been described in section 5.2.3. The capacitance is mostly determined by the width of the p–i–n layer and is therefore nearly constant. On the other hand, the real part of the microwave impedance is strongly dependent on the d.c. bias and this makes p–i–n diodes ideally suited for amplitude modulation purposes. From equation (5.7) we may conclude that the conductance of a p–i–n diode is proportional

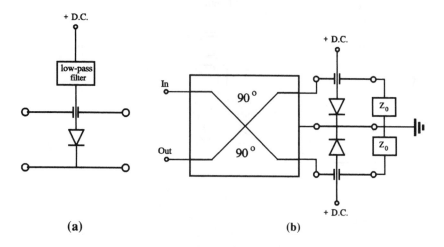

Figure 8.11   p–i–n diode (a) switch and (b) attenuator.

to the carrier density which in turn is roughly proportional to the direct current.

A switch can be made by mounting a p–i–n diode transversely in a transmission line (Fig. 8.11(a)). At zero bias its impedance will be high and the signals will pass through unattenuated. At high bias current the diode's impedance will be very low so that it presents a short circuit across the line which reflects all signals. An isolator is necessary to prevent the reflected signal from going back to the source. The same circuit can be used as a variable attenuator by choosing an intermediate bias current at which part of the signal is reflected and part is transmitted.

A more elegant solution is provided by using a 90° hybrid (section 6.4), (Fig. 8.11(b)). Ports 3 and 4 are loaded with a p–i–n diode in parallel with a matched load. At zero bias again the p–i–n diode has a very high impedance and the signals are absorbed by the matched loads. Increasing the bias means that part of the signals is reflected. Now, if a signal is fed into port 1 the signal coming back from port 4 has undergone a phase shift of 90° twice whereas the signal coming back from port 3 has zero phase shift. So these signals cancel in port 1, that is if they are of equal amplitudes. On the contrary, signals coming back into port 2 have both undergone a 90° phase shift so they add up. The result is that a signal fed into port 1 will be returned into port 2 with an attenuation determined by the bias of the p–i–n diodes and no signal will be reflected into port 1. Of course, for this principle to work it is necessary that both diodes have equal characteristics.

As switches the circuits above only switch one channel in and out. It is often necessary to switch a signal between two or more channels, e.g. in mobile communication where transmitter and receiver use the same an-

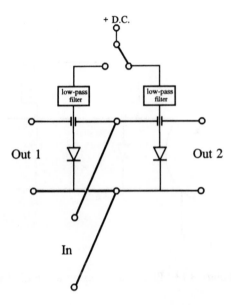

Figure 8.12 Single-pole double-throw switch.

tenna. A possible circuit for this application is sketched in Fig. 8.12(a). When diode 1 is biased it forms a short circuit and because of the quarter-wavelength line it appears as an open circuit at the junction, so signals can pass freely between lines 2 and 3, the diode in the former being unbiased. The reverse situation occurs when diode 2 is biased and diode 1 is not.

Digital phase shifters also use p–i–n diode switches. Here digital means that the phase is changed in fixed increments. An example is shown in Fig. 8.13. By biasing one of the diodes the line is short-circuited at a certain length and the reflected signal obtains a corresponding phase shift in increments of $(2 \, \Delta L/\lambda) \times 360°$.

8.2.3 Frequency modulation

Once a signal has left an oscillator or amplifier it can be modulated in amplitude but not in frequency, so frequency modulation has to be done at the oscillator itself. This type of modulation will therefore be treated in Chapter 10.

8.3 FREQUENCY MULTIPLICATION

As the frequency goes up and especially at millimeter wave frequencies it becomes increasingly difficult to build oscillators with sufficient power

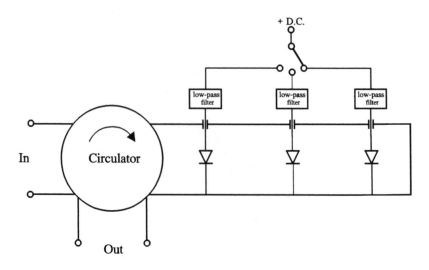

Figure 8.13 Phase shifter using p–i–n diodes.

output. Even if an oscillator is available its frequency stability may be such that one prefers to derive the signal from a lower-frequency stabilized source. In either of these cases frequency multiplication may offer a solution. If a non-linear element, be it a resistance or a capacitance, is subjected to a sinusoidal voltage then the current will contain higher harmonics of the input frequency. So in principle every non-linear element can be used as a frequency multiplier. Diodes that can be used are varactor diodes, step-recovery diodes and tunnel diodes. Devices with symmetrical (I–V or C–V) characteristics produce only odd harmonics; with asymmetric characteristics even harmonics are produced also. The principle of a diode multiplier is sketched in Fig. 8.14. Filters have been applied to keep harmonics out of the fundamental circuit and the fundamental from reaching the output. Since a multiplier is inherently a non-linear device it is important that the circuit provides the right terminations not only at the fundamental frequency and the desired harmonics but also at the other harmonics insofar as they have a significant amplitude. When the unwanted harmonics are fed back into the diode they mix again with the fundamental

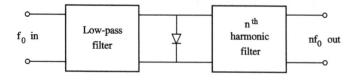

Figure 8.14 Diode frequency multiplier.

and other harmonics producing extra signal at the wanted frequency. So the operation of the multiplier can be improved by proper harmonic terminations.

Frequency doubling is applied most of the time since the efficiency of multipliers decreases rapidly with increasing harmonic number. An exception is the step-recovery diode (section 5.2.3), which is rich in higher harmonics.

In the case of varactor diodes the condition that there is no power dissipation in the device leads to the Manley–Rowe relations [2]:

$$\sum_{n=-\infty}^{\infty} \sum_{m=0}^{\infty} \frac{mP_{mn}}{m\omega_1 + n\omega_2} = 0 \qquad (8.22a)$$

$$\sum_{m=-\infty}^{\infty} \sum_{n=0}^{\infty} \frac{nP_{mn}}{m\omega_1 + n\omega_2} = 0 \qquad (8.22b)$$

where ω_1 and ω_2 are the frequencies of the signals supplied to the diode and P_{mn} is the power dissipated at a frequency $m\omega_1 + n\omega_2$. In the case of a multiplier there is no signal applied at ω_2 so only $n = 0$ applies. Then equation (8.22b) is identically zero and equation (8.22a) simplifies to

$$\sum_{1}^{\infty} P_m = 0. \qquad (8.23)$$

In other words, the total power extracted at all harmonics is equal to the power P_1 put in at the fundamental frequency. So if one manages to reflect all undesired harmonics back to the diode a varactor multiplier can have a theoretical efficiency of 100%.

Varactor diodes have been treated in section 5.3.2. The equivalent circuit is in principle the same as that of a detection diode, shown in Fig. 8.5, except that the junction resistance of a varactor is so large that it can be neglected. Hence one may conclude that it must have a cutoff frequency $f_c = 1/2\pi R_S C_J$. In multiplier applications, however, C_J is modulated between two extremes C_{Jmin} and C_{Jmax}. In this case the cutoff frequency is given by [3]

$$f_c = \frac{1}{2\pi R_S} \left(\frac{1}{C_{Jmin}} - \frac{1}{C_{Jmax}} \right). \qquad (8.24)$$

In any case, to make f_c high R_S must be kept small. For this reason varactor diodes are preferably made of GaAs because of its high electron mobility which gives a lower parasitic series resistance. Cutoff frequencies as high as 1000 GHz can be reached. The frequency dependence of varactor performance is expressed by the *quality factor* $Q = f_c/f$. Because of the finite quality factor the multiplier efficiency is always lower than 100%, especially when the output frequency is close to the cutoff frequency. As an example, let us take a simple uniformly doped Schottky varactor whose differential capacitance is given by equation (5.16). Integrating this gives the charge on the capacitor as a function of voltage:

$$Q = 2\varepsilon A[qN_D(V_D + U)]^{1/2}. \tag{8.25}$$

Suppose now that there is a time-dependent charge given by

$$Q(t) = Q_0 + Q_1 \sin(\omega t) + Q_n \sin(n\omega t). \tag{8.26}$$

The capacitive current is the time differential of this:

$$I(t) = \omega Q_1 \cos(\omega t) + n\omega Q_n \cos(n\omega t). \tag{8.27}$$

The voltage can using equation (8.26) be written as

$$
\begin{aligned}
U = -V_D + \frac{1}{qN_D(2\varepsilon A)^2}\Big(& Q_0^2 + 2Q_0Q_1 \sin(\omega t) + 2Q_0Q_n \sin(n\omega t) \\
& + Q_1Q_n\{\cos[(n+1)\omega t] - \cos[(n-1)\omega t]\} \\
& + \tfrac{1}{2}Q_1^2[1 - \cos(2\omega t)] + \tfrac{1}{2}Q_n^2[1 - \cos(2n\omega t)]\Big).
\end{aligned} \tag{8.28}
$$

The first three terms in the large parentheses contain the linear capacitive contributions which are 90° out of phase with the current so they do not dissipate or generate power. In the other terms there is none with frequency $n\omega$, however, there is one with 2ω. This implies that a varactor with this particular *C–V* characteristic cannot generate higher harmonics than the second. However, the fourth term equation of (8.28) (the one with Q_1Q_n)) shows that current at a frequency $n\omega$ produces a voltage at $(n + 1)\omega$, so higher harmonics can be generated if one allows currents to flow at the intermediate harmonics. These are called *idler currents*. Care should be taken that they are not dissipative. This can be ensured by designing the circuit to provide purely reactive terminations at the intermediate harmonics.

Like mixers, it can also be advantageous to use multidiode circuits for multipliers. A simple two-diode multiplier circuit is shown in Fig. 8.15. By symmetry considerations one can derive that in this circuit only even harmonics are produced. When one of the diodes is turned around one obtains only odd harmonics.

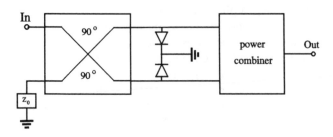

Figure 8.15 Balanced multiplier circuit.

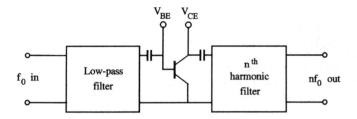

Figure 8.16 Bipolar transistor multiplier circuit.

Transistors can also be used as frequency doublers and offer amplification of the output signal as an extra bonus. The desired output frequency should be lower than the cutoff frequency otherwise the efficiency is very low. With bipolar transistors one uses the non-linear C–V characteristic of the collector–base junction. The fundamental frequency signal is fed into the emitter–base junction and produces a modulation of the collector voltage. The latter modulates the collector–base capacitance, producing the desired second harmonic. The transistor is used in a common-emitter circuit and biased in class B or C so that it draws no current in the absence of an a.c. signal. The basic circuit is shown in Fig. 8.16. The output filter rejects the fundamental frequency. It should be placed at such a distance that the resulting standing wave has a maximum at the collector.

GaAs MESFETs can be used for frequency doubling by employing the non-linearity of the I_D–V_{GS} characteristic (equation (6.12)). From this equation it can be seen that the non-linearity is strongest when $V_{GS} \approx V_T$, that is, close to pinch-off. In fact, biasing the transistor at this point, one actually uses it as a half-wave rectifier. The basic circuit for a MESFET multiplier is the same as that given for a bipolar transistor above, using the MESFET in a common-source configuration.

REFERENCES

1. Kollberg, E. (1988) In *Microwave Physics and Technique* (A. Y. Spasov ed.), World Scientific, Singapore.
2. Watson, H. A. (ed.) (1969) *Microwave Semiconductor Devices and Their Circuit Applications*, McGraw-Hill, New York.
3. Kollberg, E. (1991) *New Solid State Devices and Circuits for Millimeter Wave Applications*. Proc. Eur. Microwave Conf. 1991, MEPL, Tunbridge Wells.

FURTHER READING

1. Combes, P. F., Graffeuil, J. and Sauterau, J. F. (1989) *Microwave Components, Devices and Active Circuits*, Wiley, Chichester, New York.
2. Kollberg, E. L. (ed.) (1984) *Microwave and Millimeter Wave Mixers*, IEEE, New York.

3. Maas, S. A. (1986) *Microwave Mixers*, Artech, Dedham, MA.
4. White, J. F. (1982) *Microwave Semiconductor Engineering*, Van Nostrand Reinhold, New York.
5. Garver, R. V. (1976) *Microwave Diode Control Devices*, Artech, Dedham, MA.
6. Pengelly, R. S. (1986) *Microwave Field Effect Transistors – Theory, Design and Applications*, Wiley, Chichester, New York.
7. Soares, R., Graffeuil, J. and Obregon, J. (eds) (1983) *Applications of GaAs Mesfets*, Artech, Dedham, MA.

PROBLEMS

8.1 Calculate the current sensitivity of an ideal Schottky diode having $I_s = 1$ nA and $C_J = 1$ pF, at frequencies of 10 kHz and 10 GHz.

8.2 Calculate the voltage sensitivity of a backward diode having $V_p = 0.1$ V, $I_p = 1$ mA and $C_J = 1$ pF.

8.3 When the Schottky diode and the backward diode of the preceding problems are both terminated in a 1 kΩ resistor, which one is the more sensitive?

8.4 Calculate the mixing products of the circuit of Fig. 8.9 assuming ideal couplers and quadratic detectors.

8.5 Do the same for the circuit in Fig. 8.10(a) (Fig. 7.23 also should be referred to).

8.6 The isolation of a p–i–n diode switch in the configuration of Fig. 8.11(a) is defined as the incoming power divided by the transmitted power. For a 50 Ω line calculate the p–i–n resistance necessary to achieve an isolation of 20 dB.

8.7 Assume $n = 2$ in equation (8.26) and calculate the power at the frequencies ω and 2ω for the multiplier described by equations (8.24)–(8.28).

9

Amplifiers

9.1 INTRODUCTION

Microwave amplifiers are distinguished as small-signal and power amplifiers, according to their applications. Small-signal amplifiers are widely used in receivers as pre-amplifiers preceding the heterodyne mixer. For this type of amplifier the important requirement is a low noise figure. Often a great bandwidth is also important, in particular for pulse-modulated signals (radar). Power amplifiers are used in transmitters, repeater stations and communication satellites. In these systems, when the power is not too great, say 20 W, transistor amplifiers can be used; for greater powers vacuum tubes are still necessary. In particular, traveling wave tubes are still widely used because of their large bandwidth and high amplification.

In this chapter we will concentrate on semiconductor amplifiers. In this category we can distinguish reflection amplifiers and transmission amplifiers. In the first the active element is a diode whose real part of the impedance is negative; in the second it is a transistor.

9.2 REFLECTION AMPLIFIERS

9.2.1 Negative-resistance amplifiers

Negative-resistance amplifiers rely on the fact that the reflection coefficient of an impedance with a negative real part at the end of a transmission line is greater than 1. This we can see with the help of equation (7.13) if we substitute $Z = -R + jX$:

$$\Gamma = \frac{-R + jX - Z_0}{-R + jX + Z_0} \quad \text{so} \quad |\Gamma|^2 = \frac{(R + Z_0)^2 + X^2}{(R - Z_0)^2 + X^2} > 1. \tag{9.1}$$

Because Γ is the voltage reflection coefficient, the power amplification, being the ratio of reflected to incident power, is equal to $|\Gamma|^2$. This can be large when $R \approx Z_0$ and $X \approx 0$. In practice the values of R and X are given by the device used and Z_0 is preferably kept at a standard value, e.g. 50 Ω. Then it is necessary to insert a matching circuit (also discussed in Chapter 7)

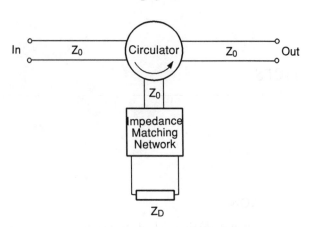

Figure 9.1 Reflection amplifier.

between Z and the transmission line which transforms Z as closely as possible to the desired value. If the matching is such that the transformed $R = Z_0$ and the transformed $X = 0$, then $|\Gamma|$ is infinite and we have an oscillator. The same matching techniques used for amplifiers can therefore also be used for oscillators. For the amplifier to be useful the incoming and reflected signals must be separated. This is done with a circulator. A complete reflection amplifier is shown schematically in Fig. 9.1. The active element in a reflection amplifier can be a tunnel diode or a Gunn diode when low noise is essential, and an IMPATT diode if high power is necessary and low noise is less important. The properties of these devices have been discussed in Chapter 4. The availability of both low-noise and low-power transistors has led to a diminishing use of diodes at frequencies below 40 GHz and this limit is shifting upwards steadily. For power amplifier at millimeter waves IMPATT diodes are still employed.

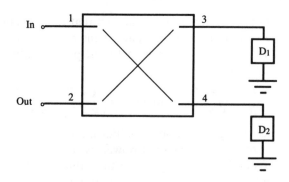

Figure 9.2 Two-diode amplifier using a 90° hybrid.

Instead of a circulator a 90° hybrid can also be used to separate incoming and outgoing signals. This is especially useful when the output of two diodes has to be combined. The principle is illustrated in Fig. 9.2. The signal entering port 1 is divided equally over ports 3 and 4 but with an extra phase delay of 90° in port 4. The signal reflected from port 3 into port 1 will now have zero phase shift and the signal reflected from port 4 into port 1 will have acquired an extra 90° phase shift, that is 180° in total. The result is that the signal returning into port 1 via port 4 is 180° phase shifted with respect to that returning via port 3 and so the two will cancel if they have equal amplitudes, that is if the diodes provide equal amplifications. In the same way we can derive that the signals returning into port 2 both have acquired 90° phase delay so they will add up. The result is that ports 1 and 2 are decoupled.

9.2.2 Parametric amplifiers

A special type of reflection amplifier is the parametric amplifier. Here the active element is a varactor diode (section 5.3.2) that is pumped with a signal of higher frequency. The principle of the parametric amplifier is sketched in Fig. 9.3. The varactor's capacitance is modulated by a signal at frequency f_1. The LCR circuit is together with the capacitance of the varactor resonant at a frequency f_3 $(<f_1)$. The whole circuit now is found to possess an impedance with a negative real part at a frequency $f_2 = f_1 - f_3$ and can therefore be used for amplification or oscillation.

To understand how this works we apply small-signal theory. Suppose the varactor is biased by a d.c. voltage V_0 and that in addition there are three a.c. components:

$$v_1(t) = V_1 \cos(\omega_1 t) \tag{9.2a}$$

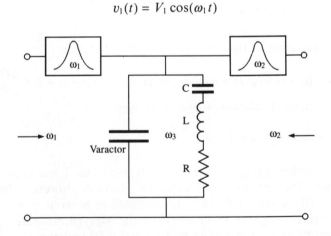

Figure 9.3 Parametric amplifier.

$$v_2(t) = V_2 \cos(\omega_2 t) \tag{9.2b}$$

$$v_3(t) = V_3 \cos(\omega_3 t + \phi) \tag{9.2c}$$

where v_1 is called the *pump voltage*, v_2 the *signal voltage* and v_3 the *idler voltage*. Furthermore $\omega_1 = \omega_2 + \omega_3$. Now the charge on the varactor Q is a function of the total voltage and, assuming that $|v_1|$, $|v_2|$ and $|v_3|$ are small compared with V_0, we can expand it in a Taylor series to second order:

$$Q(V_0 + v_1 + v_2 + v_3)$$

$$= Q(V_0) + (v_1 + v_2 + v_3)\frac{dQ}{dV} + \tfrac{1}{2}(v_1 + v_2 + v_3)^2\frac{d^2Q}{dV^2}$$

$$= Q_0 + q(t). \tag{9.3}$$

Using equations (9.2) and retaining only the components with frequencies ω_1, ω_2 and ω_3 this can be written as

$$q(t) = [V_1 \cos(\omega_1 t) + V_2 \cos(\omega_2 t) + V_3 \cos(\omega_3 t + \phi_3)]\frac{dQ}{dV}$$

$$+ [V_1 V_2 \cos(\omega_1 t) \cos(\omega_2 t) + V_1 V_3 \cos(\omega_1 t) \cos(\omega_3 t + \phi_3)$$

$$+ V_2 V_3 \cos(\omega_2 t) \cos(\omega_3 t + \phi_3)]\frac{d^2Q}{dV^2}. \tag{9.4}$$

By applying well-known trigonometric relationships this can, leaving out the terms at the sum frequencies, be reduced to

$$q(t) = [V_1 \cos(\omega_1 t) + V_2 \cos(\omega_2 t) + V_3 \cos(\omega_3 t + \phi_3)]\frac{dQ}{dV}$$

$$+ \tfrac{1}{2}[V_1 V_2 \cos(\omega_3 t) + V_1 V_3 \cos(\omega_2 t - \phi_3) \tag{9.5}$$

$$+ V_2 V_3 \cos(\omega_1 t + \phi_3)]\frac{d^2Q}{dV^2}.$$

Time differentiation of this gives the alternating current:

$$i(t) = -[\omega_1 V_1 \sin(\omega_1 t) + \omega_2 V_2 \sin(\omega_2 t) + \omega_3 V_3 \sin(\omega_3 t + \phi_3)]\frac{dQ}{dV}$$

$$- \tfrac{1}{2}[\omega_3 V_1 V_2 \sin(\omega_3 t) + \omega_2 V_1 V_3 \sin(\omega_2 t - \phi_3) \tag{9.6}$$

$$+ \omega_1 V_2 V_3 \sin(\omega_1 t + \phi_3)]\frac{d^2Q}{dV^2}.$$

The terms multiplying dQ/dV are obviously due to the linear capacitance of the varactor. They produce only capacitive current. However, the second-order terms (those with d^2Q/dV^2) can dissipate or generate power, depending on the phase angle ϕ. If we calculate the average dissipated power at each frequency we find, denoting by $i_k(t)$ and $v_k(t)$ the current and voltage components at frequency ω_k ($k = 1, 2, 3$),

$$P_1 = \overline{i_1(t)v_1(t)} = -\tfrac{1}{4}\omega_1 V_1 V_2 V_3 \frac{d^2 Q}{dV^2} \sin(\phi_3) \qquad (9.7a)$$

$$P_2 = \overline{i_2(t)v_2(t)} = +\tfrac{1}{4}\omega_2 V_1 V_2 V_3 \frac{d^2 Q}{dV^2} \sin(\phi_3) \qquad (9.7b)$$

$$P_3 = \overline{i_3(t)v_3(t)} = +\tfrac{1}{4}\omega_3 V_1 V_2 V_3 \frac{d^2 Q}{dV^2} \sin(\phi_3). \qquad (9.7c)$$

Note that $P_1/\omega_1 + P_2/\omega_2 = 0$ and $P_1/\omega_1 + P_3/\omega_3 = 0$, so the Manley–Rowe relations (section 8.3) are satisfied. Also $P_1 + P_2 + P_3 = 0$ because $\omega_1 = \omega_2 + \omega_3$ so that energy is conserved. Now the question is what determines the phase ϕ_3? Well, if the idler circuit (the series LCR circuit in Fig. 9.3) is *dissipative* then the varactor diode has to *generate* energy at the idler frequency ω_3, so $P_3 < 0$ and as a consequence of equation (9.7c) we have

$$\frac{d^2 Q}{dV^2} \sin(\phi_3) < 0.$$

Looking at equations (9.7a) and (9.7b) we immediately see that in this case $P_1 > 0$ and $P_2 < 0$, that is the pump delivers power to the varactor and at the frequency ω_2 power can be extracted so amplification is possible at this frequency. All we have to do is to take care that the idler circuit is dissipative.

The theoretical efficiency of the circuit is

$$\eta = \frac{P_2}{P_1} = \frac{\omega_2}{\omega_1} < 1. \qquad (9.8)$$

Evidently it is advantageous to choose ω_2 as close as possible to ω_1 but this cannot be driven too far since the two signals have to be kept apart by band filters.

To ensure that the idler circuit is dissipative it is not necessary to include a resistance as shown in Fig. 9.3 because the varactor itself always has some series resistance which is part of the idler circuit too. Of course power at ω_1 and ω_2 is also dissipated in this resistance so that the efficiency will be lower than the theoretical maximum given by equation (9.8).

From the simple model above we can also calculate the impedance of the circuit at the frequency ω_2. First we note that i_3 is related to v_3 by the impedance of the resonant circuit. If ω_3 is the resonant frequency of this circuit its impedance is equal to R and we can write

$$i_3(t) = -\frac{V_3 \cos(\omega_3 t + \phi_3)}{R}. \qquad (9.9a)$$

On the other hand, from equation (9.6) we have

$$i_3(t) = -\omega_3 V_3 \sin(\omega_3 t + \phi_3)\frac{dQ}{dV} - \tfrac{1}{2}\omega_3 V_1 V_2 \sin(\omega_3 t)\frac{d^2 Q}{dV^2}. \qquad (9.9b)$$

Comparing these two expressions and remembering that $V_{1,2,3}$ are defined as real quantities we find

$$\phi = -\frac{\pi}{2} + \arctan\left(\omega_3 R \frac{dQ}{dV}\right) \tag{9.10a}$$

$$V_3 = \frac{\omega_3 V_1 V_2 R \, d^2Q/dV^2}{2[1 + (\omega_3 R \, dQ/dV)^2]^{1/2}}. \tag{9.10b}$$

Substituting this in equation (9.6) the current at ω_2 can be written as

$$i_2(t) = -\omega_2 V_2 \sin(\omega_2 t) \frac{dQ}{dV} - \frac{\omega_2 \omega_3 V_1^2 V_2 R (d^2Q/dV^2)^2}{2[1 + (\omega_3 R \, dQ/dV)^2]^{1/2}} \sin(\omega_2 t - \phi_3), \tag{9.11}$$

from which we find the admittance at frequency ω_2, after some algebra, as

$$Y_2 = j\omega_2 \frac{dQ}{dV} - \frac{\omega_2 \omega_3 V_1^2 R (d^2Q/dV^2)^2}{2(1 + j\omega_3 R dQ/dV)}. \tag{9.12}$$

The first term is the linear capacitance of the varactor, and the second term is due to its non-linear behavior and contributes a negative real part as well as an extra capacitance.

The advantage of the parametric amplifier is that it contains few resistive parts so its noise is low. The main sources of noise are the varactor series resistance and the resistance of the idler circuit. Their contributions, being thermal noise, can be reduced greatly by cooling the amplifier. Disadvantages of the parametric amplifier are the necessity of an oscillator at a higher frequency (the so-called pump) and a tendency to burst spontaneously into oscillation.

9.3 TRANSMISSION AMPLIFIERS

9.3.1 Introduction

Transmission amplifiers make use of transistors or, for higher powers, of tubes. Tube amplifiers will not be discussed here. Transistor amplifiers are almost exclusively built in microstrip because connecting a device with three terminals is much simpler in this technology than in coax or waveguide. At lower frequencies, up to *ca.* 5 GHz, Si bipolar transistors are used, and from *ca.* 1 GHz GaAs MESFETs and, increasingly, HEMTs.

We distinguish narrow band amplifiers, wide band, ultrawide band and d.c.-coupled amplifiers. Narrow band amplifiers (relative bandwidth <5%) are among others used in radioastronomy and communication systems, wide band amplifiers in satellite communication systems and ultrawide band amplifiers in radars and spread spectrum communication systems. d.c.-coupled amplifiers are used to amplify pulses.

9.3.2 d.c. biasing

First of all the transistors have to be supplied with d.c. bias and the bias lines must be decoupled from the high-frequency signal. In principle this is done with series inductances and capacitances to ground. At microwave frequencies it is not so easy to use discrete inductors and capacitors so instead of a capacitance a very wide, short line is used and as a series inductance a very thin microstrip line (Fig. 9.4). The latter can be made as a meander line to increase the inductance or, when the bandwidth does not have to be large, as a quarter-wavelength line which transforms the low impedance of the capacitance to a very high impedance. At the edge of the microstrip substrate where the power line is connected, a chip capacitor can be placed. The bond wire from this capacitor to the microstrip serves as an extra series inductance.

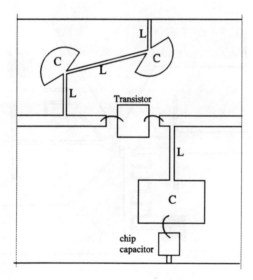

Figure 9.4 d.c. bias circuit in microstrip.

9.3.3 Impedance matching

In every amplifier matching circuits are necessary to match the transistor to the source and load. Maximum power transfer is achieved when the input impedance of the transistor is the complex conjugate of the source impedance, and the same happens at the output side. An alternative formulation is that the reflection coefficients have to be each other's complex conjugate. This is called *conjugate matching*. Different methods are used depending on the desired bandwidth.

Narrow band amplifiers can relatively easily be matched with open-ended pieces of transmission line and/or discrete capacitances and inductances. For wide band matching often a resistive feedback circuit is used, as described below. In d. c.- coupled amplifiers the matching has to be done with resistors. Because of the dissipation in the resistors the amplification per stage is low.

9.3.4 Narrow band amplifiers

When the desired bandwidth is small (5% or less) conjugate matching can be achieved by adding a reactive element to the transistor input so that the phase angle of the input impedance is the opposite of that of the source. This means that when the source is inductive one has to add a capacitor. Then the magnitudes of the two impedances are matched by the use of a quarter-wavelength transformer (Chapter 7). Instead of a discrete capacitor

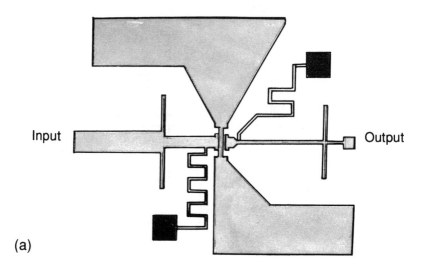

(a)

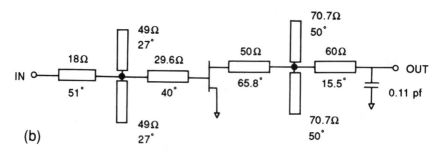

(b)

Figure 9.5 Narrow band impedance matching. Redrawn from Ferry, D. K., *Gallium Arsenide Technology*; published by Howard Sams & Co., Inc.

or inductor one can also use transmission line stubs, as explained in Chapter 7. In Fig. 9.5 an example is shown with the equivalent transmission line circuit in Fig. 9.5(a) and the actual microstrip form, which shows also the bias lines, in Fig. 9.5(b).

9.3.5 Wide band amplifiers

For larger bandwidth more complex matching circuits have to be applied. Generally the bandwidth is increased by replacing single elements by ladder networks. In microstrip these take the form of a sequence of series and parallel stubs (Fig. 9.6).

To stretch the bandwidth further towards the lower frequencies resistive coupling elements can be used (Fig. 9.7), which provide impedance

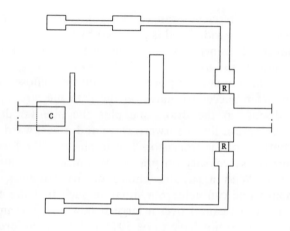

Figure 9.6 Wide band matching.

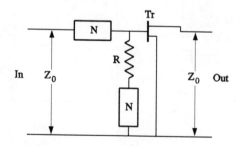

Figure 9.7 Resistive impedance matching.

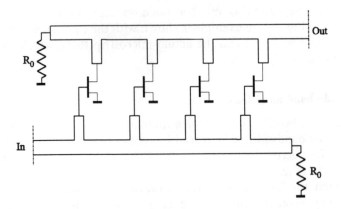

Figure 9.8 Distributed amplifier.

matching in the low-frequency range. This configuration is called a *lossy match amplifier*. A very wide band is provided by a distributed amplifier, in which a number of transistors are connected in parallel between two transmission lines (Fig. 9.8).

The lengths of the connecting pieces of line must be chosen such that the time delays between successive gates on the gate line are equal to those between the drains on the drain line plus the internal delay of each transistor. On the drain line a forward wave and a backward wave can be excited. The outputs of the transistors add in phase to the forward wave if it has the same phase velocity (in practice this means the same linewidth) as the gate line. With a suitable choice of the distances between the connection points one can achieve a situation such that the contributions to the backward wave cancel over a large bandwidth. The impedances of the connecting lines do not have to be 50 Ω and can therefore be matched to the gate and drain impedances of the transistors.

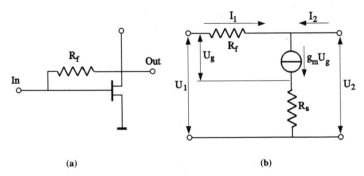

(a)   (b)

Figure 9.9 Feedback amplifier: (a) schematic; (b) low-frequency equivalent circuit.

9.3.6 Feedback amplifiers

It is well known that negative feedback improves the distortion and noise properties of amplifiers. It can also be applied to obtain wide band impedance matching. The basic circuit is shown in Fig. 9.9(a), with the equivalent circuit in Fig 9.9(b). Here the simplest possible equivalent circuit for a MESFET is used, valid only at low frequencies. The source resistance is included because it is never negligible and it also provides negative feedback. A bit of algebra then yields the admittance matrix:

$$\begin{bmatrix} I_1 \\ I_2 \end{bmatrix} = \begin{bmatrix} \dfrac{1}{R_f} & -\dfrac{1}{R_f} \\ \dfrac{g_m}{1 + g_m R_s} - \dfrac{1}{R_f} & \dfrac{1}{R_f} \end{bmatrix} \begin{bmatrix} V_1 \\ V_2 \end{bmatrix}. \tag{9.13}$$

From this the S-matrix can be calculated (Chapter 6) in a transmission line with characteristic impedance Z_0:

$$S_{11} = \frac{1}{\Delta}\left[1 - \frac{g_m Z_0^2}{R_f(1 + g_m R_s)}\right], \quad S_{12} = \frac{1}{\Delta}\frac{2Z_0}{R_f} \tag{9.14}$$

$$S_{21} = \frac{1}{\Delta}\left(\frac{2Z_0}{R_f} - \frac{2g_m Z_0}{1 + g_m R_s}\right), \quad S_{22} = \frac{1}{\Delta}\left[1 - \frac{g_m Z_0^2}{R_f(1 + g_m R_s)}\right]$$

with

$$\Delta = 1 + \frac{2Z_0}{R_f} + \frac{g_m Z_0^2}{R_f(1 + g_m R_s)}. \tag{9.15}$$

Inspection of these formulas tells us that S_{11} and S_{22} can both be zero when R_f has the correct value, given by the condition

$$g_m Z_0^2 = R_f(1 + g_m R_s). \tag{9.16}$$

When this condition is satisfied, S_{21} has the value $(Z_0 - R_f)/Z_0$. Since $|S_{21}|^2$ is the power amplification, R_f has to be greater than $2Z_0$ for any amplification to remain. This in turn means in view of equation (9.16) that g_m has to have a minimum value, equal to

$$g_m > \frac{2}{Z_0 - R_s} > \frac{2}{Z_0}. \tag{9.17}$$

So in a 50 Ω system g_m has to be at least 40 mS, otherwise we cannot achieve complete matching and amplification at the same time.

In a more realistic equivalent circuit of a MESFET (Chapter 5) there appears amongst other components a gate resistance R_g and a gate–source capacitance C_{gs} which together form a low-pass filter giving a cutoff frequency $1/2\pi R_g C_{gs}$. The frequency characteristic can be improved by

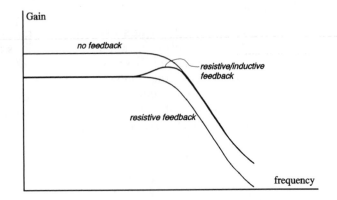

Figure 9.10   Frequency characteristic due to inductive feedback.

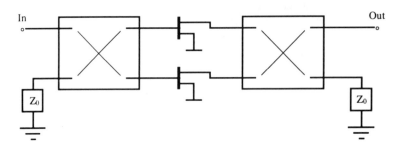

Figure 9.11   Parallel amplifier.

adding an inductance in series with the feedback resistance R_f. Its effect is to reduce the feedback, and thereby to increase the amplification, at the higher frequencies. The resulting frequency characteristic is shown in Fig. 9.10.

   To obtain more power one uses balanced amplifiers with two parallel branches, and even four or eight branches (Fig. 9.11). A great advantage of this is that if one transistor breaks down the amplifier still works, albeit with lower amplification. With a reasoning analogous to that used for the two-diode amplifier of Fig. 9.2 it can be seen that this amplifier is matched over the bandwidth of the hybrid couplers, so this is a wide band configuration too.

9.4   AMPLIFICATION AND STABILITY

9.4.1   Gain definitions

The active element in an amplifier is in general not matched to the characteristic impedance of the transmission line it is mounted in. If it is a transistor then its input and output impedances will be far from Z_0 and both S_{11} and S_{22} will be large. In this case the power gain will be low. We can improve the situation by inserting matching networks between source and input, and between output and load. The elements of the matching networks (e.g. a double-stub tuner; Chapter 6) are frequency dependent so that the gain will be maximized only for a narrow band around a certain frequency. If more bandwidth is desired one has to design the matching networks so that they provide a constant mismatch over a wider frequency band. This of course means that the gain is less than maximal. There is always a trade-off between gain and bandwidth.

Another problem is that Γ_i is a function of Γ_L and Γ_o of Γ_s. The consequence is that the two matching procedures influence each other. One has to apply an iterative process where input and output are matched alternatingly. If S_{12} is small this process quickly converges. Otherwise one has to use mathematical optimization routines that maximize the gain as a function of several variables, the variables in this case being the values of the elements of the matching networks. If for instance we use double-stub tuners (Chapter 6) there are four independent variables. A diode is easier to match since its input and output ports are the same.

An amplifier is coupled to the signal source and the load with transmission lines (Fig. 9.12). Let the source and the load have reflection coefficients Γ_s and Γ_L, respectively. Now if the amplifier is characterized by a set of S-parameters (Chapter 6) the input and output reflection coefficients Γ_i and Γ_o, respectively, are given by

$$\Gamma_i = S_{11} + \frac{S_{12}S_{21}\Gamma_L}{1 - S_{22}\Gamma_L} \tag{9.18a}$$

$$\Gamma_o = S_{22} + \frac{S_{12}S_{21}\Gamma_s}{1 - S_{11}\Gamma_s}. \tag{9.18b}$$

Different amplification factors can be defined, depending on the application. The most obvious one is the *actual power gain*, usually called *power*

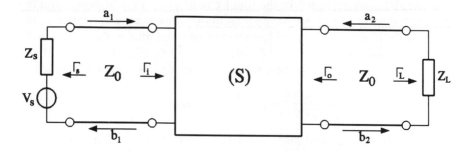

Figure 9.12 Amplifier in transmission line.

gain without further qualification, and defined as the power dissipated in the load divided by the power delivered by the source:

$$G = \frac{|b_2|^2 - |a_2|^2}{|a_1|^2 - |b_1|^2} = \frac{|S_{21}|^2(1 - |\Gamma_L|^2)}{|1 - S_{22}\Gamma_L|^2(1 - |\Gamma_i|^2)}. \tag{9.19}$$

Second, we have the *transducer gain*, defined as the power delivered to the load divided by the maximum power the source can deliver at the amplifier input (in the case that the input is matched to the source):

$$G_T = \frac{|S_{21}|^2(1 - |\Gamma_L|^2)(1 - |\Gamma_S|^2)}{|1 - S_{22}\Gamma_L|^2|1 - \Gamma_S\Gamma_i|^2}. \tag{9.20}$$

Third, we have the *available gain*, the gain that can be achieved when both the input and output are matched for maximum power transfer ($\Gamma_i = \Gamma_S^*$, $\Gamma_o = \Gamma_L^*$):

$$G_A = \frac{|S_{21}^2|(1 - |\Gamma_S|^2)}{|1 - S_{11}\Gamma_S|^2(1 - |\Gamma_o|^2)}. \tag{9.21}$$

Finally we may consider the ratio of the power delivered to the load by the amplifier to the power that would be delivered to the load if the source were directly connected to the load. This is called the *insertion gain* and it has the advantage that it can directly be measured. It is given by

$$G_I = \frac{|S_{21}|^2|1 - \Gamma_L\Gamma_S|^2}{|1 - S_{22}\Gamma_L|^2|1 - \Gamma_S\Gamma_i|^2}. \tag{9.22}$$

9.4.2 Stability

Because of the finite values of S_{12}, Γ_S and Γ_L a feedback occurs that can cause the amplifier to oscillate, which of course is not wanted. An amplifier is called *absolutely stable* if with any combination of *passive* source and load impedances it cannot oscillate. At the input side we have $b_1 = \Gamma_i a_1$ and $a_1 = \Gamma_S b_1$. This means that if $\Gamma_S\Gamma_i$ has a magnitude equal to unity the signal will circulate unattenuated in the input circuit even without power coming from the source, so that we have an oscillator. The same can happen at the output side. Since $|\Gamma_S|$ and $|\Gamma_L|$ are supposed to be smaller than unity oscillation is prevented when both $|\Gamma_i|$ and $|\Gamma_o|$ are smaller than unity. It can be shown [1] that this is the case when the *stability coefficient* K, defined as

$$K = \frac{1 + |D|^2 - |S_{11}|^2 - |S_{22}|^2}{2|S_{12}S_{21}|}, \quad D = S_{11}S_{22} - S_{12}S_{21}, \tag{9.23}$$

is greater than unity, and also $|D| > 1$. An amplifier is called *conditionally stable* when it is non-oscillating for a restricted range of passive source and load impedances.

For an unconditionally stable amplifier we define the *maximum available gain* MAG as the maximum amplification that can be obtained with optimal matching of source and load:

$$\text{MAG} = \frac{|S_{21}|}{|S_{12}|} \, [K - \text{sgn}(B)(K^2 - 1)^{1/2}] \tag{9.24}$$

$$B = 1 + |S_{11}|^2 - |S_{22}|^2 - |D|^2.$$

An amplifier that is not absolutely stable can still be matched so that it does not oscillate. In this case the amplification is called the *maximum stable gain*:

$$\text{MSG} = \frac{|S_{21}|}{|S_{12}|}. \tag{9.25}$$

The stability of an amplifier can be determined from the so-called *stability circles*. Stability means that both $|\Gamma_i|$ and $|\Gamma_o|$ are smaller than unity. Equations (9.18) show that a bilinear transformation exists between Γ_i and Γ_L, and likewise between Γ_o and Γ_s. This transformation is similar to the one that maps the impedance plane onto the Smith chart (Chapter 6) and it has the same property of transforming circles into circles. If we concentrate first on Γ_i, then the circle $|\Gamma_i| = 1$ maps onto a circle in the Γ_L plane (Fig. 9.13), so either the inside or the outside of this circle gives the Γ_L values for which the amplifier is stable. The center and radius of the circle are given by [1]

$$C_L = \frac{S_{22}^* - \Delta^* S_{11}}{|S_{22}|^2 - |\Delta|^2} \tag{9.26}$$

$$R_L = \frac{|S_{12} S_{21}|}{||S_{22}|^2 - |\Delta|^2|} \tag{9.27}$$

with

$$\Delta = S_{11} S_{22} - S_{12} S_{21}. \tag{9.28}$$

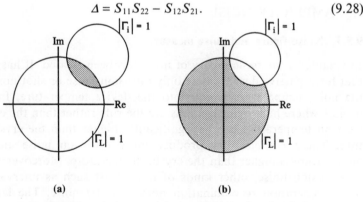

Figure 9.13 Stability circles: (a) $|S_{22}| < |\Delta|$; (b) $|S_{22}| > |\Delta|$.

When $|S_{22}| < |\Delta|$ the stable region is the inside of the circle and when $|S_{22}| > |\Delta|$ it is the outside. This can easily be seen by considering the point $\Gamma_i = 0$ which is mapped onto $\Gamma_L = S_{11}/\Delta$ (equation (9.18)). Now, this point is within the stability circle when

$$\left| \frac{S_{11}}{\Delta} - C_L \right| < R_L.$$

A little algebra reveals that this is the case when $|S_{22}| < |\Delta|$.

Another way to decide which one is the stable region is to look at the point $\Gamma_L = 0$. This obviously corresponds to $\Gamma_i = S_{11}$, so when $|S_{11}| < 1$ the point $\Gamma_L = 0$ is in a stable region and when $|S_{11}| > 1$ it is in an unstable region.

The stability circle for Γ_S, corresponding to $|\Gamma_o| < 1$, is given by similar expressions:

$$C_S = \frac{S_{11}^* - \Delta^* S_{22}}{|S_{11}|^2 - |\Delta|^2} \tag{9.29}$$

$$R_S = \frac{|S_{12}S_{21}|}{||S_{11}|^2 - |\Delta|^2|}. \tag{9.30}$$

So by plotting the inverse of equation (9.18a) or (9.18b) in the complex Γ_L plane or Γ_S plane, respectively, the stability conditions for an amplifier can be determined. This is demonstrated for the load side in Fig. 9.13. Since it may be assumed that the load is passive, Γ_L is always inside the unit circle. Then the hatched regions give the range of stable Γ_L values, on the left when $|S_{22}| < |\Delta|$, on the right for $|S_{22}| > |\Delta|$. Note that everything that has been said before refers to one frequency only. So for a wide band amplifier the procedure should be repeated for a number of frequencies across the desired band.

9.5 AMPLIFIER NOISE

9.5.1 Noise figure and noise measure

In Chapter 3 the basic theory of noise has been treated. It has been shown that for a passive device where only diffusion noise or shot noise is present the noise temperature is equal to the device temperature. In microwave devices where high field strengths are the rule rather than the exception the electron temperature can be significantly higher than the crystal temperature. These *hot electrons* produce more diffusion noise and the noise temperature is higher than the crystal temperature. Moreover, besides shot or diffusion noise, other kinds of noise exist such as intervalley transfer noise, generation–recombination noise and $1/f$ noise. The latter two are restricted to the low-frequency range, if only generation and recombination

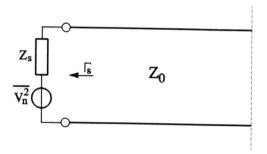

Figure 9.14 Impedance with noise.

from traps are considered. At microwave frequencies their contribution is negligible, at least in linear systems. In non-linear systems, in particular oscillators and mixers, low-frequency noise can be upconverted to the microwave frequency range. Intervalley noise in Gunn diodes and multiplication noise in IMPATT diodes are of a wide band nature and will contribute to the overall microwave noise.

On a transmission line with characteristic impedance Z_0 (Fig. 9.14) the noise source associated with Z produces a noise wave to the right that transports a power equal to

$$P = \frac{S_V Z_0}{2|Z + Z_0|^2} = \frac{2|\text{Re } Z|Z_0}{|Z + Z_0|^2} P_m = |1 - |\Gamma|^2| P_m \qquad (9.31)$$

in which Γ is the reflection coefficient of Z and P_m the maximum noise power that the device can deliver. According to equation (4.52) we can write the latter as $P_m = MkT_0B$. Note that in the case of a negative-resistance device Γ can have a magnitude larger than unity.

The noise figure F of an amplifier has been defined in Chapter 4 as the degradation of the signal-to-noise ratio at the input to that at the output under the condition that the noise at the input is produced by a matched impedance having a noise temperature equal to the standard temperature T_0. The noise power at the output of an amplifier consists of amplified input noise power GN_i (G is the power amplification) and noise power N_A produced by the amplifier itself. Normally these are assumed to be uncorrelated so they can be added simply. This gives for the noise figure

$$F = \frac{P_{in}/kT_0B}{P_{out}/N_{out}} = \frac{GN_i + N_A}{GkT_0B} = 1 + \frac{N_A}{GkT_0B}. \qquad (9.32)$$

For systems at room temperature F is a good measure of system noise. In radioastronomy and satellite communications, where one works with very low-noise receivers, F will be close to unity and, expressed in decibels, has

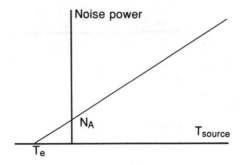

Figure 9.15 Noise of a llinear amplifier.

very small values. A handier measure in these cases is the *excess noise temperature*, defined as the temperature of a noise source at the input that would produce as much *extra* noise as the amplifier does.

When we connect the input of a linear amplifier to a noise source whose strength can be varied then the noise power at the output is a linear function of the noise temperature of the source (Fig. 9.15). Here the intersection with the vertical axis gives the added noise N_A and the extrapolation to $N = 0$ the excess noise temperature $T_e = N_A/GkB$. A comparison with equation (9.32) shows that

$$T_e = T_0(F - 1).$$

To find the relation between noise figure and noise measure let us take a look at a reflection amplifier (Fig. 9.16). The amplification of this is

$$G = |\Gamma_d'|^2$$

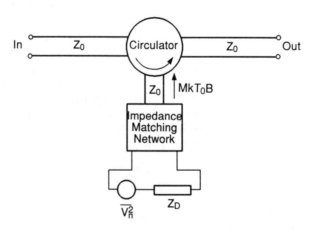

Figure 9.16 Reflection amplifier with noise.

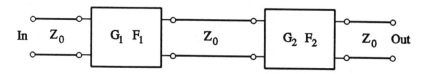

Figure 9.17 Cascade of two amplifiers.

where Γ'_d is the reflection coefficient of the negative-resistance device plus matching network, seen from the transmission line.

At the input we have a signal power P_{in} and a noise power kT_0B. At the output we have signal $P_{out} = GP_{in}$ and noise $GkT_0B + (G - 1)MkT_0B$, of which the first term represents amplified input noise and the second the contribution of the active element, according to equation (9.31). After substitution in equation (9.32) we find

$$F = 1 + M\left(1 - \frac{1}{G}\right).$$
(9.33)

This provides the link between the measurable quantity F and the device parameter M which, as was shown in section 4.4.7, can be calculated from device physics. If we have a model for the noise of an active device we can predict the noise figure of an amplifier using it, and, conversely, if we have measured the amplifier noise figure we can draw conclusions about the device physics.

Consider now a cascade of two amplifiers, (Fig. 9.17). At the input we again have signal power P_{in} and noise power kT_0B. At the input of amplifier 2 this results in power G_1P_{in} and noise $G_1F_1kT_0B$. At the output of amplifier 2 then we have $G_2G_1P_{in}$ and $G_2G_1F_1kT_0B + M_2(G_2 - 1)kT_0B$, respectively. For the noise figure of the combination we find

$$F_{1+2} = F_1 + \frac{F_2 - 1}{G_1}.$$
(9.34)

This is the famous Friis formula. It is clear that in most cases the noise figure of the first stage is the decisive quantity.

9.5.2 Noise figure circles

In Chapter 4 it has been shown that the noise behavior of a twoport can be expressed generally in two parameters, the optimal noise figure F_{opt} and the two-port noise measure $M^{(2)}$. If one now connects a twoport to an arbitrary source admittance Y_s the actual noise figure is given by [2]

$$F(Y_s) = F_{opt} + \frac{M^{(2)}|Y_s - Y_{opt}|^2}{Re(Y_{opt})\,Re(Y_s)}.$$
(9.35)

It takes a little algebra to see that lines of constant F are circles in the Y_s plane. If we now move from Y_s to its reflection coefficient Γ_s, circles will be transformed into circles, so that also in the Γ_s plane the constant noise figure loci are circles. It can be shown [3] that the center and the radius of the circle for a noise figure F_i are given by

$$C_i = \frac{\Gamma_{opt}}{N_i + 1}, \qquad R_i = \frac{[N_i(N_i + 1 - |\Gamma_{opt}|^2)]^{1/2}}{N_i + 1} \qquad (9.36)$$

where

$$N_i = \frac{F_i - F_{opt}}{4M^{(2)}} (1 - |\Gamma_{opt}|^2).$$

Note that in the cited literature these expressions are given in terms of an effective noise resistance R_n which is connected to our noise measure by

$$R_n = \frac{M^{(2)}}{\mathrm{Re}(Y_{opt})}. \qquad (9.37)$$

9.5.3 Noise measurements

From Fig. 9.15 it appears that we can measure the noise figure of an amplifier by connecting at the input a noise source with variable power (e.g. a constant noise source followed by an attenuator) and measure the output noise power.

At the output we have, at two different input noise temperatures T_1 and T_2,

$$P_1 = G[kT_1 + (F - 1)kT_0], \qquad P_2 = G[kT_2 + (F - 1)kT_0].$$

The ratio P_1/P_2 ($P_1 > P_2$) is called the Y factor and this measurement is therefore often called the Y factor measurement. For the noise figure we find

$$F = 1 + \frac{T_1 - YT_2}{(Y - 1)T_0}. \qquad (9.38)$$

For this measurement a good linear power meter is necessary. Usually this is made from a mixer stage followed by an intermediate frequency amplifier (both low noise) and a quadratic detector. If we cannot trust the linearity of the power measurement we can put another attenuator at the output and adjust this so that the output power remains constant (Fig. 9.18).

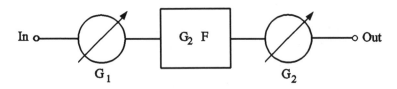

Figure 9.18 Noise measurement with two attenuators.

At the output we now have, assuming that the attenuators each contribute a noise kT_0 per unit bandwidth,

$$P_N = G_3[G_2(G_1 kT_1 + kT_0) + (F - 1)G_2 kT_0] + kT_0$$

in which G_1 and G_3 represent the 'amplification' (the inverse of the attenuation) of the attenuators and T_1 the noise temperature of the source. By doing two measurements with different settings of the attenuators and the same output power we can eliminate G_2 and calculate F. This method is especially useful in waveguide circuits since waveguide attenuators of rotary vane type can be made with high precision.

In the methods described above measurement errors can occur because of the finite reflection coefficients of the source and the power meter. An advantage of the two-attenuator method is that attenuators usually have low reflection coefficients.

9.6 POWER AMPLIFIERS

Power amplifiers are used when information-carrying signals have to be brought up to a power level of more than 100 mW, e.g. in transmitters. For this application traveling wave tubes (Chapter 2) are still much used owing to their combination of large power capacity and large bandwidth. In low to medium power applications they are increasingly replaced by transistor amplifiers.

At the lower microwave frequencies silicon bipolar transistors are used predominantly. Since the efficiency of transistors is between 25% and 60%, a great deal of power has to be conducted away as heat. Because of the good thermal conductivity of Si, transistors made of this material have excellent power capabilities. The figures range from 500 W at 500 MHz to 20 W at 2 GHz. At higher frequencies GaAs MESFETs take over. HEMTs do not perform very well in power applications owing to their limited sheet electron densities (section 6.4.3).

For power amplifiers the noise figure is not so important because they are usually preceded by either a small-signal amplifier or an oscillator, which then determines the noise figure of the combination. Of more concern are non-linear phenomena, in particular the decrease of the amplification at greater amplitudes and the occurrence of higher harmonics. In Fig. 9.19 the output power at the fundamental and higher harmonics as a function of the input power is plotted. Important characteristics of the amplifier can be read from this graph:

- the *amplification* at low input power;
- the *saturation power*;
- the *1 dB compression point*, that is the input power at which the amplification of the fundamental harmonic has dropped 1 dB, which is a measure of the power capacity;

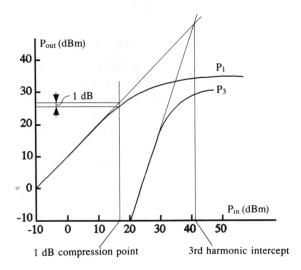

Figure 9.19 Power amplification.

- the *third-harmonic intercept*, the point where the extrapolation of the third-harmonic power curve crosses that of the first, which is a measure of the non-linearity of the amplifier.

The non-linearity gives rise to another unwanted phenomenon, *intermodulation distortion*. When two or more signals of different frequencies are amplified together, as for instance in microwave relays where many telephone channels modulated on carriers are handled, mixing products can be produced by the non-linearity of the amplifier in two ways:

- modulation sidebands can be mixed with the carrier, giving rise to low-frequency components that can be modulated again on other carriers;
- when the bandwidth is greater than one octave the mixing product of two carriers can be within the band and appear in a third channel.

Both effects produce cross-talk between channels which is of course unwanted. For these applications therefore an amplifier cannot be driven too far in the non-linear regime.

9.7 MICROWAVE MASERS

9.7.1 Introduction

Masers work on a principle completely different from the other amplifiers we have discussed so far. It is the phenomenon of *stimulated emission* of radiation. It was Einstein in 1917 who pointed out that for a correct derivation of Planck's radiation law it is necessary to assume three different processes in the interaction of radiation with matter. The first two, spontaneous emission and absorption of radiation, were already known but Einstein introduced a third one, namely a process where an incoming photon is not absorbed but releases another photon. It took until 1953 before it was realized that this process could be used for the amplification of microwave radiation. Finally Townes and coworkers realized the first *ammonia maser* based on this principle. The key papers are reprinted in ref. 4.

The word maser is an acronym of '*microwave amplification by stimulated emission of radiation*'. In 1960 followed the optical variant, the ruby laser, and in 1961 the semiconductor laser. Laser is an acronym of '*light amplification by stimulated emission of radiation*'. In spite of this name a laser is not an amplifier but an oscillator.

To have stimulated emission a system is needed that can exist in at least two stationary quantum mechanical states with different energies, Bohr's model of the atom being the paradigmatic example. The energy of an atom (or molecule) can assume only discrete values that correspond to stationary solutions of Schrödinger's equation. Non-stationary solutions (decaying exponentially in time) describe the transition from one state into another. The energy difference can be absorbed or emitted as a photon of

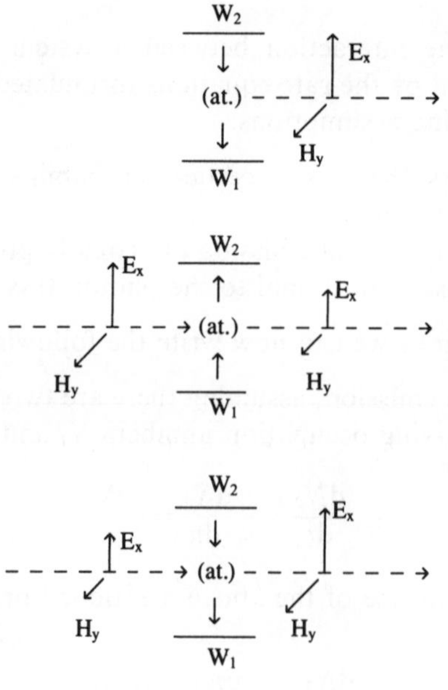

Figure 9.20 Electromagnetic interaction processes: (a) spontaneous emission; (b) absorption; (c) stimulated emission.

electromagnetic radiation. These processes are called *absorption* and *emission*, respectively.

   These transitions can be influenced by an external electromagnetic field, which exerts forces on the electrons and the atomic nucleus. This makes it plausible that an external field can induce transitions. An induced transition from a lower to a higher level is the above-mentioned *absorption*. Einstein (1917) was the first to realize that an induced transition from high to low should be equally probable. This is a third process, called *stimulated emission*. In this process the incoming photon is not absorbed but instead an extra photon is liberated. From an electromagnetic wave point of view it is important to realize that the extra photon is in phase with the incoming one. In all three processes the emitted photon is equal to the energy difference between the levels (Planck's quantum law):

$$hf = \hbar\omega = W_2 - W_1 \tag{9.39}$$

where h is Planck's original constant and $\hbar = h/2\pi$ is the modified one used mostly nowadays. In Fig. 9.20 these processes are schematically pictured. Here '(at.)' symbolically represents the atom. Besides isolated atoms there are other systems where these processes can occur: molecules, atoms embedded in a solid and the energy bands of a semiconductor.

9.7.2 Statistics

The dynamics of the interaction between a system and electromagnetic radiation is described by the rate equations formulated by Einstein in 1917. He made the following assumptions:

- for every process there is a certain probability independent of the number of particles involved;
- absorption and stimulated emission are equally probable;
- this probability is proportional to the photon flux F.

With these assumptions we can now write the following.

- For spontaneous emission, assuming there are two energy levels W_1 and W_2 ($W_1 < W_2$) having occupation numbers N_1 and N_2,

$$\frac{dN_2}{dt} = -\frac{dN_1}{dt} = -\frac{N_2}{\tau_{sp}} \tag{9.40}$$

   where τ_{sp} is the inverse of the above-mentioned probability.
- For absorption,

$$\frac{dN_2}{dt} = -\frac{dN_1}{dt} = P_{12}N_1. \tag{9.41}$$

- For stimulated emission,

$$\frac{dN_2}{dt} = -\frac{dN_1}{dt} = -P_{21}N_2. \qquad (9.42)$$

We do not use time constants in the last two cases because the probabilities are dependent on the photon flux:

$$P_{ij} = \sigma_{ij}F. \qquad (9.43)$$

If F is the number of photons per square meter per second then σ_{ij} has units of square meters and can be interpreted as a scattering cross-section. Because of the second assumption we have $P_{12} = P_{21}$ so $\sigma_{12} = \sigma_{21}$.

In equilibrium the probability that an atom has the energy W_i is proportional to $\exp(-W_i/kT)$ (Boltzmann statistics). This implies

$$\frac{N_2}{N_1} = \exp\left(\frac{W_1 - W_2}{kT}\right) = \exp\left(-\frac{hf}{kT}\right). \qquad (9.44)$$

Also, the total change of N_1 and N_2 is zero so that

$$-\frac{N_2}{\tau_{sp}} + P_{12}N_1 - P_{21}N_2 = 0, \qquad (9.45)$$

and with $P_{12} = P_{21} = \sigma_{12}F$ we obtain

$$\frac{N_2}{N_1} = \frac{\sigma_{12}\tau_{sp}F_{eq}}{\sigma_{12}\tau_{sp}F + 1}. \qquad (9.46)$$

The subscript 'eq' indicates that this refers to the equilibrium situation. If we combine equations (9.45) with (9.46) we obtain

$$F_{eq} = \frac{1/\sigma_{12}\tau_{sp}}{\exp(hf/kT) - 1}. \qquad (9.47)$$

Not accidentally, this has the same form as Planck's radiation law which gives the energy density of radiation in a frequency band Δf in equilibrium with matter at a temperature T:

$$\Delta W = \frac{8\pi hf^3}{c^3} \frac{\Delta f}{\exp(hf/kT) - 1}. \qquad (9.48)$$

Since photon flux is photon density multiplied by velocity we can write $W = hfF/c$ and compare these two equations. There is a subtle point, however, that we should keep in mind. Equation (9.48) is derived under the assumption that a large number of radiators exists covering a wide spectrum. On the other hand, in deriving equation (9.47) we have tacitly assumed that everything happens at one frequency. This of course is a simplification: in reality the energy levels W_1 and W_2 have finite widths and the radiation described by equation (9.47) also has a spectrum of finite width. Comparison with equation (9.48) allows us to relate the spectral width to the time constant of spontaneous emission:

$$\frac{1}{\sigma_{12}\tau_{sp}} = \frac{8\pi f^2}{c^2}\Delta f \tag{9.49}$$

and we can rewrite equation (9.47) as

$$F = \frac{8\pi f^2}{c^2}\frac{\Delta f}{\exp(hf/kT) - 1}. \tag{9.50}$$

The phenomenon of stimulated emission opens the possibility of amplifying an electromagnetic wave. To realize this it is necessary that there is more stimulated emission than absorption which means $N_2 > N_1$, just the opposite of what we have in the equilibrium state. Therefore this situation is called a *population inversion*, or *inversion* for short. Another way to express it is to say that the medium has a *negative temperature* because in equation (9.44) the exponential is greater than unity.

Note that in the literature often the *Einstein coefficients* A_{ij} and B_{ij} are used where $A_{21} = 1/\tau_{sp}$ and B_{ij} is defined by

$$P_{ij} = B_{ij}\frac{hfF}{c\,\Delta f}$$

so the relation between the two is

$$B_{ij} = B_{ji} = \frac{c\sigma\,\Delta f}{hf} = \frac{c^3}{8\pi hf^3}A_{21}.$$

Here we prefer to use the scattering cross-section σ because it does not contain the bandwidth Δf.

9.7.3 Pumping and inversion

The first practical realization of stimulated emission, the *ammonia maser* was made by Townes, Gordon and Zeiger in 1953 [4]. They separated a beam of ammonia (NH_3) molecules in two beams of molecules at different energy levels with the help of an inhomogeneous electric field. This is possible because the molecules in these two states have different dipole moments. The beam at the upper energy level then could be used for amplification or oscillation at a frequency of about 24 GHz.

The use of a solid with three energy levels was proposed by Bloembergen in 1956. It is then possible to produce an inversion between the upper and middle levels or between the middle and lower levels. To maintain the inversion it is necessary to bring atoms continuous from the lower level to the upper. This is called *pumping* and can be done by supplying energy at the frequency corresponding to the energy difference between these levels. It is schematically depicted in Fig. 9.21. In Fig. 9.21(a) the spontaneous decay from level 2 to 1 should be very fast and that from 3 to 2 should be

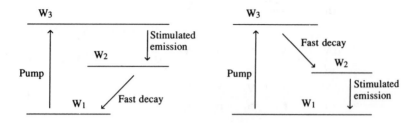

Figure 9.21 Pumping in a three-level system.

slow to maintain a high population density at level 2; in Fig. 9.21(b) it should be the other way around.

Note that there is a distinct similarity between this three-level scheme and a parametric amplifier. In both a pump is needed at a frequency higher than the signal frequency. The role of the idler is in the present case played by the transitions $2 \rightarrow 1$ and $3 \rightarrow 2$ in Figs. 9.21(a) and 9.21(b), respectively.

For microwave applications of stimulated emission one preferably uses paramagnetic atoms in a suitable host material to obtain a three-level scheme. Because of the interaction of the spin of the atom with the crystal field and an external magnetic field a fine splitting of energy levels occurs, the energy difference of which corresponds to frequencies in the microwave range. Since the separation of the levels is a function of the external field, tuning of a maser to the desired frequency is possible.

Below about 10 GHz Cr^{3+} in ruby (Al_2O_3) works well; at higher frequencies Cr^{3+} or Fe^{3+} in rutile (TiO_2) is better. The doping levels are in the range 10^{18}–10^{19}/cm^3. A possible way to make an amplifier is to put the material in a cavity and to use the latter as the negative resistance in a reflection amplifier. A better way, however, is to build a traveling wave maser (section 9.7.4).

Besides the three processes mentioned above in a solid there will be other processes (e.g. phonon emission and absorption; Chapter 3) that can induce transitions between energy levels, the so-called *non-radiative processes*. Unlike spontaneous emission phonon interactions can also cause upward transitions. Like spontaneous emission they can be represented by relaxation terms but they have much shorter time constants. Consequently the width of the spectrum is considerably increased. For the above-mentioned materials the time constants are of the order of 10^{-8}s, giving relative linewidths of about 1%.

Including non-radiative processes the rate equations for the case of Fig. 9.21(a) become

$$\frac{dN_3}{dt} = -\frac{N_3}{\tau_{32}} - \frac{N_3}{\tau_{31}} + \frac{N_2}{\tau_{23}} + \frac{N_1}{\tau_{13}}$$

$$+ \sigma_{13}F_P(N_1 - N_3) - \sigma_{32}F_S(N_3 - N_2) \quad (9.51\text{a})$$

$$\frac{dN_2}{dt} = +\frac{N_3}{\tau_{32}} - \frac{N_2}{\tau_{23}} - \frac{N_2}{\tau_{21}} + \frac{N_1}{\tau_{12}} + \sigma_{32}F_S(N_3 - N_2) \quad (9.51\text{b})$$

$$\frac{dN_1}{dt} = +\frac{N_2}{\tau_{21}} + \frac{N_3}{\tau_{31}} - \frac{N_1}{\tau_{12}} - \frac{N_1}{\tau_{13}} - \sigma_{13}F_P(N_1 - N_3). \quad (9.51\text{c})$$

Here F_P is the photon flux at the pump frequency and F_S that at the frequency of the signal we want to amplify. These equations are complemented by the conservation of the total number of particles:

$$N_1 + N_2 + N_3 = N_{\text{tot}}. \quad (9.52)$$

The whole set is now interdependent because the time derivative of equation (9.52) is equal to the sum of the equations (9.51).

In the absence of electromagnetic fields in the steady state ($\partial/\partial t \equiv 0$) the following equilibrium situation is obtained:

$$-\frac{N_3}{\tau_{32}} - \frac{N_3}{\tau_{31}} + \frac{N_2}{\tau_{23}} + \frac{N_1}{\tau_{13}} = 0 \quad (9.53\text{a})$$

$$+\frac{N_3}{\tau_{32}} - \frac{N_2}{\tau_{23}} - \frac{N_2}{\tau_{21}} + \frac{N_1}{\tau_{12}} = 0 \quad (9.53\text{b})$$

$$+\frac{N_2}{\tau_{21}} + \frac{N_3}{\tau_{31}} - \frac{N_1}{\tau_{12}} - \frac{N_1}{\tau_{13}} = 0. \quad (9.53\text{c})$$

We know that in equilibrium the ratios between the level occupancies are given by Boltzmann factors; equation (9.44). If we substitute these in equations (9.53) we find that the time constants have to be related as follows:

$$\frac{\tau_{13}}{\tau_{31}} = \exp\left(-\frac{hf_P}{kT}\right) \quad (9.54\text{a})$$

$$\frac{\tau_{23}}{\tau_{32}} = \exp\left(-\frac{hf_S}{kT}\right) \quad (9.54\text{b})$$

$$\frac{\tau_{12}}{\tau_{21}} = \exp\left[-\frac{h(f_P - f_S)}{kT}\right]. \quad (9.54\text{c})$$

This shows that the upward transitions are always less probable than the downward ones and the difference increases as the temperature goes down. This is understandable since the upward transitions need absorption of phonons and the number of available phonons decreases with temperature.

Before we proceed it will be instructive to take a little time and compare the terms with N_3 in the right-hand side of equation (9.51b) which give the contributions of spontaneous and stimulated emission at the signal frequency. They are comparable, that is the signal-to-noise ratio is about unity, when $F_S \approx 1/\sigma_{32}\tau_{sp}$. Using equation (9.49) to replace $\sigma_{32}\tau_{sp}$ and multiplying by hf to obtain the energy flux we find that for a signal frequency of 10 GHz and a spontaneous emission bandwidth $\Delta f/f$ of 10^{-3} this is the case at an energy flux of about 10^{-12} W/m^2. This is an exceedingly small number and implies that at microwave frequencies spontaneous emission is negligible for normal power levels. Including electromagnetic fields we find in the steady-state case after a little algebra

$$\frac{N_1 - N_3}{N_2} = \frac{1}{\tau_{21}\sigma_{13}F_P}. \tag{9.55}$$

This result shows that for sufficiently high pump flux the occupation of level 3 will approach that of level 1. Following the same line of reasoning as above we can conclude that the pump power needed to reach this condition is small, even considering that τ_{21} is much smaller than τ_{sp}. So we can safely assume that N_1 and N_3 are equal. For the magnitude of the inversion we find under this circumstance

$$\frac{N_3 - N_2}{N_3} = \frac{1 - \tau_{21}/\tau_{32}}{1 + \tau_{21}\sigma_{32}F_S}. \tag{9.56}$$

Evidently to obtain a strong inversion the condition $\tau_{21} \ll \tau_{32}$ must be fulfilled. It can also be seen that at high signal power levels the inversion decreases because stimulated emission depletes level 2 faster than pumping can fill it up. As a consequence the amplification decreases and the amplifier saturates. In solid-state masers this occurs at power levels of the order of 10^{-5} W.

9.7.4 Traveling wave amplification

In the previous section it was stated that a traveling wave scheme is the best way to make an amplifier based on stimulated emission. Consider a piece of material containing atoms having the energy level scheme of Fig. 9.21(a) through which an electromagnetic wave with frequency $f = (W_2 - W_1)/h$ is sent (Fig. 9.22). At the left the total photon flux is $F_S A$ (A being the cross-sectional area) photons per second. At the right it is $(F_S + \Delta F_S)A$. So $\Delta F_S A$ is the number of photons produced per second in a length Δz. From the rate equations (9.57) we have

$$\Delta F_S A = + \left.\frac{dN_2}{dt}\right|_{em} - \left.\frac{dN_2}{dt}\right|_{abs} = \sigma_{32}F_S(N_3 - N_2). \tag{9.57}$$

Converting from number N to density n and taking the limit $\Delta z \to 0$ we obtain

$$\frac{dF_S}{dz} = \sigma_{32}(n_3 - n_2)F_S. \tag{9.58}$$

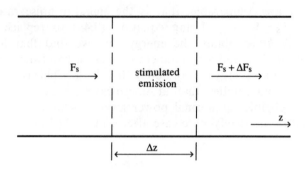

Figure 9.22 Amplification in a maser.

So the growth factor is proportional to the inversion $n_3 - n_2$:

$$F_S(z) = F_S(0) \exp\left\{\int_0^z \sigma_{32}[n_3(\zeta) - n_2(\zeta)]\,d\zeta\right\}. \tag{9.59}$$

From equation (9.56) we can see that at low signal levels $n_3 - n_2$ is constant (provided that the pump power is constant over the length of the amplifier) and the wave grows exponentially. At very high signal levels the right-hand side of equation (9.58) approaches a constant value and the wave grows only linearly. Except for stimulated emission and absorption there is also spontaneous emission which has no phase coherence with the wave and therefore manifests itself as noise. To keep the noise low we must take care that the time constant for spontaneous emission is large. In a section Δz the number of photons produced by spontaneous emission is

$$\Delta F_{\text{spont}} A = +\left.\frac{dN_2}{dt}\right|_{\text{spont}} = \frac{N_3}{\tau_{32}} = A\,\Delta z\,\frac{n_3}{\tau_{32}}. \tag{9.60}$$

Traveling to the right this will be amplified by stimulated emission so at the output we have a noise contribution

$$\Delta F_N = \Delta z\,\frac{n_3}{\tau_{32}} \exp[\sigma_{32}(n_3 - n_2)(L - z)]. \tag{9.61}$$

Since the noise contributions of the different sections are uncorrelated we can add them all up so the total noise contributed by the amplifier is

$$F_N = \int_0^L dz\,\frac{n_3}{\tau_{32}} \exp[\sigma_{32}(n_3 - n_2)(L - z)]$$

$$= \frac{1}{\sigma_{32}\tau_{32}}\frac{n_3}{n_3 - n_2}\left\{\exp[\sigma_{32}(n_3 - n_2)L] - 1\right\}. \tag{9.62}$$

This is in units of photons per square meter. To convert it to power we have to multiply it by the cross-sectional area A and by the energy of one photon (hf), with the result

$$P_N = \frac{hfA}{\sigma_{32}\tau_{32}} \frac{n_3}{n_3 - n_2} \{\exp[\sigma_{32}(n_3 - n_2)L] - 1\}. \tag{9.63}$$

In section 9.5.1 it was derived that the effective noise temperature of an amplifier is given by

$$T_e = \frac{N_A}{GkB} \tag{9.64}$$

where the added noise N_A in the present case is given by (9.63). Now the bandwidth $B = \Delta f$ of the amplifier is related to $\sigma_{32}\tau_{32}$ by equation (9.49) and the gain is given by equation (9.59) with $z = L$. Putting it all together we find

$$T_e = \frac{n_3}{n_3 - n_2} \frac{8\pi hf^3 A}{c^2 k}.$$

At 10 GHz with a cross-sectional area of 2 cm^2 (about the size of an X-band waveguide) we find $T_e \approx 0.07 n_3/(n_3 - n_2)$ K. In practice the noise temperature is always higher because of losses in the circuit. Still masers are extremely low noise amplifiers. They find applications as pre-amplifiers in radioastronomy and satellite communication.

REFERENCES

1. Pozar, D. M. (1990) *Microwave Engineering*, Addison-Wesley, Reading, MA.
2. Lange, J. (1967) Noise characterization of linear twoports in terms of invariant parameters. *IEEE J. Solid-State Circuits*, **2**, 37–40.
3. Liao, S. (1987) *Microwave Circuit Analysis and Amplifier Design*. Prentice-Hall International, Englewood Cliffs, NJ.
4. Orton, J. W., Paxman, D. H. and Walling, J. C. (1970) *The Solid-State Maser*, Pergamon, Oxford.

FURTHER READING

1. Vendelin, G. D. (1987), *Design of Amplifiers and Oscillators by the S-parameter Method*, Wiley, Chichester, New York.
2. Gonzalez, G. (1984) *Microwave Transistor Amplifiers: Analysis and Design*, Prentice-Hall, Englewood Cliffs, NJ.
3. Combes, P. F., Graffeuil, J. and Sautereau, J. -F. (1987) *Microwave Components Devices and Active Circuits*, Wiley, Chichester, New York.
4. Pengelly, R. S. (1986) *Microwave Field Effect Transistors – Theory, Design and Applications*, Wiley, Chichester, New York.
5. Soares, R., Graffeuil, J. and Obregon J. (eds) (1983) *Applications of GaAs Mesfets*, Artech, Dedham, MA.
6. Fukui, H. (ed.) (1981) *Low-Noise Microwave Transistors and Amplifiers*, IEEE, New York.

7. Decroly, J. C., Laurent, L., Lienard, J. C., Marechal, G. and Vorobeitchik, J. (1973) *Parametric Amplifiers*, Macmillan, London.
8. Matthaei, G., Young, L. and Jones, M. E. T. (1980) *Microwave Filters, Impedance Matching Networks and Coupling Structures*, Artech, Dedham, MA.

10

Oscillators

10.1 OSCILLATOR CIRCUIT THEORY

10.1.1 Introduction

Oscillators are used to generate coherent sinusoidal waveforms which may be used to transmit information or energy. In Chapter 2 we met several vacuum tube oscillators, e.g. the reflex klystron and the magnetron. Here we will discuss oscillators based on semiconductor devices.

An oscillator is by definition a non-linear circuit. It converts d.c. energy into a.c. energy and this can only be done by means of a non-linear element. We must therefore be careful with the use of concepts from linear circuit theory, such as complex impedance. Yet we will do this frequently in the following, justified by the fact that in a stable oscillating circuit a periodic signal exists in which the fundamental harmonic dominates.

In terms of circuit theory we can describe an oscillator in two ways: with a one-port model in which a resonant circuit is undamped by an impedance with a negative real part or with a two-port model where part of the output of an amplifier is fed back to the input via a resonant circuit. In both cases the resonant frequency of the circuit determines the oscillation frequency. In microwave technology the resonant circuit can be a piece of waveguide, coaxial or microstrip line, or a piece of dielectric or magnetic material (section 10.4). The description of these circuits using electromagnetic field theory often is very complicated. Therefore one tries as much as possible to model the actual structure by means of an equivalent circuit which is characterized by impedances or S-parameters. Here one is helped by the fact that microwave diodes and transistors are small compared with the wavelength, so that it is allowable to treat them as discrete circuit elements with a finite number of terminals at which voltages and currents can be defined.

It is natural to suppose that the one-port description (undamped circuit) is particularly suitable for negative-resistance diode oscillators and the other (feedback amplifier) for transistor oscillators. In fact both descriptions are equivalent and can be applied to every oscillator circuit. A negative-resistance diode can be looked upon as a reflection amplifier; on the other hand, a transistor can, with the proper feedback, appear at either of its ports as an impedance with negative real part.

The passive circuit has to perform two tasks: it determines the oscillation frequency and it matches the load to the active device. The load usually is a standard transmission line that is terminated in a reflection-free manner. Since these two functions pose different demands it would be nice if one could assign them to different parts of the circuit. In a diode oscillator this is difficult to realize but in a transistor oscillator one can place the resonator at the gate side and the load at the drain side so they can be separately optimized. Every oscillator needs a d.c. bias circuit and a high frequency output circuit. Filters have to be provided to prevent d.c. power from reaching the output and a.c. power from being lost in the bias circuit. Ways to achieve this have been described in sections 7.5.1 and 8.3.2. In the following, we will concentrate on the a.c. behavior and leave out the d.c. circuit and the filters.

10.1.2 One-port description

In the one-port description the oscillator can be represented by the equivalent circuit of Fig. 10.1. Under stationary oscillation conditions an alternating current will circulate which contains the desired oscillation frequency f_0 and higher harmonics $2f_0$, $3f_0$ etc. Going round the circuit the algebraic sum of the voltages must be zero from which follows the *oscillation condition*:

$$Z_c(n\omega_0) = -Z_d(n\omega_0), \quad n = 1, 2, \ldots. \tag{10.1}$$

The imaginary part of this equation will be satisfied at the resonant frequency of the circuit including the negative-resistance device. This resonance thus determines the oscillation frequency. The real part tells us that the losses in the circuit are compensated by the negative resistance. Usually the device behavior is such that at small signal amplitude the circuit losses are overcompensated and a signal present at the resonance frequency will grow in time. Because the diode impedance is dependent on the amplitude of the a.c. waveform an increase of the signal amplitude will lead

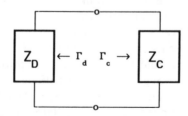

Figure 10.1 Equivalent circuit of a one-port oscillator.

to a decrease of the overcompensation until the circuit losses are exactly compensated. The oscillation can start because there is always noise present, either from the lossy parts of the circuit or from the device. This noise will always have a component at the resonance frequency.

The reason why we have phrased the above reasoning in terms of (over)compensation of losses and not simply as the negative real part of the device impedance being greater than that of the circuit will become clear in the next section.

10.1.3 Two-port description

In the two-port description of an oscillator we have the equivalent circuit of Fig. 10.2(a). Here $s(\omega)$ represents the noise, $A(\omega)$ the amplification and $H(\omega)$ the transfer characteristic of the circuit. The latter will contain the resonance that determines the oscillation frequency. Now we can describe the start-up of the oscillator as follows: initially the roundtrip gain in the loop is greater than unity. As the signal amplitude grows the amplification decreases until a stable oscillation is reached. For the signal output we then have the formula

$$S(\omega) = \frac{(1 - r)A(\omega)}{1 - rA(\omega)H(\omega)}\, s(\omega). \tag{10.2}$$

Since $s(\omega)$ is very small, we can now reformulate the oscillation condition as

$$rA(\omega)H(\omega) \approx 1. \tag{10.3}$$

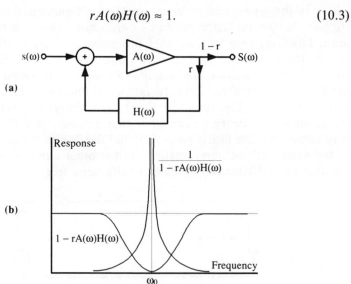

Figure 10.2 Oscillator as a feedback amplifier.

Suppose now that $A(\omega)$ is only weakly frequency dependent and $H(\omega)$ has a resonance at a frequency ω_0. Then $1-rAH$ plotted against frequency has the form of Fig. 10.2(b) and its inverse peaks very sharply at ω_0. So the frequency selectivity of the circuit is increased enormously by the feedback. This is why an oscillator with a resonant circuit of moderate frequency selectivity can nevertheless produce a signal of high spectral purity.

We can also apply this model to the circuit of Fig. 10.1. The a.c. part thereof is again drawn in Fig. 10.3, now including a noise source. Furthermore, a piece of transmission line is inserted, to make it clear that we can consider this circuit as a reflection amplifier with feedback, the feedback being caused by the reflection coefficient of the circuit impedance Z_C. If we identify $s(\omega)$ with the noise voltage of the noise source and $S(\omega)$ with the noise voltage over Z_c then we have instead of equation (10.2)

$$S(\omega) = \frac{Z_c}{Z_c + Z_d'}\, s(\omega) = \frac{(1 + \Gamma_c)(1 - \Gamma_d')}{1 - \Gamma_c\Gamma_d'}\, s(\omega) \qquad (10.4)$$

where Z_d' and Γ_c' are the diode impedance and reflection coefficient, respectively, seen from the load side of the transmission line. The condition that the numerator must be small again yields equation (10.1), now with Z_d' instead of Z_d, and alternatively

$$\Gamma_d'\Gamma_c \approx 1 \qquad (10.5)$$

which is the equivalent of equation (10.3). Equations (10.2) and (10.4) suggest that an oscillator produces only noise. This is actually true. We must, however, keep in mind that this noise is filtered with a very small bandwidth, because of the enhanced frequency selectivity discussed above. *An oscillator is a narrow band amplifier of noise.* This phrase was coined by the late H. Zaalberg van Zelst, researcher at Philips Research Laboratories and Professor of Electronics at Eindhoven University. With the help of stochastic signal theory we can show that a noise signal filtered through a very narrow band is nearly sinusoidal and the degree of coherence increases as the bandwidth becomes smaller. On the other hand, we can learn from this that an oscillator can never be totally noise free.

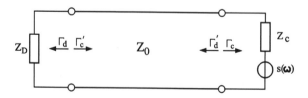

Figure 10.3 Reflection amplifier as a feedback circuit.

10.1.4 Oscillator characteristics and design

The most important characteristics of an oscillator are the oscillation frequency and the output power. Other properties to be considered are the harmonic content, the sideband noise and the d.c. power consumption. For many applications an oscillator has to be tunable over a range of frequencies. To carry information the oscillator signal has to be modulated. So the tuning and modulation characteristics are also important properties. The oscillation frequency is determined by a resonance in the circuit. Different types of resonators are used. The most important ones are described in section 10.4. Tuning can be achieved by mechanically changing the resonator or by having a bias-dependent reactance, e.g. a varactor diode, in the circuit. In the latter case one speaks about *electronic tuning*. For an electronically tunable oscillator important factors are the tuning linearity, the settling time and the post-tuning drift. The latter two are illustrated in Fig. 10.4.

At start-up, the oscillator signal is very small and small-signal conditions prevail. As the oscillation amplitude increases the oscillation frequency will change somewhat but not very much. So to determine the oscillation frequency for a given circuit a small-signal analysis can be used. Fine tuning of the frequency will nearly always be necessary after the circuit has been constructed. To this end one can use the bias dependence of the reactance of the active element, or if this is not sufficient one has to provide external means of trimming the circuit reactance. Whether or not the oscillation will start can be determined by the method of section 10.2.1. For a transistor oscillator the *stability circles* introduced in section 9.4.2 can also be used.

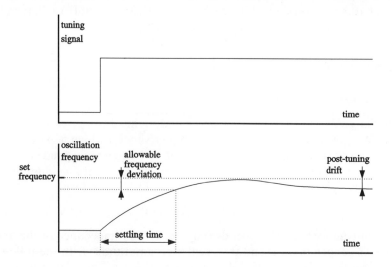

Figure 10.4 Schematic behavior of an electronically tuned oscillator.

Suppose for instance that the resonator is included in the gate (emitter) circuit and the load in the drain (collector) circuit. Then what one has to find out is whether the load reflection coefficient, transformed through the (small-signal) S-parameters of the transistor, is in an unstable region. This evaluation has to be done at the resonant frequency of the resonator of course.

For a large-signal analysis two methods are available. Because the oscillator signal is usually reasonably sinusoidal, S-parameter measurements as a function of signal amplitude (*large-signal S-parameters*) can be made of the active element and these can be used to determine the amplitude at which the energy delivered by the active device matches the losses in the circuit plus external load. This method of course gives no information about the higher harmonic content and is unfit for devices which produce a large amount of higher harmonics, notably Gunn and IMPATT diodes. The other method takes full account of the non-linearity of the active device. Either one makes a full time-dependent analysis of the circuit, using the non-linear device characteristics, or one applies the so-called *harmonic balance method*, in which the device is treated by a non-linear analysis and the circuit (which is assumed to be linear) by complex calculus for each harmonic. Usually not more than five harmonics are included. The full time-dependent analysis can, if it is possible to describe the circuit in terms of lumped elements, be done using a circuit analysis program such as SPICE®. For transmission line circuits there are commercial programs available, such as COMPACT®, TOUCHSTONE® or MDS®, which can perform non-linear analysis of microwave circuits.

Noise analysis of oscillators can be very complicated, because of the up- and downconversion of noise between baseband and oscillator sidebands. Oscillator noise is treated in section 10.3.2.

10.2 STABILITY

10.2.1 Conditions for oscillation

A general analysis of the stability of oscillators has been given by Kurokawa [1]. We give it here in a somewhat simpler version which is based on the assumption that we can extend the concept of frequency-dependent impedance $Z(\omega)$ by analytic continuation to complex frequency ω. Suppose the current through the high-frequency (h.f.) circuit has the form

$$i(t) = \text{Re}[A \exp(j\omega t)]. \qquad (10.6)$$

The circuit impedance Z_c depends only on ω. The impedance of the active element Z_d depends on ω and the oscillation amplitude A. In the following we assume that Z_d depends only on A, since the ω dependence is usually comparatively weak.

We now write the oscillation condition as

$$Z_c(\omega) + Z_d(A) = 0 \qquad (10.7)$$

and we assume that this holds also for complex ω. A double Taylor series development around ω and A ($\omega = \omega_0 + \delta\omega$, $A = A_0 + \delta A$) then yields:

$$Z_c(\omega_0) + Z_d(A_0) + \delta\omega \frac{dZ_c}{d\omega_0} + \delta A \frac{dZ_d}{dA_0} = 0. \qquad (10.8)$$

According to the oscillation condition the first two terms add up to zero, so that we can write, with the short-hand notation $dZ_c/d\omega_0 \Rightarrow \mathscr{Z}_{c\omega}$ etc.,

$$\delta\omega = -\frac{\mathscr{Z}_{dA}}{\mathscr{Z}_{c\omega}} \delta A. \qquad (10.9)$$

Splitting this into real and imaginary parts we obtain

$$\mathrm{Re}(\delta\omega) = -\frac{\mathscr{R}_{dA} \cdot \mathscr{R}_{c\omega} + \mathscr{X}_{dA} \cdot \mathscr{X}_{c\omega}}{|\mathscr{Z}_{c\omega}|^2} \delta A \qquad (10.10a)$$

$$\mathrm{Im}(\delta\omega) = -\frac{\mathscr{X}_{dA} \cdot \mathscr{R}_{c\omega} - \mathscr{R}_{dA} \cdot \mathscr{X}_{c\omega}}{|\mathscr{Z}_{c\omega}|^2} \delta A. \qquad (10.10b)$$

Now $\mathrm{Im}(\delta\omega) > 0$ means that the oscillation amplitude decreases and $\mathrm{Im}(\delta\omega) < 0$ that it increases. Stable oscillation therefore demands that, if $\delta A > 0$, $\mathrm{Im}(\delta\omega) > 0$, and, if $\delta A < 0$, $\mathrm{Im}(\delta\omega)$ must be negative. It then follows from equation (10.10b) that

$$\mathscr{X}_{dA} \cdot \mathscr{R}_{c\omega} - \mathscr{R}_{dA} \cdot \mathscr{X}_{c\omega} < 0 \qquad (10.11)$$

or, after dividing by $\mathscr{R}_{c\omega} \cdot \mathscr{R}_{dA}$,

$$\frac{\mathscr{X}_{c\omega}}{\mathscr{R}_{c\omega}} > \frac{\mathscr{X}_{dA}}{\mathscr{R}_{dA}} \qquad \text{when } \mathscr{R}_{c\omega} \cdot \mathscr{R}_{dA} > 0 \qquad (10.12a)$$

$$\frac{\mathscr{X}_{c\omega}}{\mathscr{R}_{c\omega}} < \frac{\mathscr{X}_{dA}}{\mathscr{R}_{dA}} \qquad \text{when } \mathscr{R}_{c\omega} \cdot \mathscr{R}_{dA} < 0. \qquad (10.12b)$$

Given two possibilities for the signs of $\mathscr{R}_{dA}$ and $\mathscr{X}_{dA}$ we can distinguish four cases which are depicted in Fig. 10.5. The arrow $\mathscr{Z}_{dA}$ indicates the direction in which Z_d moves in the complex plane when A increases. Inspection of the relations (10.12) yields the result that condition (10.12a) is fulfilled when the arrow $\mathscr{Z}_{cw}$ representing the movement of Z_c as a function of frequency is in the sector marked 'a' and condition (10.12b) in the sector marked 'b'. In all cases sectors a and b add up to a semicircle having the same orientation with respect to $\mathscr{Z}_{dA}$. The conclusion from these figures is that for all four cases stable oscillation results when when $\mathscr{Z}_{c\omega}$ makes a counterclockwise angle between 0° and 180° with respect to $\mathscr{Z}_{dA}$.

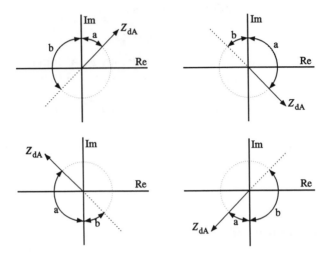

Figure 10.5 Graphical analysis of oscillator stability.

This result allows a graphical analysis of oscillator stability. If $-Z_d(A)$ and $Z_c(\omega)$ are plotted in the complex plane, the curves will intersect in a number of points and these intersections give possible points of oscillation. Let arrows give the directions in which Z_d and Z_c move when A and ω increase. The intersection points will be stable when going clockwise from the arrow connected with $-Z_d$ to the one connected with Z_c an angle between $0°$ and $180°$ is found.

Note that the terminology may be somewhat confusing: 'stable' here means oscillating steadily and unstable means a decaying or growing oscillation. In another context (amplifiers for instance) stable would mean no oscillation.

We can perhaps make this rather abstract theory come alive by applying it to two simple examples, a parallel and a series resonant circuit (Fig. 10.6). For simplicity we assume that Z_d is real and negative and its magnitude decreases with increasing amplitude of the a.c. voltage.

For the parallel circuit the impedance can be written as

$$Z_C = \frac{j\omega L}{1 - \omega^2 LC + j\omega L/R}. \tag{10.13}$$

In the complex plane this describes a circle which is followed clockwise when ω increases (Fig. 10.7(a)). On the other hand, $-R_D$ moves over the real axis from right to left when the oscillation amplitude increases. So the angle between the two is greater than $180°$ and the intersection is an unstable point, meaning that any oscillation in this point will be damped.

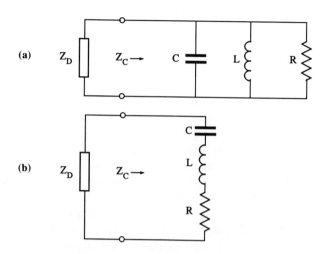

Figure 10.6 Two simple oscillator circuits.

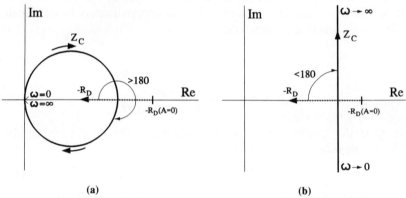

Figure 10.7 Stability diagrams for the circuits of Fig. 10.6.

For the series circuit the impedance is

$$Z_C = j\omega L + R + \frac{1}{j\omega C}. \tag{10.14}$$

The locus of this when ω changes is a vertical line (Fig. 10.7(b)) and the intersection angle now is smaller than 180° meaning that this circuit will oscillate stably.

It has been pointed out [2] that the same results can be derived by applying Nyquist stability theory to equation (10.4). This theory says that when the locus of $\Gamma_c\Gamma_d$ for ω going from $-\infty$ to $+\infty$ encircles the point 1 the circuit will oscillate. For the two examples above it is easily seen that this means that the series circuit will oscillate whereas the parallel circuit will not.

10.2.2 Amplitude and frequency stability

Once an oscillator is operating one becomes interested in how stable its frequency and amplitude really are. The frequency stability is particularly important in microwave systems. The short-term stability is influenced by variations in bias current and ambient temperature. This is called (*current or temperature*) *pushing*. The oscillation frequency can also be influenced by variations of the load impedance. This effect is called *load pulling*.

The long-term stability depends largely on the aging of the components the oscillator is made of, in particular the semiconductor devices. Other factors of influence are oxidation of metal parts and absorption of water vapour by dielectrics. Even adsorption of water on free semiconductor surfaces can affect the oscillation frequency.

Suppose the passive part of the oscillator contains a load impedance which satisfies the oscillation condition (10.1). To fix the oscillation frequency this impedance will be connected to the active device through a two-port which often is some kind of resonator (Fig. 10.8). Here the resonator is a hollow cavity which is coupled to coaxial lines by the protruding ends of their center conductors. This twoport can approximately be described by the following set of S-parameters:

$$S_{12} = S_{21} = \frac{j/Q}{\omega_0/\omega - \omega/\omega_0 + j/Q} \tag{10.15}$$

$$S_{11} = S_{22} = \frac{\omega_0/\omega - \omega/\omega_0}{\omega_0/\omega - \omega/\omega_0 + j/Q}. \tag{10.16}$$

Here Q is the quality factor of the resonator. The oscillation condition (10.5) can now be rearranged to give

$$S_{12}S_{21}\Gamma_d\Gamma_L = (1 - S_{11}\Gamma_d)(1 - S_{22}\Gamma_L). \tag{10.17}$$

Around the resonant frequency ω_0, S_{11} and S_{22} are very small so that this equation can be approximated by

$$-\frac{1/Q^2}{(\omega_0/\omega - \omega/\omega_0 + j/Q)^2}\Gamma_d\Gamma_L = 1. \tag{10.18}$$

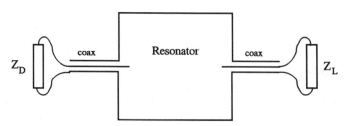

Figure 10.8 Oscillator configuration with resonant cavity.

To this equation we can apply a variation ΔI to assess current pushing or a variation $\Delta \Gamma_L$ for load pulling, either of which will lead to a frequency deviation $\Delta \omega$ and an amplitude deviation ΔA. If we evaluate the results at $\omega = \omega_0$ we obtain

$$-\frac{4jQ}{\omega_0} \Gamma_d \Gamma_L \Delta \omega + \Gamma_L \frac{d\Gamma_d}{dA} \Delta A + \Gamma_L \frac{d\Gamma_d}{dI} \Delta I = 0 \qquad (10.19)$$

$$-\frac{4jQ}{\omega_0} \Gamma_d \Gamma_L \Delta \omega + \Gamma_d \Delta \Gamma_L + \Gamma_L \frac{d\Gamma_d}{dA} \Delta A = 0. \qquad (10.20)$$

As before we can split these in real and imaginary parts to find $\Delta \omega$ and ΔA but even without doing this it can be seen that $\Delta \omega$ will be smaller the greater Q is. This shows the importance of the quality factor for the frequency stability.

Another consideration is that the negative resistance of a device always goes through an extremum for some bias current. This is the point where we expect most power from the device and, since in the point $\mathcal{R}_{dI} = 0$ we may expect $d\Gamma_d/dI$ also to be small so it follows that this is also a good bias point as far as stability is concerned.

10.3 MODULATION AND NOISE

10.3.1 Tuning and modulation

For many applications an oscillator has to be tuned, that is the oscillation frequency has to be changed from one predetermined value to another. For applications where information has to be transmitted the carrier wave generated by the oscillator has to be modulated. Modulation of an oscillator can be done in amplitude, frequency or phase. In the microwave field frequency modulation (FM) and phase modulation (PM) are preferred for several reasons, (FM) for analog signals and PM for digital signals. Tuning and frequency modulation differ only in the speed with which the frequency changes. Tuning is possible by mechanically changing the size of the resonator, otherwise the same techniques can be used for tuning and FM. Since Z_d depends on the d.c. bias current I_0 it is in principle possible to modulate an oscillator via its bias current. This is called *bias tuning (or modulation)*. Although this method looks attractive because of its simplicty it has disadvantages too: usually AM and FM modulation occur at the same time, the modulation characteristics can be non-linear and the greater the modulation sensitivity the greater also the sensitivity for noise in the bias current. So, amplitude modulation is best done externally with a p–i–n diode modulator as described in Chapter 7. For FM and PM one can use modulation of X_c by including a varactor diode or a YIG resonator (section 10.4.3) in the circuit. At high microwave frequencies, however, bias tuning is still preferred because of its simplicity.

10.3.2 Oscillator noise

No oscillator is a perfect sinusoidal generator, as already has been shown
in section 10.1.3. Besides higher harmonics there are also stochastic fluctuations in amplitude and frequency, which are called noise. Because of the
noise the spectrum of an oscillator around the oscillation frequency always
has sidebands as shown in Fig. 10.9(a). We can describe the signal with
noise by the following expression:

$$V(t) = [A_0 + a(t)] \cos[\omega_c t + \phi(t)] \qquad (10.21)$$

where $a(t)$ and $\phi(t)$ are random signals. $a(t)$ describes the *AM noise*, $\phi(t)$ is
called *phase noise* and $d\phi/dt$ *FM noise*. The use of the term phase noise is
only meaningful when the maximum phase deviation does not exceed
2π rad. In a vector diagram (Fig. 10.9(b)) we can represent the noise-free
carrier by a vector of constant length which rotates with a constant velocity
and the noise by an extra vector with varying amplitude and rotation velocity.
The component of the latter which lies in the direction of the carrier vector
describes the AM noise and the orthogonal component the FM noise.
Usually it is assumed that the total noise power is small compared with the
power at the central frequency f_c. A common measure for the noise then is
the power in the bandwidth B at a distance f_m from f_c, divided by the
bandwidth and in dB relative to the carrier wave power (dBc), or

$$\text{oscillator noise figure} = \frac{P_d(f_m, B)}{P_c(f_c)B} \text{ dBc/Hz.} \qquad (10.22)$$

This is specified for AM and FM separately. Usually P_n is measured in one
sideband so it should correctly be called the *single-sideband oscillator noise
figure*. For FM noise one often uses another measure, the root mean square
frequency deviation Δf_{RMS}, that is the root of the average quadratic
frequency deviation in a specified bandwidth divided by that bandwidth. A
similar definition holds for $\Delta \phi_{RMS}$. Here too the distance from the carrier
frequency must be specified. The relation with the FM noise power is

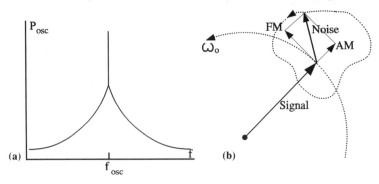

Figure 10.9 Oscillator noise: (a) spectrum; (b) vector diagram.

$$(\Delta\phi_{RMS})^2 = \left(\frac{\Delta f_{RMS}}{f_m}\right)^2 = 2\,\frac{P_{FM}}{P_c}.$$ (10.23)

For one-port oscillators it can be related to the device noise measure, defined in Chapter 7 [3, 4]:

$$(\Delta\phi_{RMS})^2 = \frac{1}{Q_{ext}^2}\,\frac{MkT_0B}{P_c}$$ (10.24)

where Q_{ext} is the loaded quality factor (section 10.4.1) of the resonator.

Sources of noise in oscillators are as follows:

- noise in the active element;
- noise in the bias current;
- noise in the lossy parts of the passive circuit.

Not only the noise around the oscillation frequency but also the low frequency noise are important. The latter is, owing to the non-linear properties of the active element, upconverted to the sidebands of the oscillator. The upconversion is related to the modulation sensitivity of the oscillator, that is the extent to which the oscillation frequency depends on the bias current. It will be clear that an oscillator which is modulated via the bias current is also sensitive to the noise in the bias current. The $1/f$ noise in the bias current and the active element are therefore important quantities. Because Gunn diodes have much $1/f$ noise their noise performance close to the carrier is worse than that of IMPATT diodes. Further away from the carrier IMPATT diodes are noisier because here the multiplication noise predominates. Besides upconversion there is also downconversion, i.e. sideband noise is downconverted to baseband noise. Because of this effect amplitude modulation noise can be converted to frequency modulation noise and vice versa. The noise of oscillators can be measured in the following ways.

- For amplitude noise one has to rectify the oscillator signal. This gives the function $A_0 + a(t)$ of equation (10.21). The noise component of this can be determined by low-frequency measurement techniques.
- For FM noise the situation is more complicated. Essentially one first has to convert it to AM noise which can be measured as described above, subtracting the original AM noise. A method to do this has been described by Ondria [5].

10.3.3 Injection locking

In section 10.1.3 it has been shown that an oscillator is really a narrow band amplifier of the noise that is always present in the circuit. Instead of noise we can also amplify the signal of another oscillator which can be injected at any point of the feedback loop. This is called *injection locking* and is used to reduce the noise of the oscillator. Of course the frequency of the injected signal must be within the bandwidth of the oscillator, or more precisely, the amplified injected signal must be larger than the amplified noise. With reference to equation (10.2),

$$\left| \frac{(1-r)A(\omega_i)}{1 - rA(\omega_i)H(\omega_i)} S(\omega_i) \right| > \left| \frac{(1-r)A(\omega_0)}{1 - rA(\omega_0)H(\omega_0)} s(\omega_0) \right| \qquad (10.25)$$

where $S(\omega_i)$ is the injected signal at frequency ω_i and $s(\omega_0)$ the noise at the frequency ω_0 where the denominator has its minimum. Since the fraction

$$\frac{(1-r)A(\omega)}{1 - rA(\omega)H(\omega)}$$

is a very sharply peaked function of ω, ω_i cannot be much different from ω_0. The allowable deviation is greater the stronger the injected signal is. The frequency range of ω_i where condition (10.25) is satisfied is called the *locking range*. An extensive theoretical treatment of injection locking has been given by Kurokawa [6].

10.3.4 Back rectification

Oscillator diodes are by definition non-linear elements and when they are subjected to a sinusoidal voltage the current will contain higher harmonics but also a d.c. component, especially when the voltage amplitude is large. Compare this with the shift of the bias point of a detection diode (section 8.1.2). As a consequence in an oscillating diode an extra direct current will flow and the bias point will be shifted according to the d.c. load line. This effect is called *back rectification* or *d.c. restoration*. An unpleasant consequence of this is that devices that normally have a d.c. characteristic with a positive

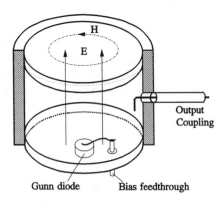

Figure 10.10 TM$_{010}$ resonator.

slope can acquire a negative slope which can lead to bias circuit oscillations at low frequencies. Back rectification is not all bad, however. It can be used to build very simple Doppler radar systems. A Doppler radar sends out a continuous wave. When this is reflected by a moving object the frequency is changed and the reflected wave can be mixed with the original signal to obtain the difference frequency which is directly proportional to the speed of the reflecting object. The back rectification effect makes it possible to use the oscillator diode itself as the mixer, leading to an extremely simple circuit. It has been found that Gunn and BARITT diodes are especially well suited for this application.

10.4 RESONATORS

10.4.1 Transmission line resonators

Any structure that supports standing waves is a resonator. In particular, a section of transmission line, be it coaxial, waveguide or microstrip, between two discontinuities constitutes a resonator and can be used to determine the frequency of an oscillator. Every resonator has an infinite number of resonant frequencies and when designing an oscillator one should take care that the other resonances are far away from the desired one. An important quantity is the *quality factor Q* referred to already in section 10.2.2 which is defined generally as

$$Q = 2\pi \times \frac{\text{energy stored in the resonator}}{\text{energy dissipated in one signal period}}.$$

When the resonator is completely closed off from the outside world the dissipation is in the cavity walls and the medium filling it. This is called the *internal* or *unloaded Q*. When the cavity is connected to the outside the dissipated energy includes energy given off to the external load. In this case one speaks of the *external* or *loaded Q*.

A popular resonator for Gunn diode oscillators is the cylindrical cavity whose fundamental mode is the TM_{010} mode (Fig. 10.10).

IMPATT diodes are often mounted under a post in a waveguide. The post can be provided with a cap as in Fig. 10.11. The space under the cap is an open resonator and at the same time provides impedance matching. The circuit is completed by a moveable short at about a quarter-wavelength from the post, allowing fine tuning.

A simple coaxial resonator is made with a moveable sleeve (Fig. 10.12). The sleeve gives locally a lower characteristic impedance and when it is a quarter-wavelength long it acts as an impedance transformer. The space between the sleeve and the end is the resonator. In practice often three sleeves are used to increase the range over which impedance matching is

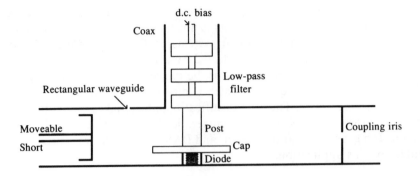

Figure 10.11 Post-and-cap circuit in waveguide.

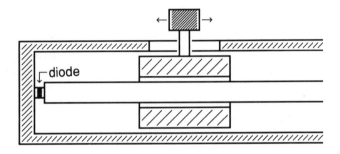

Figure 10.12 Coaxial resonator with moveable sleeve.

possible. This circuit is handy when a quick evaluation has to be made of the oscillating properties of a new batch of diodes.

Microstrip resonators have been discussed in section 6.2. Their quality factors are low in general so if one wants a stable microstrip oscillator a high-Q cavity, e.g. a dielectric resonator (section 10.4.2) should be added.

10.4.2 Dielectric resonators

A piece of dielectric material with a larger permittivity than the surrounding air also behaves as a resonator. It is called an *open resonator* because part of the electromagnetic field is always in the space outside the dielectric. The confinement of the field is greater and the size is smaller at a given resonant frequency for a greater difference in dielectric constant between the material and air. The field pattern is similar to that of a closed metal box of the same form. Preferably cylindrical resonators are used (Fig. 10.13), with the field pattern of the TE_{01m} mode in which the magnetic field lines are in the direction of the axis. The field lines that extend outside the resonator can be used for coupling with a microstrip line.

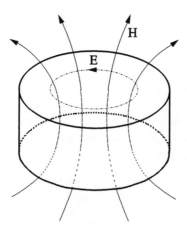

Figure 10.13 Dielectric resonator.

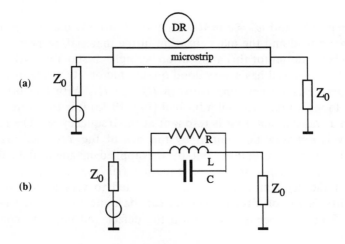

Figure 10.14 (a) Microstrip mounting and (b) equivalent circuit of DR.

Figure 10.14 shows how a dielectric resonator is mounted close to a microstrip line and also the equivalent circuit. In the latter the transformer ratio is a decreasing function of the distance to the line.

The materials which are used for dielectric resonators come from the family of $BaTiO_3$ and have a high dielectric permittivity ε_r which makes it possible to keep the dimensions small. These resonators are therefore excellent for use in microstrip circuits at frequencies from 2 GHz upwards. The quality factor is higher than that of cavity resonators so that stable and low noise oscillators are realized. The temperature variation is also less than that of metal cavities. A DRO (*dielectric resonator oscillator*) therefore suffers less from frequency pushing than a CSO (*cavity-stabilized oscillator*).

10.4.3 YIG resonators

Another possibility is to make the resonator of a magnetic material in which case the magnetic permeability is greater then that of air. An extra degree of freedom can be obtained by subjecting the material to a d.c. magnetization. Now the permeability μ assumes the form of a tensor, which without losses has the following form:

$$\mu = \mu_0 \begin{bmatrix} 1 + \dfrac{\omega_m \omega_h}{\omega_h^2 - \omega^2} & j \dfrac{\omega_m \omega}{\omega_h^2 - \omega^2} & 0 \\[3ex] -j \dfrac{\omega_m \omega}{\omega_h^2 - \omega^2} & 1 + \dfrac{\omega_m \omega_h}{\omega_h^2 - \omega^2} & 0 \\[3ex] 0 & 0 & 1 + \dfrac{\omega_m}{\omega_h} \end{bmatrix} \tag{10.26}$$

with $\omega_h = e/m\,H_0$ and $\omega_m = e/m\,M_0$. H_0 and M_0 are the d.c. components of the magnetic field and the magnetization in the material, respectively.

A popular material for this application is single-crystal YIG (yttrium iron garnet). This material has a very good quality factor. By virtue of equation (10.26) there exists a coupling between H_x and H_y. With two orthogonal coupling loops around a small YIG ball (Fig. 10.15) it is therefore possible to obtain a transmission that is resonant at the frequency ω_h. The resonant frequency is not now determined by the size of the YIG ball, as in the case of a dielectric resonator, but by the applied magnetic field through the denominator of μ_{xy} and μ_{yx} in equation (10.26). The ball is made so small that the geometric resonances are shifted to very high frequencies. The quality factor of YIG resonators can be quite high, in the range of 10^4 at 4 GHz. It depends strongly on the defect and impurity content of

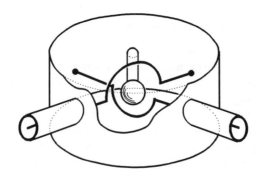

Figure 10.15 YIG resonator filter with coupling loops.

the material. For this reason single-crystal material is always used. Another factor of influence is the surface roughness so careful polishing is necessary.

The great advantage of the YIG resonator is that the resonant frequency can be tuned over a wide range by varying the magnetic bias which is generated by a field coil. Since ω_h is proportional to the magnetic field the tuning curve can be extremely linear, with deviations of less than 0.1%. The temperature variation depends on the orientation of the crystal with respect to the magnetic bias field. This opens the possibility to compensate for the temperature dependence of the other parts of the circuit. Tunable YIG filters are also used for other applications, e.g. in spectrum analyzers.

10.5 DIODE OSCILLATORS

10.5.1 Introduction

Diode oscillators, or more generally negative-resistance oscillators, employ one of the devices treated in Chapter 4 having an impedance with a negative real part.

The negative resistance of tunnel diodes and Gunn diodes is very broad band, for tunnel diodes from zero frequency to an upper limit which is about 1000 GHz. In the case of Gunn diodes it depends on the d.c. characteristic. If this has a negative slope the negative resistance will exist from d.c. conditions to around 100 GHz; if it has a positive slope every-where there will be a negative resistance in a frequency band around the inverse of the transit time, but this band can be very wide, an octave or more. IMPATT diodes have in principle only a negative resistance around the transit-time frequency, but owing to their non-linear properties the d.c. characteristic of an oscillating diode can acquire a negative slope. So with all three the bias circuit must be designed carefully to have no resonance which would excite parasitic oscillations at low frequencies.

10.5.2 Tunnel diode oscillators

Tunnel diode oscillators are characterized by low output power and low noise. The maximally attainable frequency is very high. With DBRT diodes oscillation frequencies up to 700 GHz have been reached, albeit with negligible output power. The most important application is as local oscil-lators in low noise receivers for small signals, e.g. in radio-astronomy and satellite communication. In these applications often superconducting mixers are applied which need very low local oscillator power.

10.5.3 Gunn diode oscillators

We have seen in Chapter 4 that in Gunn diodes of sufficient length domains can form consisting of an accumulation layer and a depletion layer. These domains will also form in Gunn oscillators, at least in those for low frequencies, say a few gigahertz. Their behavior will, however, be strongly influenced by the fact that the diode voltage in an oscillator consists of a d.c. and a comparatively large a.c. component. A domain will form when the total voltage rises above a certain value and will subsequently grow and travel to the anode. When the frequency is lower than the inverse of the domain transit time the domain will reach the anode before or at the time the voltage goes below the critical value necessary to sustain a domain. Since the voltage keeps going down the formation of a new domain will be delayed until the voltage rises again. This is called the *delayed-domain mode*. In the case that the period is shorter than the inverse domain propagation time the voltage will be going down before the domain has reached the anode and the domain will be dissipated. This is called the *quenched-domain mode*. At high a.c. amplitudes the so-called *LSA* (limited space-charge accumulation) *mode* can occur. Here domain formation is largely prevented because for the greatest part of the cycle the average electric field is in a range where the differential mobility is negative but small, so that the domain formation goes slowly, and in the lowest part of the cycle the field drops below the threshold value so that any trace of domain formation will be wiped out. Which mode will be operative is largely determined by the product of doping and n region length, as well as the frequency. This can be visualized by a diagram in which the product of

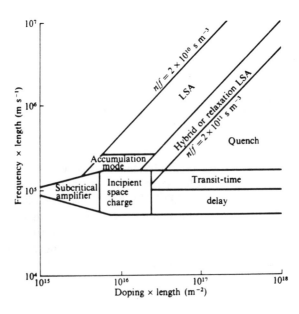

Figure 10.16 Mode diagram for Gunn oscillators. Reproduced with permission from Hobson, G. S., *The Gunn Effect*; published by Oxford University Press, 1974.

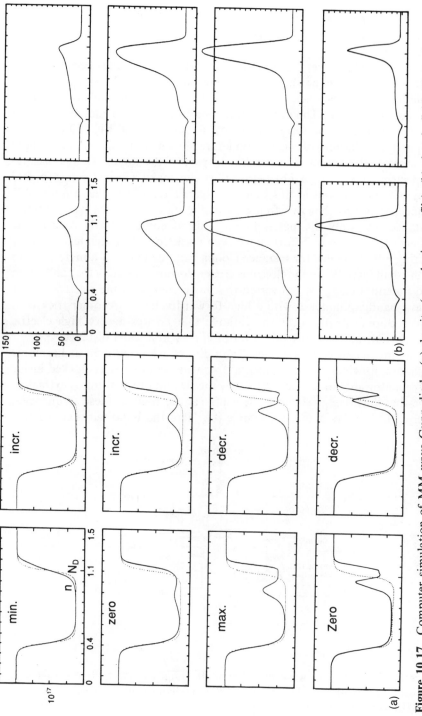

Figure 10.17 Computer simulation of MM wave Gunn diode: (a) density (. . ., doping profile); (b) electric field. The points are taken at intervals of $\pi/4$ rad, starting at the minimum of the diode voltage.

doping and length is plotted on one axis and that of frequency and length on the other (Fig. 10.16) Reference [7] provides a detailed discussion of these modes.

All these modes have little practical importance nowadays because at the lower microwave frequencies Gunn diodes are replaced by transistors and for millimeter waves where they are still used the dimensions are too small for domains to form. The dominant mode of operation is something called *traveling accumulation layer mode* and is sketched in Fig. 10.17 where computer simulations of such a diode are shown. It is interesting to note that most of the action is in the second part of the diode. This is because the electrons need to get hot before they can make a transfer to a higher valley (Chapter 3) and in this time they have already traveled a noticeable distance. It means that this first part of the diode acts as a passive series resistance that only dissipates power. This of course greatly reduces the efficiency of the diode. Efforts have been made to reduce the dead zone by injecting hot electrons, for instance from a cathode made of a material with a higher bandgap. When the electrons cross a downward step in the conduction band potential energy is converted to kinetic energy so the electrons enter the low bandgap material with a high forward velocity. Another trick is to make a doping profile near the cathode that creates a very high electric field, so that the electrons gain a lot of energy in a short distance.

At millimeter wave frequencies Gunn diodes are often used as harmonic oscillators, in which the fundamental frequency component is locked inside the resonator and an harmonic, usually the second or the third, is extracted. Somewhat surprisingly lumped elements can be used at these frequencies. Figure 10.18 shows a recent example [8]. Here the bond wires connecting

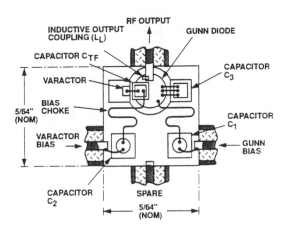

Figure 10.18 Millimeter wave third harmonic Gunn oscillator. Reproduced with permission from Cohen, L. D. (1991) *A Millimeter-Wave Third Harmonic Gunn VCO* . . ., **MTT-S** Digest, 937–8. © 1991 IEEE.

the device act as inductances. Not shown in the figure is the output coupling of the coaxial line into a waveguide whose cutoff frequency is above the fundamental frequency. The frequency can be varactor tuned between 70 and 90 GHz.

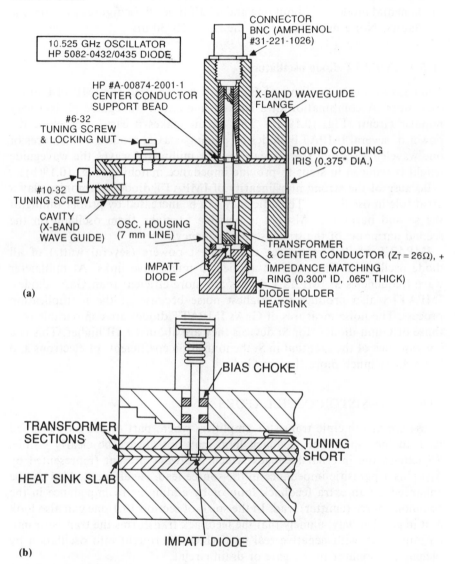

Figure 10.19 IMPATT diode oscillator circuits: (a) X band (8–12 GHz) © copyright (1971) Hewlett-Packard Company, Reproduced with permission. (b) Q band (30–50 GHz). Redrawn from Milford, T. A. and Bernick, R. L. (1979) *Millimeter-Wave CW Impact Diodes and Oscillators*, IEEE Trans. **MTT-27**, 483–92. © 1979 IEEE.

Gunn diode oscillators have medium output powers, from 1 W at the lower frequencies to 10 mW at 100 GHz. They find use, in addition to local oscillators, in small transmitters, especially for portable apparatus (police radar). GaAs is usable to about 80 GHz and InP up to 170 GHz in the fundamental mode and about one and a half times these figures as harmonic oscillators. Noise measures are in the range 20–30 dB.

10.5.4 IMPATT diode oscillators

The coaxial line with sleeve of Fig. 10.12 can be used for IMPATT diode oscillators. A combination of a coaxial line and a waveguide [9] is a very popular circuit (Fig. 10.19(a)). This circuit makes it easy to combine the power of several IMPATT diodes by placing coaxial mounts at distances of one wavelength along the waveguide. For millimeter waves the waveguide height is reduced in steps to provide impedance matching (Fig. 10.19(b)).

Because of the strong non-linearity of IMPATT diodes harmonics play a great role in oscillators. The efficiency can be increased by proper tuning of the second harmonic. Also, it is possible to make them oscillate on the second harmonic of the transit-time frequency.

IMPATT oscillators deliver the greatest powers (several watts) of all diode oscillators. They can be applied in microwave links. At millimeter wave frequencies Si IMPATT diodes are more efficient than GaAs diodes. IMPATTS also produce the highest noise because of the multiplication process. The noise measures of GaAs IMPATT diodes are comparable with those of Gunn diodes; for Si devices they are about 10 dB higher. This is a consequence of the fact that in Si the ionization coefficients of electrons and holes differ much more than in GaAs.

10.6 TRANSISTOR OSCILLATORS

These are in principle transistor amplifiers where part of the output is fed back to the input. The feedback loop may not always be clearly visible. Sometimes the internal feedback which is always present (represented by S_{12}) plus a parasitic impedance in the source lead, is sufficient. This can be enhanced by an extra feedback loop or by adding extra impedance in the common source (emitter) lead. In the spirit of section 10.1 one can also look at it in another way, namely that the feedback transforms the transistor into an impedance with negative real part which is brought into oscillation by placing a resonator in the gate or drain circuit.

A quick look at the feedback circuit of Fig. 9.9 and its S-parameters shows that with these purely resistive feedbacks $|S_{11}|$ and $|S_{22}|$ can never become greater than unity. However, when both resistances are replaced by complex impedances it will be possible and, likewise, when the gate–source and gate–drain capacitances are not neglected.

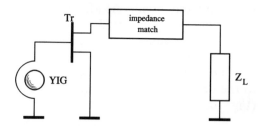

Figure 10.20 YIG-tuned transistor oscillator.

At frequencies below 5 GHz Si bipolar transistors are used frequently in circuits that are essentially variations on the well-known Colpitts or Clapp oscillator circuits.

At higher frequencies often the negative-resistance concept is used but in a more refined way than the simple diode oscillator circuit. The fact that the transistor is a twoport makes it possible to place the frequency-determining resonator at one side and the load at the other, as in Fig. 10.20. This separation makes it easier to fulfill simultaneously the conditions for frequency tuning and impedance matching. Oscillators which are tuned with YIG resonators (section 10.4.3) are often constructed this way. If the resonant circuit is a YIG filter the oscillation frequency is determined by the magnetic field and is electronically tunable when the field is produced

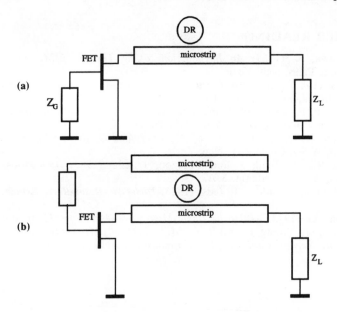

Figure 10.21 Dielectric resonator oscillator (DRO) types.

by a current coil. In this way oscillators are made that can be tuned from 1 to 18 GHz or from 26 to 40 GHz.

In oscillators with fixed frequency in microstrip technology nowadays ever more dielectric resonators (section 10.4.2) are used. The dielectric resonator can be coupled to the gate lead, which gives a circuit similar to Fig. 10.20, or to the drain lead (Fig. 10.21(a)). It can also be part of the feedback loop (Fig. 10.21(b)).

REFERENCES

1. Kurokawa, K. (1969) Some basic characteristics of broadband negative resistance oscillators. *Bell Syst. Tech. J.*, **48**, 1937–55.
2. Jackson, R. W. (1992) Criteria for the onset of oscillation in microwave circuits. *IEEE Trans. Microwave Theory Tech.*, **40**, 566–9.
3. Edson, W. A. (1960) Noise in oscillators. *Proc. IRE*, **48**, 1454–66.
4. Josenhans, J. (1966) Noise spectra of Read diode and Gunn oscillators. *Proc. IEEE*, **54**, 1478–9.
5. Ondria, J. G. (1968) A microwave system for measurement of AM and FM noise spectra. *IEEE Trans. Microwave Theory Tech.*, **16**, 767–81.
6. Kurokawa, K. (1973) Injection locking of microwave solid-state oscillators. *Proc. IEEE*, **61**, 1386–410.
7. Carroll, J. A. (1970) *Hot Electron Microwave Generators*, Arnold, London.
8. Cohen, L. D. (1991) A millimeter wave, third harmonic, Gunn VCO with ultrawideband tuning. *1991 IEEE MTT-S Dig*. 937–8.
9. Magalhaes, F. M. and Kurokawa, K. (1970) A single-tuned oscillator for Impatt characterization. *Proc. IEEE*, **58**, 831–2.

FURTHER READING

1. Kurokawa, K. (1969) *An Introduction to the Theory of Microwave Circuits*, Academic Press, New York.
2. Liao, S. (1987) *Microwave Circuit Analysis and Amplifier Design*, Prentice Hall International, Englewood Cliffs, NJ.
3. Combes, P. F., Graffeuil, J. and Sautereau, J. -F. (1987) *Microwave Components, Devices and Active Circuits*, Wiley, Chichester.
4. Vendelin, G. D. (1987) *Design of Amplifiers and Oscillators by the S-parameter Method*, Wiley, Chichester, New York.
5. Soares, R., Graffeuil, J. and Obregon, J. (eds) (1983) *Applications of GaAs Mesfets*, Artech, Dedham, MA.
6. Kaifez, D. and Guillon, P. (eds) (1986) *Dielectric Resonators*, Artech, Dedham, MA.
7. Gibbons, G. (1973) *Avalanche Diode Microwave Oscillators*, Clarendon, Oxford.
8. Matthaei, G., Young, L. and Jones, M. E. T. (1980) *Microwave Filters, Impedance Matching Networks and Coupling Structures*, Artech, Dedham, MA.

11

Monolithic microwave integrated circuits

11.1 INTRODUCTION

11.1.1 Hybrid and monolithic microwave ICs

The essence of a monolithic IC is that all components, both passive and active, and their interconnections are made on the same semiconductor substrate. The microwave monolithic ICs (MMICs) are direct descendants of the hybrid microstrip integrated circuits (MICs) in which the active and some of the passive components are mounted as discrete devices on dielectric substrates. These have been treated in Chapters 8, 9 and 10. In MMICs the substrate has to be semi-insulating to accommodate the necessary transmission lines and passive components. The active components are made in isolated conducting regions of the semiconductor material. For the connections (microstrip lines) and passive components it is important that the substrate forms a dielectric with low losses. For this purpose GaAs is better owing to its higher resistivity in undoped (or compensated) state (Chapter 3). For the active components GaAs is also better than Si because of its greater electron velocity.

Soon after the fabrication of the first GaAs MESFETs the idea arose of integrating these with passive components on the same substrate. We distinguish three kinds of monolithic integrated circuits.

- Analog (microwave) circuits are designed for a limited frequency band and sinusoidal signals; these are commonly designated as MMICs (monolithic microwave integrated circuits).
- Digital circuits must be broad band and are suited for pulse signals of large amplitude; these are sometimes grouped under the name gigabit electronics. Gigabit digital ICs look very much like their low-frequency counterparts: they contain transistors, diodes, resistors and capacitors but no inductors, and the interconnections are kept as short as possible.
- Optoelectronic integrated circuits (OEICs) have optical emitters or detectors which are combined with one of the two above; besides these

one also talks about integrated optical circuits (IOCs), in which all the components are optical.

11.1.2 Why monolithic?

Compared with hybrid MICs, analog MMICs have a number of (potential) advantages:

- smaller size;
- higher reliability;
- potentially lower cost at large production numbers;
- lower parasitics (no bond wires) and better reproducibility giving better performance, which especially at high frequencies is very important.

There are disadvantages too:

- at low production numbers the cost can be higher because of higher substrate consumption and lower yield;
- higher losses and noise;
- more cross-talk;
- no possibility of adjustment after fabrication;
- limited range of values for components such as capacitors and inductors;
- poorer heatsinking.

To illustrate the cost problem we can make the following rough calculation. Suppose we start from 3 in wafers with a usable area of about 25 cm^2 and the chip size is 10 mm^2, then we have about 250 chips per wafer. To have the design made by a GaAs foundry will cost about \$50,000. A yield of 40% then gives a unit price of \$500. If we apply the same calculation to discrete transistors with about 0.5 mm^2 chip area and a yield of 60%, the unit price is \$13. So if we want to make a circuit containing only a few transistors it will be cheaper to do it in hybrid technology. However, in mass production the cost per processed wafer will be much less than the \$50,000 quoted above for a single run and the MMIC method becomes more competitive.

Another reason that MMICs have been slow in replacing MICs is that comparable noise figures are more difficult to realize in MMICs than in MICs owing to higher losses in the passive components and the interconnections. Experience of the last five years, however, shows that the advantages of MMICs are slowly overcoming the disadvantages and markets are opening up in many areas of which some are listed as follows:
military (low volume, high unit price), e.g.

- phased array radars,
- telecom systems,
- surveillance and countermeasures, and
- smart weapons;

civil (low volume, high unit price), e.g.

- satellite communications,
- optical fiber communication,
- instrumentation, and
- supercomputers;

civil (high volume, low unit price), e.g.

- direct broadcast satellite (DBS) television,
- global positioning system (GPS),
- mobile telephone,
- high definition television (HDTV), and
- automobile radar.

Most of these applications require analog circuits, some ask for digital circuits and some need a mix of both. In optical fiber communication systems OEICs and IOCs are required. Another application of OEICs could be as interconnections between circuit boards in computer systems instead of backplane connections.

For digital ICs the question is not hybrid vs monolithic. Nobody would think nowadays of building even the smallest digital circuit in hybrid technology. The competition here is between different technologies, mainly between the old established cheap and reliable silicon technologies and the newly emerging GaAs technologies which are faster but more expensive, and which cannot reach the high levels of integration that silicon has achieved.

11.1.3 Available technologies

Several technologies are now available for integrated circuits with microwave bandwidths:

- silicon bipolar emitter-coupled logic (ECL), which is the only Si technology that can reach gigahertz speeds;
- GaAs MESFET technology, in which there are several logic families which will be described further on;
- AlGaAs/GaAs HEMT technology, which is the previous one with the MESFETs replaced by faster and less power-consuming HEMTs.

Besides these there are new technologies emerging which in the future may replace the above-mentioned ones:

- SiGe bipolar technology where the classical bipolar transistors are replaced by heterojunction bipolar transistors (HBT) with SiGe base;
- AlGaAs/GaAs and strained layer AlGaAs/InGaAs/GaAs HBT technology;

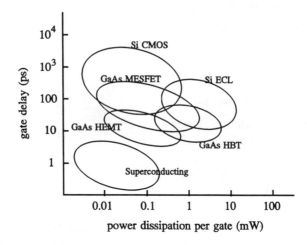

Figure 11.1 Comparison of digital technologies.

- InAlAs/InGaAs HEMT or HBT on InP substrates;
- superconducting Josephson junction technology.

An early review of digital technologies that is still instructive to read because of its attention to system considerations was given by Bosch [1]. Often the different technologies for digital ICs are compared in a so-called speed–power diagram, where the delay of a logic gate is plotted against the power dissipation per gate (Fig. 11.1). This type of diagram is often used because it gives a comparison at a glance on the two most important points. Speed is important for obvious reasons, and power dissipation per gate because it largely determines how many gates can be packed on one chip. Diagrams such as this should, however, not be taken too seriously. Besides being time dependent a comparison of this nature can only be indicative as long as one does not specify a circuit. An honest comparison can only be made if one compares similar circuits, e.g. memories of the same size. Before a choice can be made for a specific application other factors have to be considered: cost, reliability, compatibility with existing circuitry etc.

11.2 BIPOLAR TECHNOLOGIES

11.2.1 Introduction

High-speed digital technologies based on bipolar transistors all use emitter-coupled logic (ECL) as their basic circuitry. Sometimes the name current-

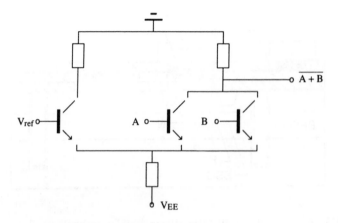

Figure 11.2 Basic ECL NOR gate.

mode logic (CML) is used which is actually a wider class of circuits encompassing ECL. The basic form of an ECL NOR gate is shown in Fig. 11.2.

The voltage levels representing logic 0 and 1 are below and above the reference voltage V_{ref} which itself is about halfway between zero and V_{EE}.

11.2.2 Silicon bipolar and heterobipolar technologies

Base thickness and emitter widths of bipolars have been steadily reduced over the years and with conventional technologies base widths of 0.1 µm and emitter widths of 2 µm are feasible, leading to bit rates in the low gigabit per second range. 'Conventional' means that ion implantation is used for the base and emitter dopings.

A less conventional technique that has recently been developed is to diffuse the base from a locally deposited solid source of borosilicate glass (BSG), which consists of silicon dioxide with a few percent of boron trioxide. In the BSG then a window is etched in which n-type polycrystalline silicon is deposited as an emitter. By this method very thin bases can be made and by also keeping the emitter area very small (0.5 µm × 2 µm) cutoff frequencies of 40 GHz have been reached [1]. A cross-section of this structure is shown in Fig. 11.3.

Further improvements are made with new technologies such as MBE which enable the base to be made of silicon–germanium (Chapter 6). Improvements from SiGe are also expected in CMOS technology which is hampered by the slowness of the p channel transistors. With SiGe technology a kind of silicon HEMT can be made in which the channel consists of an SiGe layer with 20–40% Ge. Because of the Ge-content and the effects

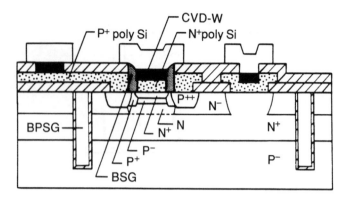

Figure 11.3 An example of 40 GHz silicon bipolar technology. Relabelled from Sugiyama, M., Takemura, H., Ogawa, C. *et al.* (1989) *A 40 GHz f_T Si Bipolar Transistor LSI Technology*, Int. Electron Devices Meeting 1989, 221–4. © 1989 IEEE Press.

of strain the hole mobility in this layer is much higher than in pure Si, so faster p channel transistors are expected.

11.2.3   GaAs heterobipolar technology

The GaAs heterojunction bipolar transistor has been treated in Chapter 6. Like its silicon counterpart, it is well suited for monolithic integration. Fig. 11.4

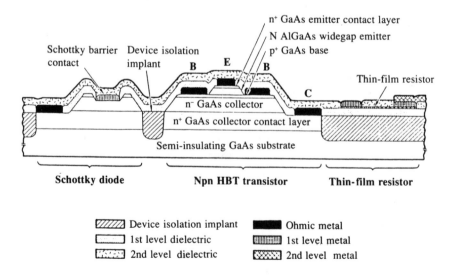

Figure 11.4   An example of GaAs heterobipolar IC technology.

shows some components of a GaAs HBT IC technology. GaAs HBTs can be used for analog (mainly power amplifiers) as well as digital applications, e.g. frequency dividers.

11.3 GaAs FET TECHNOLOGIES

11.3.1 Introduction

Analog MMICs are almost exclusively made in GaAs technology using MESFETs and increasingly HEMTs as the active devices. Here a short overview of the most important features of the current technologies is given. There are several good books which treat GaAs technology in more detail [2–6]. The idea of replacing the dielectric substrates of MICs by GaAs and integrating passive and active components came quite naturally. The great bandwidth of the GaAs MESFET was of course also an invitation to digital circuit designers to try to develop digital ICs based on this device. The first GaAs ICs appeared around 1975. Considering that the MESFET itself really took off around 1970 this is a quick development. It is in part due to the fact that GaAs technology could profit from the technological developments made by the people working with silicon, notably the photolithographic techniques and also ion implantation which could be copied almost without change. In other areas such as epitaxy, etching and contacting the technology had to be developed afresh and much of the research effort of the past years has been spent in these areas. With gatelengths of 0.5 and 0.25 µm becoming the standard now speeds of over 10 Gbit/s can be obtained. Recently the HEMT promises even higher speeds at lower power densities.

11.3.2 Active device technology

The most used active component is the MESFET (Chapter 6). It is suited for many analog applications: amplification, frequency conversion, oscillation, phase shifting and switching. The Schottky-barrier diode is used as a detector, mixer or varactor diode for tuning of resonant circuits. Because MESFETs and Schottky diodes need different doping concentrations attempts have often been made to have the diode's functions performed by MESFETs too, to save processing steps. There is a trade-off then between processing complexity and circuit performance. In digital circuits the MESFET functions as a switch and as an inverting amplifier. It can also be used instead of a resistor as a passive load with the gate connected to the source. Diodes can be used as switches and as level shifters. The MESFET needs an n-type GaAs layer for the channel and an n^+ layer for the source and drain contacts. The n^+ contact layer is etched away in the gate region when the gate recess is made. These layers can be epitaxially grown on the semi-insulating GaAs substrate. In the past, when GaAs substrates were Cr-doped to make them semi-insulating, it was necessary to grow an undoped buffer layer first since these substrates contained many impurities which can lead to increased noise, backgating and other unpleasant effects. Nowadays semi-insulating substrates can be grown with low impurity content so the quality is much better. Sometimes, however, a p-type buffer is grown to prevent substrate currents. The epitaxial growth is nowadays done by MBE or MOVPE (Chapter 3). In production MOVPE seems to be the preferred technique.

Another technology is ion implantation. Here too it is advantageous to grow a buffer layer first. An advantage of ion implantation over epitaxial growth is that it can be done locally using a mask of photoresist or a dielectric giving the possibility of realizing different doping concentrations on the same substrate. This is important when there are MESFETs and Schottky diodes on the same chip. Implanted n-type layers have to be annealed at a fairly high temperature (700–800 °C) to repair the lattice damage that has been caused by the ion bombardment. If no precautions are taken arsenic will evaporate from the wafer, leaving gallium droplets on the surface. To prevent this either the wafer is capped with a layer of Si_3N_4 or the anneal furnace is filled with AsH_3 under pressure. Care has also to be taken that the results of the previous processing steps are not destroyed by the high-temperature process. Implantation should therefore be done as early in the process as possible. Besides there is a strong tendency nowadays to keep the high-temperature processes as short as possible using rapid thermal annealing (Chapter 3).

Device isolation can be accomplished by etching away the n layers around the devices or by implantation with protons or boron ions. These render the layers non-conducting. Boron is preferred because it diffuses much less easily than hydrogen. The advantage of implantation is that the surface remains planar. Isolation implants do not have to be annealed. In the case that the n layers are made by ion implantation isolation can be done as above but also by implanting the n dopes only locally.

Important with the small gate lengths that are necessary to get the required speed is the use of self-aligned gate techniques. There are two main processes which are illustrated by Fig. 11.5. The first, called the SAGFET (self-aligned gate FET) process, uses the gate itself as a mask for the ion implant that forms the source and drain contact layers. To this end the gate has to have a T-structure, so it is built up from two metals. An added benefit of the T-shape is that it allows a small gate length and yet a low gate resistance. In this process the gate must be able to withstand the high-temperature annealing step that is necessary after the implantation. This is why these gates are made of so-called refractory metals, e.g. tungsten or titanium, or alloys such as WSi. The other process, called SAINT (self-aligned implanted n layer technique) uses a dummy gate of photoresist for a mask. This is used

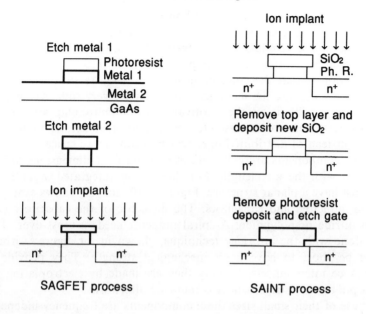

Figure 11.5 Self-aligned gate technologies.

for the implantation and then to deposit SiO$_2$. After the photoresist is removed the gate metal is deposited in the hole in the SiO$_2$ layer.

One problem that has long hampered the development of large-scale digital GaAs ICs is the control of the uniformity of the threshold voltages. This is especially critical when enhancement MESFETs or HEMTs are used. Across a wafer the standard deviation of the threshold voltages should be well below 100 mV, otherwise the yield in working circuits will be very low. To realize what this means we can make a simple calculation. At threshold the voltage drop across the depletion layer of a MESFET is about 2 V. A look at equation 6.5a tells us that a variation of 0.1 V on 2 V will be caused by a doping variation of 5% or a variation in layer width of 2.5%. Keeping the threshold uniformity well below 0.1 V then means that the doping uniformity should be 1% or 2% and the thickness uniformity less than 1%, or about 2 nm, that is seven atomic layers. These clearly are quite stringent demands and they form besides the high cost another reason why GaAs LSI technology has been slow in conquering a place on the market.

11.3.3 Passive components and interconnections for analog MMICs

In section 11.1.1 it has been stated that analog MMICs are in principle copies of hybrid MICs as they are made on alumina or plastic substrates.

There is a difference in approach, however. The use of available area is more critical in MMICs. Since the cost of processing a wafer is independent of the number of circuits on the wafer, the price of a chip is directly proportional to its area so it is important to minimize this area. Another consideration is that the connecting lines on a GaAs substrate are more lossy, mainly because the metallization layers are very thin. Keeping them as short as possible is another motivation to minimize chip area. For this reason one prefers to use discrete components (capacitors, inductors) in MMICs instead of distributed ones which occupy more space.

Passive components are called discrete when their dimensions are small compared with the wavelength. For these to be integrated monolithically they must have a planar structure. Figures 7.10 and 7.11 shows examples of discrete inductors and capacitors. The dielectric of the capacitors can be silicon nitride or polyimide. A spiral inductor needs a cross-over. This is often done with the *airbridge* technique, shown in the figure. Airbridges should be short to have good mechanical strength. They provide low-capacitance interconnects. Usually they are made by electroplating metal over a photoresist stripe, which is removed afterwards.

Because of their small sizes these components are frequency-independent up to a certain frequency. At higher frequencies their parasitics start to play a role. It is shown in Chapter 7 how the parasitics can be represented. One should realize that this too is approximate. The true frequency dependence of these components can only be obtained from a full electromagnetic field analysis. The parasitics have to be taken into account in the circuit design. Since accurate models are not available an iterative procedure has to be followed, consisting of a first design using estimated circuit parameters, making and measuring the circuit and adapting the parameters to the measured results, followed by a redesign. This method converges reasonably fast. A problem with the discrete components is their low Q-value because of the thin metallization layer.

The connections in GaAs ICs are mostly made of gold. This should be thick to limit losses so often the evaporated gold layers are thickened by electroplating. In analog ICs they should have the form of microstrip lines which implies that the substrate must be polished and gold-plated. When connecting lines have to cross each other an airbridge is made, as shown in Fig. 11.6.

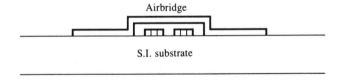

Figure 11.6 Metal airbridge interconnect.

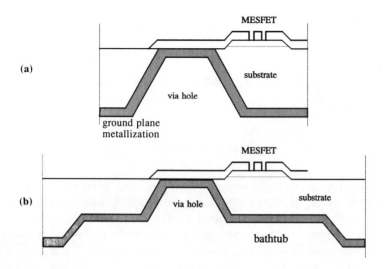

Figure 11.7 (a) Via hole and (b) bathtub structure.

Important in analog ICs for microwave frequencies is the low- inductance grounding of common-source FETs. The best solution is grounding via a hole that is etched through the substrate (a so-called *via hole*) from the ground plane and metallized together with the latter (Fig. 11.7(a)). Note that this necessitates making a photoresist pattern on the back side of the wafer that has to be aligned with the pattern on the front side. This can be done by shining infrared light through the wafer. When the wavelength is longer than about 0.9 µm the photon energy is less than the bandgap and the light will not be absorbed in the GaAs. Of course it cannot be seen by the naked eye but has to be monitored via an infrared-sensitive TV camera.

Both the alignment and the etching are easier the thinner the substrate is. Furthermore, a thin substrate provides a better heat sink for the FETs (remember that GaAs is a bad heat conductor). On the other hand, a thin substrate causes greater losses in the microstrip lines, especially at high frequencies. A solution for this problem is the bathtub structure of Fig. 11.7(b). Here the substrate is made thinner only where the FETs are to be placed.

Via holes and bathtubs are usually etched by wet chemical techniques, since the etch depth is rather large: the substrate is first lapped down from the back side to a thickness of about 150 µm and through this thickness the holes are etched. Because of the undercutting that goes with wet etching the hole is much wider at the substrate side than at the top side. A recent modification of the via hole principle entails etching a hole from the front side and depositing the source pad metal in this hole. Since the use of wet etching

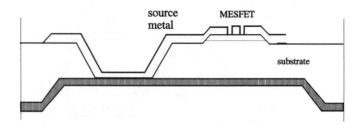

Figure 11.8   Front side via hole.

would produce a big hole, plasma etching is used, a technique that allows holes with almost vertical walls under the proper conditions (Fig. 11.8).

An example of a fairly complicated MMIC is shown in Fig. 11.9. This chip measures 1.8 mm × 5.8 mm and contains a digitally controlled (6-bit) phase shifter, five amplifiers and two transmit–receive switches [7]. It gives a good impression of the compactness that can be achieved using discrete components.

11.3.4   Digital technologies

Gigabit electronics is the nickname for ultrafast digital electronics with bit rates that exceed 1 Gbit/s. These circuits put new demands on design as well as devices. In the frequency range over 1 GHz one has to take account of the microwave character of the circuits. Propagation delays and reflection of signals on transmission lines have to be reckoned with and devices have to have high cutoff frequencies.

Contrary to analog MMICs the interconnections are not made as microstrip lines. The propagation delays would be too long and the capacitances to ground would cause unnecessary loading of the logic gates. Instead one tries to keep the interconnects as short as possible.

The speed of digital circuits is limited by the following:

- intrinsic response of the switching transistors (10–100 ps);
- RC time constants of input and output circuits (10–1000 ps);
- signal propagation times on the chip (10–100 ps).

Clearly it is important to keep the RC time constants short. The output load of a logic gate consists mostly of capacitances, partly from the input of the next gate and partly from the interconnects. This capacitance has to be charged by the transistor at the output of the logic gate. So transistors for fast digital circuits should fulfill the following requirements:

- short switching times, where both the rise time of the output pulse and its delay with respect to the input pulse are important;
- sufficient current capability for the charging and discharging of circuit capacitances.

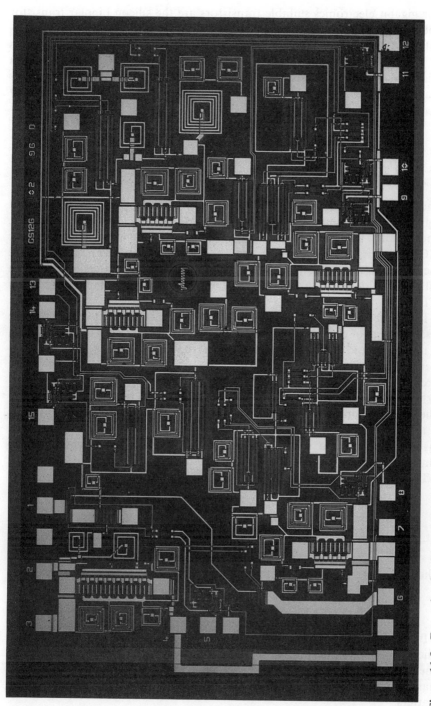

Figure 11.9 Example of a GaAs MESFET MMIC. By courtesy of Dr F. A. Myers, GEC–Marconi.

For gigabit electronics suitable transistors are the silicon bipolar transistor (including HBT) and the GaAs MESFET and HBT. The Si CMOS technology does not yet reach gigabit rates mainly because of the slowness of the p channel transistors although steady progress is being made in this field. Bipolar transistors have inherently better current capabilities than FETs but FETs have smaller input capacitances and higher cutoff frequencies.

11.4 GaAs LOGIC FAMILIES

11.4.1 Introduction

The GaAs MESFET is the most frequently used transistor in gigabit logic circuits although the use of the HEMT and the HBT is increasing. The circuits using HBTs belong to the family of emitter-coupled logic (ECL) circuits, well known from the silicon bipolar field. HEMTs are used in the same types of circuits as MESFETs so in the following we will give a description of the most important logic families based on MESFETs.

The depletion MESFET is the standard device. It is easy to make but the disadvantage of circuits with only depletion transistors is that level shifting is necessary and two supply voltages are required. Circuits with enhancement MESFETs on the contrary need only one supply voltage. These circuits, however, have small logic swings (about 0.5 V) because the Schottky barrier gate cannot be biased more than 0.8 V in the forward direction. Their fabrication therefore needs very tight control of parameters. Extremely thin and uniform active layers must be made. JFETs with a p^+ gate can sustain 1.1 V in the forward direction so that the logic swing can be somewhat larger.

The most well-known MESFET logic families are

- buffered FET logic (BFL),
- directly coupled FET logic (DCFL),
- Schottky diode FET logic (SDFL), and
- source-coupled FET logic (SCFL).

Then there are variations on these themes, e.g.

- capacitor diode FET logic (CDFL), an offspring of BFL, and
- source follower FET logic (SFFL), a variant of SCFL.

In the following the most important of these families will be discussed. There is no generally accepted standard process, although DCFL is the one used by the greatest number of manufacturers. On the whole every manufacturer seems to prefer his own process, probably because of patent matters.

11.4.2 Buffered FET logic

This is the first technology in which GaAs digital ICs were made. It uses only depletion-mode transistors. A logical 0 is represented by a voltage level around − 2 V and a logical 1 by a voltage around zero. The circuit for a NOR gate is given in Fig. 11.10. This type of circuit needs a negative voltage on the gate to switch the transistor off and a positive drain voltage. As a consequence the output voltage level has to be shifted so that input and output levels correspond. An advantage of BFL is that its techno-logy is simple. The first GaAs digital ICs were made in BFL using mesa-isolated epitaxial n-layer MESFETs. The design can also be implemented using HEMTs. Each of the forward-biased diodes gives a voltage drop of about 0.8 V. These diodes lead to a high dissipation and a large consumption of chip area. NOR gates are realized with parallel FETs as in the figure and NAND gates with series FETs. Logic functions of a higher order are made by combinations of parallel and series transistors. In a NAND function the number of FETs in series is limited because with more than three FETs the total on resistance causes too high a voltage drop.

BFL has a low output impedance (source follower) and can therefore charge and discharge capacitances quickly. This makes this family fast, but at the cost of large dissipation.

Note that in this case and others shown below transistors are used as loads instead of resistors. This has several advantages: transistors are smaller than resistors and the non-linear characteristic of the transistor gives a faster rise time and a better noise margin.

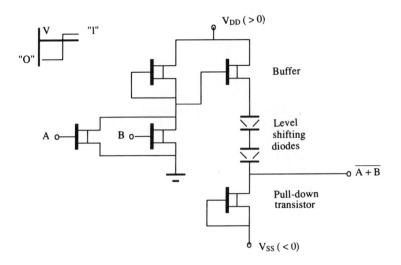

Figure 11.10 Buffered FET logic NOR gate.

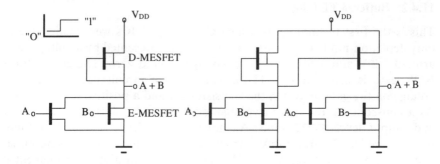

Figure 11.11 NOR gates in direct-coupled FET logic: (a) unbuffered; (b) buffered (SBFL).

11.4.3 Directly coupled FET logic

This logic family uses enhancement-mode transistors (normally off) as switches and depletion-mode transistors (normally on) as loads. This makes it possible, with 0 V as a logical 0 and a positive voltage as a logical 1, to use only one power supply.

In Fig. 11.11 two NOR gates are drawn as examples of this kind of logic. The one in Fig. 11.11(b) is also called SBFL (super buffered DCFL).

11.4.4 Schottky diode FET logic

Like BFL this logic uses depletion transistors only. The most important differences between BFL and SDFL are as follows: the logic OR function

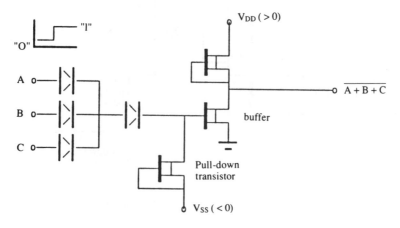

Figure 11.12 Schottky diode FET logic.

is formed by Schottky diodes (hence the name Schottky diode FET logic) and level shifting is realized by the switching diodes and an extra diode at the input.

SDFL retains the high speed of BFL but has much less power dissipation and consumes less chip area per gate. An example of a NOR gate with this logic is depicted in Fig. 11.12. The diodes need to draw only little current and can therefore be small. The greatest power saving is achieved by having the level shifting done not at the output but at the input side where the current is much smaller. SDFL leads to very dense circuits because switching diodes take much less space than FETs. $1 \times 2\,\mu m^2$ is necessary for a switching diode and $3 \times 3\,\mu m^2$ for a level shifting diode. Moreover, diodes do not need a connection to a power supply.

SDFL logic cannot be implemented using only standard MESFET techniques because the diodes need much different doping profiles. The doping must be lower and deeper for a low depletion capacitance. SDFL uses local implantation in selected areas.

11.4.5 Capacitor diode FET logic

This logic has in common with BFL that a stack of two (or three) forward-biased diodes takes care of the level shifting. A reverse-biased diode with a large area is now placed parallel to these diodes (Fig. 11.13). This diode passes the leading edge of the pulse capacitively and thereby speeds up the circuit. Because the switching transient is capacitive it does not dissipate power so the overall dissipation is less than in BFL.

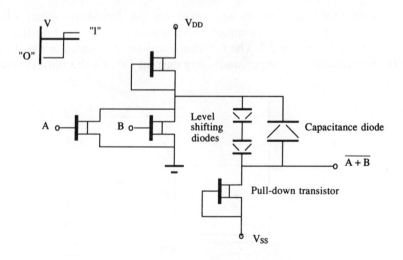

Figure 11.13 Capacitor diode FET logic (CDFL).

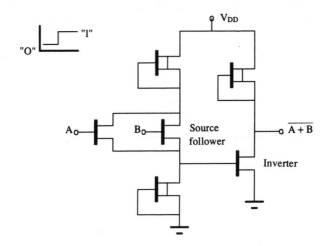

Figure 11.14 Source follower FET logic.

11.4.6 Source follower FET logic

This consists of an E-MESFET source follower and an inverter (Fig. 11.14). As loads D-MESFETs are used. Another name for this family is *source-coupled FET logic* (SCFL). The main advantage of SFFL is its good noise margin. This is obtained at the cost of sacrificing some speed and power.

11.5 SUPERCONDUCTIVE TECHNOLOGIES

Superconducting logic circuits are based on the Josephson effect which occurs in a superconductor–insulator–superconductor (SIS) junction that has been described in section 5.7. The $I–V$ characteristic is repeated in Fig. 11.15. It has two branches: the superconducting current (S) and the normal tunnel

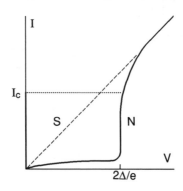

Figure 11.15 Bistable $I–V$ characteristic of an SIS junction.

current (N). If the device is biased from a current source we have an almost horizontal loadline that crosses both branches and it is possible to switch between two stable bias points. This is the basis of superconductive logic.

The superconducting current cannot exceed a critical value I_C. This value is reduced by the application of a magnetic field. When bias is applied the device will come up in the zero-voltage state. Switching to the finite-voltage state now can occur either by a magnetic field that reduces the critical current below the bias current or by injecting an extra current so that the total exceeds the critical value. When the magnetic field or the injected current is taken away the device remains on the N branch. Switching back can occur only by reducing the bias current to zero and increasing it again. This means that the bias current has to be pulsed. The preferred material for logic applications is niobium with aluminum oxide as the insulator. The junctions are made by sputtering Nb, then Al which is oxidized in an atmosphere of 90% Ar and 10% O_2 and then sputtering the second Nb layer. As a substrate material silicon is used. Figure 11.16 gives a cross-section of a superconducting logic circuit.

In practice the basic switching circuit consists of two Josephson junctions in parallel, a so-called SQUID (superconducting quantum interference

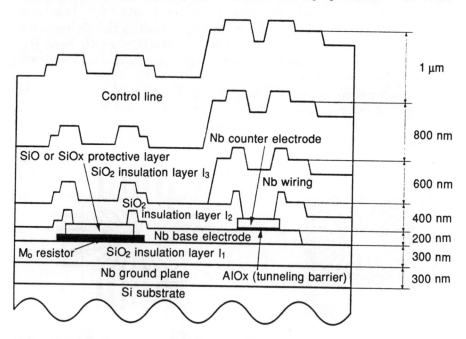

Figure 11.16 Cross-section of a Josephson junction logic circuit. Reproduced with permission from Hasuo, S., Josephson Devices for Computer Applications in *International Journal of High Speed Electronics*, **3**(1); published by World Scientific Company, 1992.

device). The name derives from the fact that all superconducting electron pairs in a superconductor have the same wavefunction and if the superconductor splits into two branches there has to be phase coherence of the wavefunction in the two branches. This coherence can be destroyed by a magnetic field or by injecting extra current in one branch. If this happens the supercurrent stops and the device switches to the normal branch. Because of the phase coherence condition the SQUID is much more sensitive than a single junction.

There are many logic families possible in this technology. Based on the way switching is accomplished they can be classified into three groups: magnetic, current injection and mixed. The last one is the most complicated but it can combine the advantages of the two others. A recent review is given in ref. 8.

11.6 INTEGRATED CIRCUIT PACKAGING

Integrated circuits, like diodes and transistors, can be used in packaged or unpackaged form. Analog microwave ICs (MMICs) are often mounted as chips on alumina microstrip circuits which contain bias circuits, filters, circulators and other parts that cannot or do not have to be integrated monolithically. The casing of the microstrip circuit, which is normally closed, serves as a package for the whole. MMICs, which have few connections, can also be mounted in packages like those used for power transistors (section 6.8).

Digital integrated circuits are usually sold in packaged form. They are characterized by their large number of connections, and thus special IC packages are necessary. The usual dual-in-line packages, however, are unfit for high-speed digital ICs, because they have rather long leads which

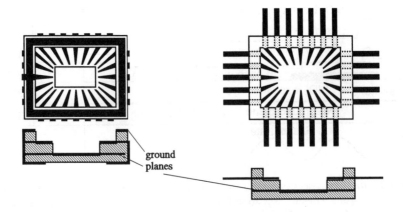

Figure 11.17 Surface mount package suitable for high-speed ICs.

contribute intolerably great inductances. Better suited are the so-called surface mount packages (Fig. 11.17(a)), which have much lower parasitics. What one tries to achieve in high-speed packages is that the leads together with the metal bottom of the package form transmission lines with a well-defined characteristic impedance of 50 or 100 Ω (Fig. 11.17(b)). Another possibility is to alternate the leads with ground planes, so that they acquire the character of coplanar transmission lines (section 7.3). This also provides isolation between the leads. Also, it is not possible to use one size of package for all ICs, since this would have to fit the largest IC and then the smaller ICs would have to be connected by unnecessarily long bond wires. Thus, in principle for every IC a fitting package should be designed. Often packages are designed to contain more ICs, so that a complete subsystem can be mounted in one package (information about high-speed IC packaging can be found in ref. 9).

REFERENCES

1. Bosch, B. G. (1979) Gigabit Electronics – A Review. *IEEE Proc.*, **67**, 340–79.
2. Thomas, H., Morgan, D. V., Thomas, B., Aubrey, J. E. and Morgan, G. B. (eds) (1986) *Gallium Arsenide for Devices and Integrated Circuits*, IEE Electronic Materials and Devices Series 3, Peregrinus, London.
3. Williams, R. (1990) *Modern GaAs Processing Methods*, Artech, Boston, MA, London.
4. Howes, M. J. and Morgan, D. V. (eds) (1985) *Gallium Arsenide, Materials, Devices, and Circuits*, Wiley, New York.
5. Goyal, R. (ed.) (1989) *Monolithic Microwave Integrated Circuits: Technology and Design*, Artech, Norwood, MA.
6. Ferry, D. K. (ed.) (1985) *Gallium Arsenide Technology*, Sams, Indianapolis, IN.
7. Lane, A. A., Jenkins, J. A., Green, C. R. and Myers, F. A. (1989) S and C band multifunction MMICs for phased array radar. Proc. GaAs IC Symp. 1989, IEEE New York, pp. 259–62.
8. Hasuo, S. (1992) Josephson devices for computer applications. *Int. J. High-Speed Electron.*, **3**, 13–52.
9. IEEE (1985–9) Proc. GaAs Symp., IEEE, New York.

FURTHER READING

1. Ladbrooke, P. H. (1989) *MMIC Design: GaAs FETs and HEMTs*, Artech, Boston, MA.
2. Pucel, R. A. (ed.) (1985) *Monolithic Microwave Integrated Circuits*, IEEE, New York.
3. Rose-Innes, A. C. and Rhoderick, E. H. (1987) *Introduction to Superconductivity*, 2nd edn, Pergamon, Oxford.
4. Sugiyama, M., Takemura, H., Ogawa, C., Tashiro, T., Morikawa, T. and Nakamae, M. (1989) A 40 GHz f_T Si bipolar transistor LSI technology. Proc. Int. Electron Devices Meet. 1989, IEEE, New York, pp. 221–4.

Index